A SHORT PRIMER ON
Why Cancer Still Sucks

D1073728

DAVID J. STEWART, MD

Tellwell Talent
www.tellwell.ca

ISBN
978-0-2288-7199-6 (Hardcover)
978-0-2288-7198-9 (Paperback)
978-0-2288-7200-9 (eBook)

Dedication

To my patients and their loved ones, whose need and suffering drive me, whose thanks have richly rewarded me, and whose courage constantly inspires me.

Table of Contents

Foreword

Dr. David Stewart and I met soon after my wife Annie was diagnosed with a rare lung cancer that tends to afflict fit women who don't smoke. When she was diagnosed, it had already spread to a few of her organs and lymphatic system. She was given three to four weeks to live. We met Dave about a week and a half into those four weeks. It had taken that amount of time to determine the specific kind of cancer she had – an extraordinarily short time. Weeks of diagnostic delay were the norm in Canada in those days.

I knew that delay was the enemy, especially in Annie's case where the cancer had already metastasized throughout her body. One of the chapters in this book explains why. I knew because I am a cancer survivor. It had taken almost five months from detecting my first lump to getting a sufficient diagnosis to start treatment. In that time, one lump had become four lumps. Luckily, the metastases were confined to a local region, and therefore the cancer was curable. My wife's case was different. She could ill afford any further metastatic spread.

We had bulldozed through the system to speed up diagnostic and testing processes. We needed a physician who would work with us as a partner to act quickly and decisively as an advocate. Dave was our man. He is responsive, kind, knowledgeable and innovative. Working together, he and I got Annie on an experimental drug within two weeks, and eventually another after that one failed. She lived a "simple, happy life" (her words) for fifteen and a half months, not four weeks.

This book is important for several reasons. It is written by an internationally respected oncologist with a huge heart and remarkable competence. He explains why, despite progress, cancer is a scourge of our times. Almost half of us will develop cancer at some point in our lives. If you're lucky, you'll be one of the survivors. Or, you'll have the emotionally and physically challenging honor to care for a loved one who has cancer... or maybe, like me, you'll experience both.

This book is also a toolbox. At one level, it is a primer about cancer, suitable for all of us. At another, it is a textbook for the more advanced, who want to know in more depth and detail. At either level, it will inform you about emerging diagnostic tests, causes of cancer, preventative measures, promising and existing treatments, and likely progress in the near and far term. You'll learn about surgery, radiotherapy, immunotherapy, precision medicines, conventional chemotherapy, alternative therapies and more.

For me, though, the most important message in this book is about the North American healthcare system. The US and Canadian systems each have strengths and weakness that help and hinder cancer prevention, diagnosis, and treatment. You'll learn about them in the pages that follow. For example, what can be done to prevent cancers? What is it that needs to change in drug approval processes and clinical trial protocols? How can we make optimal use of advanced, rapidly emerging tests, medicines, and other interventions? What are the systemic and process barriers to timely, effective prevention and care? How can they be overcome?

You'll read about how critical it is to get cancer patients faster access to better diagnoses and treatments.

Before she died, Annie asked me to ensure that other patients with lethal diseases benefit from the kind of advocacy Dave and I provided. So, we started LSTN, The Life-Saving Therapies Network (www.lifesavingtherapies.com). In many ways, the chapters in this book about the healthcare system, drug approvals and clinical trial protocols are its intellectual and moral underpinnings.

Enjoy this book. I hope that it informs you, makes you think and maybe even impels you to speak out about urgently needed reforms.

John-Peter Bradford, Ph.D., FCMC

Preface

My objectives for this book

I had three main objectives in writing this book. The first objective was to give cancer patients, their families, other members of the public, healthcare trainees and non-experts a better understanding of cancer. What causes it? Why is it so common? What are the limitations of screening? How does cancer cause symptoms? How do therapies work and how do they cause side effects? Why might they fail?

I tell trainees that if we can't cure a patient, we can at least help them understand why we cannot. Patients repeatedly tell me that uncertainty can be worse than bad news.

Particularly early in a patient's journey, there are several sources of uncertainty. Not knowing if they have cancer. Are they curable or treatable? Why is the therapy failing? Are there other therapy options available? Understanding the enemy gives people strength. If a person understands what is happening, they are better equipped to make important choices in how to proceed.

I also tell trainees that you can be an excellent oncologist if all you know are the results of the clinical trials that guide evidence-based care. But if you also understand a few of the basics of cancer biology, it can help you make more rational clinical decisions in areas where the clinical trials shed imperfect light.

My second very important objective in writing this book was to raise public awareness of "systems" obstacles we face in the fight against cancer. We need allies in the struggle to make everything happen faster. To speed up the development and approval of effective new treatments, the funding for those therapies, the tests a patient must undergo to diagnose and characterize a cancer, and the initiation of a patient's therapy.

There are too many impediments. It doesn't have to be this way. We can and must find more effective approaches. Even if you read little else in this book, please at least look at the overview sections of Chapters 11 to 14. As stated in Chapter 12, this is a call to action. Our leaders need to get the message that we must do better. The rapid government response to COVID-19 demonstrates clearly what concerted action can achieve. When it comes to cancer, there are many <u>reasons</u> why things proceed at a much slower pace than with COVID, but there are no valid <u>excuses</u>.

The third objective was to give the family and friends of oncologists some insight into what drives many of us—why most oncologists love what they are doing and why our workdays can be so long.

The past

Following his victory at the Battle of Zela in 47 BC, Julius Cesar triumphantly declared, "Veni, vidi, vici." "I came, I saw, I conquered." In the war on cancer, we have made sustained, slow progress, but there are very few areas in which we have truly conquered.

In 1976, I began my training in oncology. Since that time, all my patients have had the common experience of having had cancer. Caring for patients with cancer has been challenging and difficult, but it has also been exceptionally gratifying, rewarding, interesting and exciting. In this book, I will share some of the things I have learned during my journey.

The late 1970s and early 1980s were heady times in oncology. The pioneering work of Emil J. Freireich and others meant that children with acute lymphoblastic leukemia had gone from facing certain death just a few years earlier to having a 65–70% possibility of being cured. Several other malignancies had also become potentially curable even in patients with widespread disease, including Hodgkin's disease, some other types of lymphomas and leukemia, testicular cancer, gestational choriocarcinoma,[i] and some childhood cancers such as rhabdomyosarcoma, Wilms' tumor, and others. The cure rate of localized osteogenic sarcoma had increased from 10% with surgery alone to 60% if chemotherapy was administered after surgery.

---

[i]   Gestational choriocarcinoma is a malignancy that arises from a pregnancy. Before the availability of modern chemotherapy, it was highly lethal, but it can often now be permanently cured, even when widely metastatic.

The present

When I started my oncology training, I brashly predicted that within a very few years, we would conquer all malignancies with new chemotherapy approaches. But like the ancient Roman failure to subjugate the northern British Isles, we have slowed in our attempt to advance past the Hadrian's Wall of common adult malignancies such as cancers of the breast, lung, prostate, colon, and pancreas. We cannot cure any of these once they have metastasized widely.

Unequivocally, we have made substantial progress. Modern radiotherapy techniques are much more effective and less toxic than those of just a few decades ago. For most malignancies, systemic therapies like chemotherapy, targeted therapies, and immunotherapy can prolong life expectancy and can alleviate suffering in at least some patients with advanced disease. Radiotherapy or a systemic therapy that shrinks a patient's cancer is often a more potent analgesic than morphine because shrinking the cancer takes painful pressure off surrounding structures.

The most effective way to help patients regain an appetite that has been suppressed by metastatic cancer is to shrink tumors. In lung cancer, therapies that shrink tumors may eliminate a patient's need for round-the-clock oxygen supplementation. These therapies may also quiet the incessant, hacking cough that robs patients of their sleep.

But even if we can make metastatic cancers shrink, we usually cannot cure them. After a period of control, the cancer recovers. The inevitable emergence of resistant cells enables it to grow back.

So why are we in our current situation? This book is about giving people insight into why things are the way they are.

The book's structure

In each chapter, I start with a ***Short Primer*** section. This is the basis of the book's name. This section is a brief overview for readers with no medical background. Each Short Primer section is followed by a ***Further Details and References*** section that is somewhat more technical and includes supporting documentation. My friend, frequent collaborator, and literary adviser John-Peter Bradford refers to this as the "nerdy section" of each chapter.

In the early chapters, I discuss why cancer is now so common and how it causes distressing symptoms. I explain why screening tests like mammography and colonoscopy may work in some patients but fail to detect cancers early enough to cure other patients. I then briefly discuss how surgery, radiotherapy, chemotherapy, hormonal therapies, new targeted therapies, and immunotherapy work, why they may fail, and a few of their more prominent unpleasant side effects. I examine the fallacies of alternative therapies and how current oncology dogma, myths, and legends hold us back.

In Chapters 11 to 14, I address systems issues. In Chapters 11 and 12, I discuss why progress against cancer has been so agonizingly slow, the price we pay for this slow progress, and why cancer therapies are so expensive. In this story, there are no true villains. Just about everyone is trying to do the right thing, but in the process, we repeatedly have been tripping each other up.

Contributing substantially to the problem are the different perspectives and priorities that obscure the big picture and impede proceeding together with a common focus. We currently have no one leadership body that can drive rapid progress. In this book, I will present a few thoughts on a better path forward.

In Chapter 13, I delve further into the high cost of cancer therapy and what we can do about it. In Chapter 14, I discuss how both the American and Canadian healthcare systems fail in different ways.

Finally, in Chapter 15, I will talk briefly about how cancer care in the future may be much different than it is now. Despite all the things that slow us down, we <u>are</u> making progress, and tomorrow will be much brighter than today.

Disclaimer: In this book I present numerous established cancer facts and I will present my own opinions. However, there is substantial uncertainty in several areas, and there may be other reasonable ways of looking at things rather than the way in which I interpret them. I find that I learn most when thoughtful people disagree with me.

In the book, I will also talk about things I tell my patients. However, nothing in here should be interpreted as being medical advice from me to you as an individual reader. If you have cancer, discuss your own situation with your healthcare providers and seek other sources of information.

Finally, the views and opinions expressed in this book are my own. They do not represent those of the University of Ottawa or The Ottawa Hospital.

Disclosure of potential conflicts of interest: Most of my income is from care of cancer patients. However, each year an average of about 1.5% of my income comes from pharmaceutical companies or government agencies that ask my opinion on questions such as how we might make faster progress or how specific therapies might be made more effective. The highest that this has reached in any one year is less than 3% of my income. I have found that if I charge someone for my opinion, they are more likely to listen to it. I also own a 3% share of a patent on a test to predict benefit from an experimental gene therapy. In the past ten years, I have earned a bit less than $3,000 from this patent.

I have tried to ensure that none of these financial interests has colored my perspective.

Book Proceeds: A portion of the net proceeds from sale of this book will be offered to several deserving groups,[ii] including the Life Saving Therapies Network, Lung Cancer Canada, The Ottawa Hospital Foundation, the Ottawa Regional Cancer Foundation, Queen's University Meds '74 Fund, the University of Texas MD Anderson Cancer Center, the Canadian Cancer Society, Main Street Community Services, and others.

[ii]   At the time of publication, I had not discussed this with any of these groups. Some might possibly have policies that would preclude accepting the donation.

— 1 —

Dealing with the Deluge: Why is Cancer so Common?

The first reason that cancer still sucks is that it is everywhere. You can contract it no matter how healthy your lifestyle or how pure your family history. The same goes for your friends, neighbors, and loved ones.

Short Primer

Current projections are that 39% of all males and 38% of all females in the United States[1] and 49% of all males and 45% of all females in Canada[2] will develop cancer at some point in their lives. Overall, cancer causes about 22% of all deaths in the United States (compared to 23% for heart disease)[1] and it causes 30% of all deaths in Canada (compared to 19% for heart disease).[2]

While cancer is much more common in older people than in younger people, it is second only to injury as a cause of death in children and young adults.[3] Researchers use a measure called "potential life-years lost" or "years of life lost" to describe the impact of a disease on premature deaths.[iii] Cancer is our leading cause of potential years of life lost.[4,5]

Over the years, numerous celebrities, magazine and newspaper articles, and books have made wide-ranging claims about things that will increase your risk of getting cancer or that will protect you from it. Some of these

---

iii   Number of "life-years lost" for an individual is calculated by subtracting the age at which the person died from the age at which an average person would have died.

claims have merit and some do not. One source will report something to be protective and another will report this same factor to be harmful. Why does this happen? It happens because how a researcher asks a question, and what other factors they take into consideration will affect the answer they get. Some celebrities and journalists may also have personal views that are at best weakly backed by science.

My wife, Lesley, and many of our friends have repeatedly bemoaned this conflicting advice. In this chapter, I will outline my perspective about some of these protective or aggravating factors.

Age: It has been darkly observed that the most effective way to avoid ever developing cancer is to die young of something else. For many of you, the single biggest factor driving your risk of developing cancer is simply getting another day older. Every day of your life, billions of cells in your body divide to replace old or damaged cells, and every cell division carries with it a small risk of a cancer-causing mutation. Part of the reason that cancer is an even bigger problem in Canada than in the US may be that Canadians tend to live longer than Americans (by about three years for females and 4.5 years for males).[6] With every year of additional life, the risk of developing cancer rises.

Tobacco: Tobacco is the next largest cause of cancer in North America. It plays a role in about 30% of all cancer cases. Smoking is most strongly associated with lung cancer, but also increases the risk of cancers of the bladder, head and neck, colon, breast, and several other malignancies.

We know far more now about the risks of smoking than we used to. When I was young, my father smoked up to three packs of cigarettes per day. As 6- or 7-year-olds, my friends and I would sneak an occasional smoke to try to figure out why cigarettes were so appealing to our parents.

Some heavy smokers never develop smoking-related cancers. However, when people ask me, "Why did Uncle Joe never develop cancer despite smoking heavily until he died cancer-free at age 99?" I tell them, "Some soldiers who go to war manage to survive it. But that does not mean that going to war is good for your health." The same is true for smoking.

If you are currently a smoker, you should quit. As soon as you quit, your risk of developing a smoking-related cancer begins to go down, compared to if you continued to smoke.[iv] If you have been a heavy smoker for many years, quitting smoking will not bring your cancer risk down to the same level as if you never smoked, but it will reduce your risk substantially compared to your friends who keep on smoking.

Because there are so many people who used to smoke compared to the number who are still smoking, more of the patients we now see with lung cancer are former smokers rather than being current smokers. However, the current number of North Americans with lung cancer would now be much higher if fewer smokers had quit smoking.

Even if you have already developed lung cancer, you should still quit smoking. Continuing to smoke makes your cancer more resistant to cancer treatments and decreases their benefits.

Quitting smoking can be very difficult. Smokers often say that a cigarette helps them think more clearly and makes them feel good. They are right. When brain cells are exposed to the nicotine in tobacco smoke, the nicotine induces increased production by the brain cells of nicotine-binding chemical groups called nicotine receptors.

Nicotine levels in the blood and brain fall gradually over a few hours after a smoker's last cigarette. If the brain's nicotine receptors are not receiving their regular "feed" of nicotine, the unhappy receptors start making brain cells malfunction. The person cannot think as clearly, and they can feel irritable and unwell. Smoking another cigarette supplies the nicotine receptors with nicotine, and the smoker can then temporarily think as clearly and feel as well as a nonsmoker. They don't feel better than a nonsmoker when they smoke, but they no longer feel a lot worse than a nonsmoker.

I have been told that this is the reason that the cockpit of commercial planes was the last part of the plane in which smoking was banned. A pilot who smoked might be fine when the plane took off, but if he had not had any nicotine, he might be cognitively impaired four hours later when it was time to land the plane.

The only way to fix this nicotine dependency long term is to stop supplying the nicotine receptors with nicotine. They then gradually go away over a period of weeks.

[iv]   Your risk of a heart attack also decreases rapidly if you quit smoking.

I like using the analogy of feeding the bears. Our farmhouse retreat is in a wilderness area surrounded by dense bush. There are a lot of bears around, but they typically stay far from the house. So, they don't bother us. However, if someone fed the bears, more and more would show up at the house, and they would be very unhappy and dangerous if no one fed them. The only way to stop them from coming around would be to stop feeding them. But it could take weeks before they stopped showing up.

When a smoker smokes, they are feeding the nicotine-addicted bears in their brain. Some might consider shooting a bothersome bear, but you can't shoot the bear if it is living in your head. The only way to eventually get rid of it is to stop feeding it. Don't feed the bears!

Weight: Being overweight is another major contributor to developing cancer. Too much weight increases the level of inflammation in your body, and inflammation increases your risk of cancer. If you are overweight, you will also have high levels in your body of various hormones and growth factors that drive cells to divide more frequently. If more of your cells are dividing, then more of your cells will develop the mutations that occur in the process of cell division. The more mutations, the higher the risk of bad mutations. All it takes is a few bad mutations to start you down the road to developing a cancer.

You should concentrate on losing excess weight and not gaining it back. That is often easier said than done, but your risk of developing some cancers can depend on it. However, many of us are prone to "weight cycling" (also referred to as "yo-yo dieting" or "the rhythm method of girth control"), working hard to lose weight only to have it rapidly reaccumulate. While there are some reasons that weight cycling could harm your health, most large studies do not support this concern. So, my advice is to keep on working at taking off the extra pounds and try to make this the time that you are finally successful in keeping them off.

Physical Exercise: The flip side: put a high priority on being active. Regular exercise reduces your risk of developing cancer. For me, that means regularly walking the 4 km round trip to work even on the hottest Ottawa summer days and the coldest Ottawa winter days.[v] It also means spending as much weekend time as I can out in the bush at our farm, clearing ski trails in the winter and pursuing other projects in the summer.[vi]

Nutrition: What you eat makes a difference. My wife Lesley ensures that I now eat far less beef and far more salads than I did when I was young and carefree. This is a good idea for you too. Eating a lot of red meat (particularly beef and lamb) and processed foods[vii] is associated with an increased risk of cancer. Eating more fresh fruits and vegetables (particularly green-yellow cruciferous vegetables like cabbage, broccoli, cauliflower, and Brussel sprouts) is associated with a reduced cancer risk. Eating more whole grains, nuts, skim milk, yogurt, and soy foods may also be associated with reduced risk. Chicken, fish, and pork appear to convey minimal risk, unlike beef and lamb.

We hoped that vitamin supplements might help, but this is unequivocally not the case. Whatever it is in fresh fruits and vegetables that protects from cancer, it is not just the vitamins. It is probably an entourage effect whereby the whole plant and not simply extracts from it provides protection. For people who can eat a normal diet, vitamin supplements do not reduce the risk of cancer. Some studies have even shown that certain vitamin supplements (particularly some antioxidants) may increase the risk of developing cancer. Some people need supplements because of a vitamin deficiency, because of pregnancy, or because they are unable to eat or absorb their food properly, but you should generally not take vitamin supplements unless your physician recommends that you do.

Alcohol: While alcohol in moderation appears to reduce the risk of heart attacks and strokes, drinking even small amounts of any type of

---

[v]   For many years, I ran an average of 10 km per day. I eventually stopped running when the hip and knee damage caused by it caught up to me, and when people who were walking briskly began to pass me as I "ran." I was never very fast and slowed down further as my joints wore out.

[vi]   In the summer of 2021, my major activity was focused on battling a huge gypsy moth infestation!

[vii]   Processed foods include meats or vegetables that have been salted, cured, smoked, pickled, etc.

alcohol increases the risk of developing cancer. Red wine may not be quite as bad as other alcohol beverages, but even it carries some risk.

To me, there are few things more refreshing after a full day working outside in the sun than an ice-cold Moosehead beer. For a special dinner, I love a medium rare rib eye steak (not just meat, but red meat!) accompanied by a full-bodied Shiraz,[viii] followed by a desert of 3-year-old Balderson cheddar cheese with a tawny port. For both unhealthy foods and alcohol, moderation helps. Even with moderation, you are increasing your risk. But the less you consume, the less you increase your risk.

Sun: Excess sun exposure (particularly sunburns during childhood) is the major cause of common skin cancers and malignant melanoma. While we need sun exposure for our skin to produce vitamin D, it is important to limit sun exposure and to use sunscreens and protective clothing. When we go on tropical vacations, Lesley and our friends lie on the beach in the sun and taunt me for sitting in the shade, wearing sunscreen and a hat (and sometimes even a long-sleeved shirt), but I do not let their scorn deter me. We will see who gets the last laugh on this![ix]

Radiation: We now have a much greater appreciation of the risks of x-rays than we once did. I remember as a child going to a shoe store and having the salesperson x-ray my shoed feet to try to convince my parents that my old shoes were too small for me.[x] I also remember at the fall fair how a chiropractor x-rayed my spine to try to convince my parents that I needed spine manipulations. Decades ago, radiation was used by some physicians to treat acne, enlarged tonsils, ring worm, plantar warts, and unwanted facial hair. This no longer happens.

While x-rays and scans are essential for diagnosis and management of many health conditions, caution should also be exercised here. Any one x-ray or scan carries only a very small risk, but the risk is higher for young people than for older people.

Diagnostic x-rays are not the only source of radiation you may be exposed to. Flying exposes you to increased levels of cosmic radiation from the sun. The radioactive gas radon is also a hazard. It can seep into

---

viii   Although my wife Lesley prefers a buttery Chardonnay.

ix   Lesley says, "At least <u>we</u> enjoyed ourselves!"

x   The salesperson was right, and I have the bunions to prove it.

your basement from surrounding rock and soil, particularly if you live in an area with a lot of granite in the local rock, as is the case in Ottawa. At our home in Ottawa, I had our basement tested for radon,[xi] but the levels were fortunately very low. However, if we had found high levels, I could have easily managed this by simply improving the ventilation. If you live in an area where radon levels can be high, you might not want to have your child's bedroom or playroom in your basement until you have confirmed that radon levels are low.

Chronic infections: Human papilloma virus (HPV) infections (most acquired through sexual contact) can increase the risk of cancers of the cervix, head and neck, and some other types. Hepatitis B increases the risk of liver cancer. Vaccination against these viruses can reduce infection risk. You should consider discussing vaccination with your physician.

A chronic bacterial stomach infection with the organism helicobacter pylori increases the risk of stomach ulcers but also increases the risk of stomach cancer. This infection can generally be easily eradicated with the appropriate treatment. Any other process that causes chronic inflammation may also increase cancer risk.

Sleep deprivation and shift work: Does sleep deprivation increase the risk of cancer? When I was an internal medicine resident at Montreal's Royal Victoria Hospital in the mid-1970s, I was on call every second night, and would usually be up all night on those nights. I would then have to put in a full day's work after being up all night.[xii] Some weekends on call meant working 60 hours straight, with barely enough time to even sit down. However, as unpleasant as sleep deprivation can be, there is little evidence that it increases the risk of cancer. This is certainly good news for parents

---

[xi]   There was no need for this in Houston since there are very few basements there. If a Houston house had a basement, it would be at high risk of flooding in the next tropical storm to hit the city.

[xii]   Thankfully, such practices are no longer permitted. Now, if a trainee is on call for the night, they are given the next day off to recuperate. This is much better for trainees and is also probably safer for patients.

of young children everywhere![xiii] Shift work is associated with a slight increase in risk of some cancers, but this does not appear to be linked to sleep deprivation.

Stress: Stress is an intrinsic part of life. In fact, the stress response was important in our evolution as a species. It helped us recognize and deal with situations that were life-threatening and could have been species-ending. The problem is that in modern life, the stress response is no longer always helpful. So, we tend to see saber-toothed tigers where there are none and we worry too much. Does stress cause cancer? There is no clear answer to this question. Some studies suggest that it might, while others have found no association.

Heredity: Some people inherit factors that put them at increased risk of developing cancer. An example is a decreased ability to repair the mutations that may cause cancer. If several of your relatives have developed cancer at a very early age (in their twenties or younger), you might consider discussing with your physician the pros and cons of having yourself tested for one of these hereditary factors. Your risk may also be increased if several "first degree" relatives (i.e., your parent, sibling, or child) developed related types of cancer (particularly breast and ovarian cancer, or colon cancer). For example, when actor Angelina Jolie lost her mother, aunt, and grandmother to cancer, she had herself tested and found that she had inherited a mutation of a gene called *BRCA1*. Based on this information, she decided to undergo the preventative removal of both breasts.

Other stuff: There are also several other things that can increase your risk of cancer such as exposure to asbestos, air pollution, herbicides, pesticides, and workplace dust and fumes in factories and mines.

Don't worry, be prudent: Cancer is everywhere and so is the risk of contracting it. You may reduce your risk by changing your approach to life, but it is impossible to completely avoid risk. And as Lesley keeps on reminding me, it is important to be prudent, but it is equally important that we remember to still have a good time and live a full life.

[xiii] When my twins were born during my first-year fellowship at MD Anderson Cancer Center, I often joked that the only nights I got any sleep were my nights on call at the hospital.

Further Details and References

Below I will go into further detail and have added some supporting documentation.

The biological link between normal aging and cancer: Cancer develops due to mutations and related gene changes in your cells. These alterations occur as cells divide.

What are mutations, and why are they important? A mutation is a faulty copy of a gene. Each cell has approximately 19,000 to 20,000 genes, and every gene is the blueprint (or "code") for a different protein.[7] Each of the genes comprise an average of about 26,000 pairs of deoxyribonucleotide acid (DNA) "bases."[xiv] [8] Cellular machinery "prints" additional protein molecules. The arrangement or pattern of DNA bases in a gene tells it what the proteins should look like. When cells divide, they make a second copy of each of their genes to pass on to the new "daughter" cell. If a cell supplies a daughter cell with a mutated, faulty copy of the gene, then the daughter cell will pass this faulty gene to its daughter cells for the rest of one's life. The protein produced by this faulty gene may be a faulty protein. Many of these faulty proteins do not cause a problem, but some can contribute to the development of a cancer.

[xiv] DNA bases are the four molecules adenine, cytosine, guanine, and thymine. They are attached to each other end-to-end like a long string. The order in which they are added to the gene forms a code that cellular machinery then uses to "build" a specific protein. This cellular machinery creates proteins by stringing together molecular substances called amino acids, and if one wrong amino acid is added, it may markedly change the function and characteristics of the protein to which it is added. Three DNA bases in a row code for one amino acid. For example, if the next three DNA bases in the gene are guanine-guanine-thymine, the protein building machinery in the cell will bring in glycine as the next amino acid to be added as the protein is built. However, if there is a mutation such that the next three DNA bases are instead guanine-adenine-thymine, the protein building machinery will instead bring in the amino acid aspartate. If it is mutated to guanine-thymine-thymine, it will instead bring in a valine. If it is mutated to thymine-guanine-thymine, it will instead bring in cysteine. For example, a mutation in the *KRAS* gene can be a driving factor in several cancers. In some lung cancers, a single glycine in *KRAS* is replaced by a cysteine, while in some cancers of the pancreas and colon this single glycine may be replaced by an aspartate or valine. This simple change helps drive the cancer.

Cancer is uncontrolled cell growth. The more mutations you have in your body, the higher the probability that you have a mutation that can lead to the development of cancer.

You have approximately 37 trillion cells in your body[9] and about 100 billion of them divide each day to replace aging and damaged cells.[10] Every time that a cell in your body divides, an average of three new mutations occur in that cell.[11] That means that every day you are alive, you experience about 300 billion new mutations. Given these numbers, it is incredible that we do not all develop cancer at an early age.

Happily, many mutations are repaired by your body's defense systems,[11] or cause death of the cell (through a process called apoptosis) or permanent loss of its ability to divide (through a process called senescence).[12] However, at least some of the mutations will persist. The longer you have lived, the higher the number of persistent mutations you will have. Some of these persistent mutations might eventually lead to the development of cancer.

Oncogenes and uncontrolled cell division: Protein changes that enable cells to divide in an uncontrolled way drive development of cancer. Ordinarily, cell division is highly controlled. Several proteins in a cell are responsible for making a cell "decide" to divide to produce a daughter cell or else to continue as it is, without dividing. But these proteins can only cause cell division if specific events "activate" them.[xv] Factors in the cell's environment interact with molecules on the cell surface to tell the

---

[xv] Activation of these proteins is typically an outcome of a change in shape of the protein that results from another molecule binding to it in a highly specific way, like a key going into a lock. The molecule binding to the protein can, for example, be another protein whose shape and chemical structure permits it to fit precisely into a specific "pocket" in the target protein or it can be a phosphorus molecule. When this activating protein or phosphorus attaches to the target protein, it causes a change in shape of the target protein, and this in turn gives the target protein properties it did not have in its initial shape. For example, the change may enable it to add phosphorus to other nearby proteins, thereby changing their shape. This creates a chain reaction, with one protein changing another which in turn changes yet another in a chain reaction that reaches all the way down to the cell nucleus. In the cell nucleus, this chain reaction ultimately initiates the process of cell division. Normally, cell division cannot occur unless this chain reaction is started by very specific, controlled interactions between initiating molecules that may be at the cell surface.

cell that it is time to divide. For example, a molecule in the environment might tell the cell that there are sufficient nutrients available to support cell growth. Or another molecule might indicate tissue damage that requires cell division to heal a wound.

However, one of the proteins in the cell division chain may mutate in a way that it does not require anything to activate it. It is turned on continuously. In this scenario, cell division keeps on going nonstop without an outside stimulus to tell the cell that division is needed. These mutated genes that drive uncontrolled cell division are called oncogenes.

We currently know of several potential oncogenes, and others keep on being discovered. For example, lung adenocarcinoma may be driven by alterations of genes for one of several different growth factors. Examples of potential lung adenocarcinoma oncogenes include *EGFR, ALK, KRAS, BRAF, RET, MET, ROS1, HER2, RET and NTRK*. About half of malignant melanomas are driven by *BRAF* mutations. While specific oncogenes are known to be important in some malignancies, there are other malignancies for which specific driving oncogenes have not yet been discovered.

Tumor suppressor genes: Another important class of genes are "tumor suppressor genes." Unless the body tells them otherwise, the protein products of these genes ordinarily block cell division. If they are altered, cells can divide in an uncontrolled manner.

Tumor suppressor gene proteins also identify damaged DNA. If they detect damaged DNA, they have three options. They can force the cell to stop dividing until the DNA damage is repaired. Alternatively, they can induce apoptosis (leading to cell death) or senescence (resulting in a cell that is still alive but can no longer divide). These latter two processes prevent a cell with unrepaired mutations from producing daughter cells with mutations.

Tumor suppressor genes may be lost or inactivated through mutations. They can also be inactivated through what are called epigenetic changes.[13] These are changes brought about by modification of gene expression, rather than by alteration of the DNA genetic code. Epigenetic events are responsible for giving different cells their own unique functions, despite all the cells in your body having the same genes. This is the reason that cells in your liver (for example) perform functions that are different from those performed by the cells in your kidneys or brain.

Epigenetic processes can also turn off tumor suppressor genes that are not supposed to be turned off. These abnormal epigenetic changes can be permanent and can be passed from a cell to its daughter cell during cell division.

The need for several mutations to create a cancer cell: In most cases, several mutations and epigenetic changes are needed in a given cell for it to become malignant.[14] If this were not the case and it only took one or two alterations, then we would all develop cancer at a very early age.

Other impacts of aging: Normally, cells in an organism compete. As cells divide, younger "more fit" cells induce the death or removal of older, less fit cells.[15] A mutated cell may be less fit than other normal cells around it. Generally, normal cells tend to eliminate mutant cells. However, as you age, your cells become less fit in general, and they may no longer be able to eliminate mutant cells. This may permit mutant cells to survive, divide, and give rise to a malignancy.[15]

Other factors associated with cancer risk: The number one cause of cancer is mutations that occur during normal cell division as you age. However, several other factors can increase or decrease the risk of development of cancer. Broadly speaking, these factors may drive cell growth and increase the number of cell divisions in an individual (thereby increasing the opportunity for mutations to occur). Alternatively, they may increase the amount of DNA damage (thereby increasing the number of mutations). Finally, they may decrease the ability to repair mutations or increase the probability of survival of mutated cells.

As a rule, modest exposure to these factors carries minimal risk, but risk rises with increasing exposure. For most of these factors, trying to eliminate all exposure would be prohibitively expensive, would create problems in other areas, or would be an excessive infringement on personal freedoms. The important thing is for you to know about the issue and to manage your personal risk accordingly.

Tobacco: Tobacco smoke contains at least 60 carcinogens (i.e., chemicals that can cause cancer by increasing mutations).[16] In North America, smoking is the leading preventable cause of cancer. It plays a role in about 30% of all cases of cancer.[17] Smoking is responsible for about 85% of all lung cancers,[18] but the carcinogens from tobacco smoke can also pass down through the mouth and throat, esophagus and stomach, colon and rectum, contributing to development of cancers in these organs.[19,20] The carcinogens are also absorbed into the blood stream and passed out through the urine, contributing to the development of cancers of the kidney and bladder.[19] Smoking also increases the risk of acute myelogenous leukemia[21] and cancers of the pancreas,[22] liver,[23] cervix,[24] and breast.[25]

Compared to cigarette smoking, smoking a pipe or cigar involves less inhalation of smoke deep into the lungs. Pipe and cigar tobaccos are more alkaline than cigarette tobacco.[26] Consequently, nicotine from pipe and cigar smoking is easily absorbed into the blood stream right from the wall of the mouth. On the other hand, cigarette smoke must be inhaled deep into the lungs for much of the nicotine to be absorbed.[xvi] Pipe and cigar smoking is associated with development of cancers of the lung, head and neck, esophagus, pancreas, and liver.[27,28]

While exposure to someone else's tobacco smoke is not nearly as bad as personally being a smoker, "secondhand smoke" (also called "passive smoking" or "environmental tobacco exposure") may nevertheless increase your risk of cancer of the lung,[29] bladder,[30] colon,[31] kidney,[32] cervix,[32] or breast.[25]

No matter how much you have smoked, there are benefits to quitting. If you are a heavy smoker and you quit smoking, your risk of developing lung cancer or another smoking-associated cancer will gradually decrease over a period of years,[22,33] although your risk would remain higher than the cancer risk of a person who had never smoked.[33]

Former smokers may develop a smoking-associated cancer decades after they quit.[33] This happens since several mutations are needed within one cell to make it malignant. For example, you may have acquired all except one required mutation while you were still smoking. Then, many years later, you might develop that one final required mutation as a simple result of aging. The bad news is that you would now have developed that last mutation required for a cancer to develop. The good news is that if you had not stopped smoking when you did, this last required mutation and the onset of your cancer might have happened many years earlier.

[xvi] Some tobacco companies changed the pH (acidity/alkalinity) of cigarette tobacco to further decrease nicotine absorption so that the unwitting smoker had to smoke more cigarettes in a day to feed their nicotine receptors.

Carcinogens are not the only way that tobacco smoke contributes to development of lung cancer. Smoking damages lung tissue, leading to the development of chronic obstructive pulmonary disease or COPD (i.e., emphysema or chronic bronchitis). COPD makes the lungs more susceptible to infection, and chronic infection leads to chronic inflammation. Chronic inflammation by itself contributes to the development of cancer through several different mechanisms.[34,35] Smokers who develop COPD are more likely to develop lung cancer than are smokers without COPD.[36]

I strongly advise those who have already developed cancer to quit smoking. For a variety of malignancies, survival is better for cancer patients who quit smoking than if they continue to smoke.[37-41] For patients with incurable metastatic cancers, toxicity of chemotherapy may be increased and effectiveness is decreased in patients who continue to smoke.[39,42] Continuing to smoke also increases the risk of developing a second smoking-associated cancer in those who are cured of their first cancer.[39,43]

Electronic cigarettes: What about use of electronic cigarette or vaping? While the jury is still out on some specifics, vaping cannot be regarded as being safe. Many experts feel that it is probably safer than cigarette smoking since it avoids most of the carcinogens present in tobacco smoke. On the other hand, e-cigarettes introduce other chemicals, many of which could increase the risk of lung damage or cancer.[44,45] In any case, nicotine itself can drive tumor cell growth in the laboratory, so any source of nicotine might carry some risk.[45] In addition, vaping can lead to nicotine addiction. This can lead to initiation of cigarette smoking in individuals who had previously been non-smokers.[46]

Cannabis: Epidemiological research suggests that regular smoking of cannabis could increase your risk of lung cancer. One study suggests that your risk from one joint per day is roughly equivalent to that of smoking one pack of cigarettes per day.[47] However, at this point, the overall evidence around whether cannabis use increases the risk of cancer is inconclusive.[48]

Ingestion by mouth of cannabis derivatives (cannabinoids) such as oils might avoid the lung toxicity caused by smoking cannabis. In the laboratory some cannabinoids have anticancer activities or can make chemotherapy more effective. However, they also have actions that might promote cancer growth. While cannabinoids may help with the pain, insomnia and decreased appetite caused by your cancer, it is not known at this time whether they might be beneficial or harmful in controlling cancer.[48,49] If you do not have cancer, my advice is to avoid at least smoking cannabis. For my patients with cancer, I tell them that I do not personally prescribe cannabis products, but do not have an objection to patients using them if they find that they help their pain, nausea, or other symptoms.

Sun exposure: Skin damage from sun exposure is by far the major cause of skin cancers.[50] Tanning lamps are also hazardous. Severe sunburn at a young age is particularly important for development of malignant melanoma.[51] Regular use by adults of sunscreens reduces the risk of developing squamous cell carcinoma of the skin by 40%. Sunscreen use by adults has a more modest impact on probability of development of basal cell carcinomas of the skin.[52] It has no obvious impact on the risk of development of malignant melanomas.[53]

Sunscreen use by adults has relatively little impact on skin cancer risk since the important skin-damaging sun exposure occurs during childhood.[54] Furthermore, people using sunscreens often stay in the sun longer than the sunscreen is effective and longer than people not using sunscreen. This prolonged exposure can reduce or eliminate the beneficial impact of sunscreen use.[55]

In children, sunscreen use reduces the risk of moles called melanocytic nevi that increase the risk of melanoma later in life.[56] Sunscreen use during childhood reduces the risk of later development of melanoma by about 40% (although this protection is reduced for children who spend longer amounts of time in the sun).[57] It is estimated that childhood and adolescence use of sunscreens would reduce the life-time risk of squamous and basal cell skin cancers by almost 80%.[58] Children and adults should use sunscreen and wear protective clothing, while limiting time spent in the sun.

High body weight: Diet and physical activity are also important. Up to 20% of all cancers may be related to obesity,[59] making it the second leading cause (after smoking) of preventable cancers. One measure of obesity, the Body Mass Index or BMI is the weight in kilograms divided by the square of the height in meters. For example, if your weight is 100 kg and your height is 1.80 meters, then your BMI would be 100/ (1.80 x 1.80) = 30.9.[xvii]

[xvii] If using weight in pounds and height in inches, instead of kilograms and meters, the formula for BMI is:

730 x Weight (pounds)/(height [inches] x height [inches])

15

As BMI increases above 25, your risk of dying from cancer increases. The higher the BMI, the higher the risk.[60] For both sexes, excess body weight is associated with an increased risk of cancers of the esophagus, thyroid, colon, kidney, leukemia, multiple myeloma, and lymphoma. Overweight males also have an increased risk of cancer of the rectum and malignant melanoma. Obesity is not clearly linked to risk of prostate cancer, but is associated with development of a particularly aggressive form of prostate cancer.[61] Overweight females have an increased risk of cancers of the endometrium, gallbladder, breast, and pancreas.[62,63]

Consuming too many calories can stimulate cell growth and division, in part by increasing insulin production.[64,65] Remember, more cell divisions mean more mutations. High fat diets also increase inflammation which may play a role in developing several types of cancer.[66]

Abdominal obesity: Increased fat within the abdomen (called visceral obesity, central obesity, or abdominal obesity) is characterized by an increased waist size and in an increased ratio of waist to hip size. Abdominal obesity is associated with an increased risk of death from cancer in both males and females.[67] It alters the way in which your fat cells respond to insulin, causing resistance to insulin.[68] This results in both higher insulin levels and higher total body inflammation.[69] High insulin levels, high inflammation, and foods that increase inflammation[70] can all increase your risk of cancer.

The high insulin levels and greater inflammation associated with visceral obesity are part of the "metabolic syndrome." The metabolic syndrome includes not just high insulin levels, diabetes and inflammation, but also high blood pressure, high serum triglycerides, deficiency of heart-protective high density lipoproteins, and increased risk of blood clots and liver disease.[71]

Physical activity and weight loss: Regular exercise reduces visceral obesity.[72,73] Physical activity reduces your risk of developing acute leukemia, multiple myeloma, and cancers of the colon, rectum, breast, endometrium, ovary, prostate, bladder, kidney, stomach, esophagus, liver, head and neck, and lung.[74-77] The beneficial impact of physical activity is due in part to it reducing excessive sex hormones, metabolic hormones, and inflammation.[78]

Cancer risk may also be reduced by weight loss. For example, cancer risk is decreased by stomach bypass (bariatric surgery) in morbidly obese patients.[79,80] The diabetic drug metformin is associated with reduced cancer risk in adults with diabetes.[81]

Repeated episodes of weight loss followed by regaining the lost weight is called "weight cycling." Some early studies suggested that weight cycling was associated with an increased risk of death from cancer and other causes. However, more recent studies have detected no major negative or positive effects of weight cycling.[82-85]

Generalized obesity, hormones, and breast cancer risk: Fat beneath the skin (subcutaneous fat) is associated with the risk of developing breast cancer. The risk from subcutaneous fat does not appear to be as great as the risk associated with abdominal fat.[86] However, the location of the fat could impact the type of breast cancer. Women with increased subcutaneous fat are more likely to develop breast cancers that are driven by the hormone estrogen (called estrogen receptor positive breast cancer). Women with predominantly abdominal obesity are more likely to develop breast cancers that are not driven by estrogen (called estrogen receptor negative breast cancer).[87]

The increased risk of estrogen receptor positive breast cancer with subcutaneous obesity may be because high subcutaneous fat content results in higher levels of estrogen and related hormones.[88] Just as high calorie levels can increase cancer risk by increasing the number of mutations, high estrogen levels could increase mutations by increasing the rate of cell division. There are also several other mechanisms by which estrogen may increase the risk of breast cancer.[89]

Diet: What you eat is important. Canada's current food guide[90] is a good place to start. Consuming more fruits and vegetables is associated with a reduction in risk for cancers in general,[91] and with some individual types of cancer, particularly of the colon[92,93] and breast.[94] As for vegetables, green-yellow vegetables and cruciferous vegetables such as cabbage, broccoli, cauliflower, and Brussel sprouts are strongly associated with reduced cancer risk.[91]

Diets rich in fruits and vegetables may be protective since there are protective substances in these foods. Alternatively, these diets may be protective simply because they are associated with lower calorie intake and less obesity. People who eat a lot of fruits and vegetables might also have more healthy lifestyles in general (e.g., more exercise, less smoking). Or all three possibilities could be at work.

Questions like this are often best answered by randomized controlled trials. In these trials, the equivalent of a flip of a coin decides whether an individual participant does or does not receive a therapy. This random selection process helps ensure that the two groups in the study are balanced for other factors (such as total calorie intake, exercise, smoking) that might affect the risk of developing cancer. For example, antioxidants in fresh fruits and vegetables were thought to be responsible for a reduced risk of cancer. However, randomized controlled trials have been done in which one group received antioxidants and the other group received a placebo.[xviii] These studies have consistently failed to show that antioxidants protect from cancer. In fact, the risk of developing cancer was increased by the antioxidant supplementation in some of these studies. Negative studies have included beta carotene,[95,96] selenium,[97,98] and vitamins A,[96] C,[99] D,[100,101] and E.[95,98,99] It is unknown why these studies were negative. One hypothesis is that the antioxidants might help cancer cells survive instead of preventing them from forming.

Researchers have not yet been able to identify the factors in fruits and vegetables that reduce cancer risk. Several elements in fruits and vegetables might play a role. For example, many of them are rich in compounds called flavonoids that are associated with reduced inflammation.[102]

Many other dietary components are associated with a reduced risk of cancer. High consumption of whole grains (e.g., whole wheat) is associated with reduction of risk of cancers in general[103] and with reduced risk of cancers of the colon[92,93,104] and breast[104] in particular. On the other hand, consumption of refined grains (e.g., white bread) and white rice is not associated with reduced cancer risk.[103] High dietary fiber consumption may be associated with reduced risk of cancers of the colon, breast,[104,105] ovary,[106] esophagus, stomach, and pancreas.[105] High consumption of nuts is

---

xviii These are "double-blind" studies, with a code number assigned to each patient. Neither the patient nor the investigator knows which patients received the antioxidant and which received placebo until the assessment for new cancers is completed. The code is broken at the end of the trial so that the number of cancers can be calculated for each of the two groups.

also associated with a reduced risk of dying from cancer.[107] Dairy products in general are associated with a reduced risk of colon cancer,[92,93] while skim milk (but not whole milk), yogurt, and soy foods are associated with reduced risk of breast cancer. [108]

Meat consumption: "Red meat" and "processed meat" are associated with an increased risk of several types of cancer. Red meat includes beef, veal, pork, and lamb while processed meat consists of meat that has been salted, cured, fermented, or smoked.[109] When red meat is cooked, a variety of carcinogens such as chemicals called heterocyclic aromatic amines and polycyclic aromatic hydrocarbons are produced, particularly when using high temperatures by frying, grilling, or barbecuing.[109] Processed meats have increased levels of the polycyclic hydrocarbons found in cooked red meat, but also have other carcinogens called nitrites and nitrosamines.[109] The processing helps preserve meat longer and reduces the risk of food poisoning, but it comes at the price of a higher cancer risk.

High red meat consumption is associated with increased risk of cancers of the colon,[92,110,111] lung,[110] esophagus,[110] stomach,[110] and breast.[108] The strongest association may be between beef consumption and risk of colon cancer.[111] High lamb consumption also appears to increase the risk of colon cancer.[111] On the other hand, there is no strong association between high pork consumption and cancer.[110,111]

Poultry[110-112] and fish[93,108,112-116] consumption showed either no association with an increased risk, or were associated with a reduced risk of development for some cancers.[93,110,112-116]

High consumption of processed meats has been associated with an increase in several cancers, including cancers of the colon,[92,110] breast,[94,108] esophagus,[110] stomach,[110,117] and bladder.[110] Preserved or pickled fish and vegetables are also associated with an increased risk of stomach cancer.[117,118] High consumption of saturated fats is associated with an increased risk of breast cancer.[94]

The Mediterranean diet: The Mediterranean diet is associated with a reduced risk of cancer.[119] This diet, common in the Mediterranean region, is characterized by high consumption of vegetables, fruits, nuts, legumes, unprocessed cereals, and olive oil. Also, moderate amounts of fish and wine and very little meat and dairy products (apart from some cheeses) are consumed as part of this diet.[120]

Alcohol: Individuals who drink at least some alcohol live somewhat longer, on average, than people who drink no alcohol at all. Overall, lowest death rates occur in men and women who drink modestly (i.e., at least some alcohol, but no more than two drinks per day for men and no more than one drink per day for women).[121]

However, drinking is associated in different ways with different diseases. Alcohol is associated with reduced death rates from heart disease and strokes, but even relatively light drinking (in some cases even an average of less than one drink per day) may increase the risk of dying from cancers of the head and neck region, esophagus, liver, and breast.[xix] [121-124] Alcohol consumption has also been associated with an increased risk of cancers of the colon,[121,125,126] stomach,[127,128] lung,[129] prostate,[130,131] endometrium,[132] skin,[133-135] and brain.[136] Beer and liquor differ somewhat from each other in their association with specific cancers. To muddy the water a bit further, alcohol consumption may be associated with reduced risk of lymphomas[137] and cancers of the kidney[138] and bladder.[139]

Wine consumption is less strongly associated with cancer risk than drinking beer and liquor. But wine is nevertheless associated with an increased risk of cancers of the breast[140] and head and neck.[123,124] White— but not red—wine consumption is also associated with an increased risk of prostate cancer,[131] malignant melanoma,[134] and squamous[135] and basal cell carcinomas of the skin.[133] Red wine may actually reduce the risk of prostate cancer.[131,141]

The bottom line is that no matter the type of alcohol, cancer risk is increased. However, the risk appears to be lower with red wine than with other alcohol choices. Moderation reduces risk, and despite the cancer-related negatives, alcohol consumption is associated with a reduced probability of death from heart disease and strokes. Alcohol consumption is similar in the US and Canada (slightly higher in Canada than in the US in 2010, but slightly lower in 2016),[142] so this does not explain differences between the two countries in cancer death rates.

Factors that impact hormone-related cancers: In a previous section, I discussed how obesity might drive hormone-related cancers by increasing

---

[xix] Of interest, high alcohol consumption is associated with an increased risk of death from "injuries and other causes" in men, but with a reduced risk in women!

the body's production of factors like estrogen. Other factors related to hormones may also impact cancer risk. Women who have their first menstrual period at an early age (early menarche) or have a later than usual age for menopause onset have a higher number of years of estrogen exposure. This is associated with increased risk of breast cancer.[143] Early menarche is also associated with an increased risk of cancers of the ovary[144] and endometrium.[145] Early age of first pregnancy and having multiple pregnancies reduces the long-term risk of breast cancer due to reduced estrogen production during pregnancy.[143] Pregnancy also reduces the risk of cancers of the ovary and endometrium.[146]

Hormone replacement therapy after menopause: On average, the benefits of hormone replacement therapy greatly outweigh the risks for post-menopausal women. There may be an increased risk of cancer but a reduced risk of heart disease and osteoporosis.[89,147] If estrogen is taken alone, it increases the risk of endometrial cancer.[148] Adding progestin to the estrogen reduces the risk of endometrial cancer[148] but increases the risk of breast cancer if the progestin is taken for a prolonged period of time.[147,149] Consequently, women who have not had a hysterectomy typically receive hormone replacement therapy consisting of estrogen and progestin combined, but typically only receive the replacement therapy for three to five years to reduce the risk that it might cause breast cancer.

On the other hand, women who have had a hysterectomy have no risk of endometrial cancer and hence do not need progestin with the estrogen. In these women, estrogen replacement alone either has minimal impact on breast cancer risk[147] or can actually reduce breast cancer risk.[149] The estrogen is typically stopped after about ten years since the risks may start to outweigh the benefits by that point.[147]

Hormone replacement therapy is also associated with an increased risk of ovarian cancer,[150] but with a reduced risk of developing colon cancer.[151]

Birth control pills: Use of birth control pills is associated with a modest increase in the risk of breast cancer and cancer of the cervix, but with a reduced risk of development of cancers of the endometrium, ovary, and colon.[152]

Impact of miscellaneous other therapies on cancer risk: Drugs that were intended for other purposes may favorably impact the risk of developing cancer. For example, some studies have reported a reduced risk of colon cancer in individuals who frequently take aspirin to reduce the risk of heart attacks or for other reasons.[153,154] However, randomized controlled trials of low dose aspirin have yielded conflicting results, so I cannot at this time recommend that you take low dose aspirin to try to reduce your cancer risk.[155,156]

Other medications that have been associated with a reduction in cancer risk are metformin in the treatment of diabetes,[157] statins in the treatment of high cholesterol,[158] finasteride for the treatment of prostate enlargement,[159] and bisphosphonates for osteoporosis.[160] For each of these examples, it is not known if the medication actually reduces the risk of developing cancer or whether the association is due to some unrelated factor.

Drugs suppressing the immune system: People who have undergone organ transplants and are on chronic immunosuppressive medications to prevent rejection of their new organ are at increased risk of development of skin cancers, lymphomas, virus-related cancers,[xx] and other malignancies.[161] There is suggestive[162] but inconclusive[163] evidence that cancer risk is increased by effective new immunosuppressive monoclonal antibodies and related agents that target tumor necrosis factor (TNF), etc., for the treatment of psoriasis, Crohn's disease, and rheumatoid arthritis. If there is a risk associated with these monoclonals, it appears to be low.[164,165] Any apparent association may be related to concurrent use of other immunosuppressive agents like thiopurine.[166] These monoclonal antibodies do not appear to increase the risk of the recurrence of preexisting cancers.[167]

I further discuss the overall impact of the immune system on risk of cancer development in Chapter 7.

Radiation exposure: Canada has a lot of granite, and granite can contain uranium. Radon is generated by the decay of naturally occurring uranium in soil and bedrock and can permeate up into the basement of a home. Radon inhalation is estimated to cause about 2% of all cancers in North America. It increases risk of lung and kidney cancers, melanoma, leukemia, and some childhood cancers.[168] It is easy to test for radon. If high levels are found, they can be reduced inexpensively by improving the basement's ventilation.[168]

[xx] Virus-related cancers include those caused by the human papilloma virus or the hepatitis B virus.

Those beautiful granite counter tops which we all love can also be impregnated with uranium which releases radiation, although the radiation dose is generally not high enough to carry much risk.[169] Other sources of low dose radiation exposure include gypsum board and concrete.[169]

The benefits of diagnostic x-rays and scans substantially outweigh their risks, but they are not completely risk-free. These procedures account for an estimated 14% of an average individual's total exposure to radiation from all sources, and cause about 0.6% of all cancers.[170] Scans carry a higher risk for the young than for the old. Risk may also vary with gender and type of procedure. For example, a female child who undergoes a CAT scan of the upper spine would have an estimated 0.7% risk of eventually developing a cancer-related to the procedure. Conversely, a 70 year-old would only have a 0.01% risk of later developing a cancer-related to a CAT scan of the head.[171] Your risk of later developing a scan-related malignancy would also be increased if you or your family members had inherited a defect in the cellular machinery that repairs DNA damage.[172,173]

Atomic bomb survivors have had an increased risk of a wide range of cancers, even if radiation exposure was relatively low.[172] People exposed to radiation from the Chernobyl accident had an increased risk of thyroid cancer and leukemia.[174]

The earth is constantly bombarded by cosmic rays from the sun and galaxies.[175] Your cells can repair a lot of the damage from this cosmic radiation. In the laboratory, exposing cells to very low dose radiation improves their ability to repair gene damage from later higher doses.[176]

The atmosphere filters out quite a bit of the cosmic radiation. During air travel, there is less atmosphere to protect you, so you are exposed to a higher radiation dose. The earth's geomagnetic fields also help protect us from cosmic rays, but the amount of protection decreases as you get closer to the earth's north and south poles. Hence, people who fly a lot have increased exposure to these cosmic rays, particularly when in flights at higher altitudes and at higher northern latitudes.[175] There is an elevated incidence of breast cancer and melanoma in flight crews, but no study has yet conclusively determined whether this is related to radiation exposure or to unrelated lifestyle factors and reproductive history.[177]

Radiotherapy can be of substantial benefit in the treatment of cancers, but a small proportion of patients will go on to develop a new malignancy due to the radiotherapy.[178,179] For example, an estimated one out of every 125 lung cancers are caused by radiation exposure years earlier as a result of radiotherapy for breast cancer or another malignancy.

As I pointed out in the earlier summary section of this chapter, the risks of radiation exposure are now well understood. In the past, radiation was used to treat benign conditions like acne, hirsutism (i.e., excess hair growth), tinea capitis (i.e., ring worm of the scalp), warts, and enlarged tonsils. Children and young adults who received radiation for these conditions are at increased risk of developing malignancies later in life.[180-182]

Electromagnetic fields: Electromagnetic fields (e.g., related to microwave ovens, Wi-Fi, radio and television signals, and electrical power lines) have not been shown to be associated with cancer risk. The one caveat is a possible increase in risk of leukemia in children who have had high exposure to electromagnetic fields from high tension power lines.[183] Cell phone use has not consistently been associated with risk of any type of cancer. Some studies, but not others, have suggested a weak association of high cell phone use with risk of developing a brain tumor.[184]

Sleep deprivation and shift work: Shift work is associated with a slightly increased risk of prostate cancer in men[185] and breast cancer in women.[186] However, any impact of shift work may not be mediated by sleep deprivation. Short sleep duration (generally less than seven hours per night) does not appear to be associated with an increased risk of cancer, but long sleep duration (generally more than nine hours per night) is.[187] Among cancer survivors, longer sleep duration is actually associated with a higher risk of death from cancer.[188] At this point, we don't know if there is any causative association or whether the association is indirect.

Stress: While some assessments have suggested a weak association between stressful life events and the development of breast cancer,[189] others have shown no major association.[190-192] Similarly, some assessments suggest an association between workplace stress and development of lung, colorectal, and other cancers,[193] while others do not.[190] Again, it is not possible to say whether there is any causal association between stress and cancer.

Inflammatory diseases: As noted earlier, chronic inflammation related to emphysema or abdominal obesity increases the risk of cancer. Part of the negative impact of both alcohol and tobacco is related to the chronic inflammation that they cause. There are also several illnesses causing chronic inflammation that are associated with an increased risk of cancer. For example, chronic inflammation of the bowel from ulcerative colitis or Crohn's disease carries an increased risk of colon cancer.[194] Chronic helicobacter pylori infections may cause stomach cancer.[195] The risk of lung cancer is increased by chronic inflammation from COPD,[36] asbestos exposure,[196] mining and other occupations associated with high dust exposure,[197] or other conditions causing lung scarring[198] or fibrosis.[199] The risk of cancers in the mouth is increased by chronic inflammation from tooth loss and periodontal disease.[200] Liver cancers may be caused by inflammation from obesity-associated fatty liver, alcohol-induced cirrhosis, chronic viral hepatitis B or C infections, or toxins from food molds.[201] Psoriasis is associated with increased risk of lymphomas and cancers of the skin and lung.[202] The autoimmune disease scleroderma is associated with an increased risk of breast cancer, lung cancer, and hematologic malignancies.[203]

Other infections associated with cancer: Overall, about 15% of cancers are caused by viral infections.[204] In addition to an association of liver cancers with chronic hepatitis infections, some other viruses are also known to be associated with cancer. For example, the human papilloma virus (HPV) is particularly likely to be transmitted during sexual intercourse or during oral-genital contact. HPV causes mutations in cells. These mutations can lead to development of cancers of the cervix, vagina, vulva, anus, head and neck, and penis.[204] Currently available vaccines against hepatitis B and HPV now hold the promise of reducing cancers related to them. Most people should be vaccinated against hepatitis B. In both Canada and the USA, HPV vaccine is recommended for both males and females ages 9 to 26, although there are some minor differences between the countries in recommendation details.

A small proportion of those who have been exposed to the Epstein-Barr virus (which causes infectious mononucleosis) will eventually develop virus-related Hodgkin's disease, other lymphomas or nasopharyngeal carcinomas (a malignancy that starts high up at the top of the nasal cavity).[205] Virus-related nasopharyngeal carcinoma is more common in the Asian population and Inuit than in other populations.[205]

AIDS patients have an increased risk of cancers caused by the Epstein-Barr and human papilloma viruses, and may also develop a cancer called Kaposi's sarcoma caused by a herpes infection.[206]

Heredity and cancer: Unfortunately, one may also inherit an increased susceptibility to cancer. There are, however, other explanations for a high occurrence of cancer in a family. These include exposure to similar cancer-causing factors, such as smoking, and other environmental influences. Also, since cancer is so common anyway, it may look like a particular family has an unusually high rate of cancer, but this is often due to chance alone, rather than being due to heredity.

About 5% of all cancers are strongly linked to heredity. There is a good chance that heredity is a factor if several members of a family develop related types of cancer at an early age (childhood, teens or twenties).[207] Heredity may in some situations also play a role in development of cancers later in life.[208]

One example of a "hereditary cancer syndrome" is an inherited mutation in the RB gene. The RB mutation greatly increases the risk of young children developing retinoblastoma, a cancer of the eye.[209] If there is a strong family history of colon cancer developing at an early age, this increases the probability of an inherited predisposition to colon cancer such as Gardner's Syndrome (also known as "familial polyposis") or Lynch syndrome.[210] If multiple family members develop breast or ovarian cancer (particularly at an early age), this increases the probability of an inherited mutation in the BRCA1 or BRCA2 genes that are important in DNA repair.[211] Inheritance of a BRCA mutation also increases the risk of melanomas and of cancers of the fallopian tube, prostate, stomach, and pancreas.[211] There are also a number of other inherited cancer syndromes such as Li-Fraumeni syndrome, ataxia telangiectasia, Cowden syndrome, Von Hippel-Lindau Disease and Multiple endocrine neoplasia.[207,212]

Cancer-associated mutations like these are referred to as dichotomous factors. A dichotomous factor is one that is either present (thereby giving an individual a substantially increased risk of developing cancer, compared to those without the factor) or absent. If your family history strongly suggests an inherited predisposition to cancer, it might be reasonable for you to consider undergoing genetic testing. On the other hand, genetic testing would unlikely be helpful if you do not have a strong family history, since there would be only a very small possibility that a relevant mutation would be found.

In addition, if tests are done in low-risk individuals, something called Bayes' theorem comes into play.[213] There is no test that is perfect, and every time a test is done, it carries a risk of a "false positive"- i.e., the test will be reported as positive due to a lab error rather than the abnormality really being present. The risk of a false positive remains the same across most patients, while the risk of a truly abnormal test increases in patients whose family history is suggestive. Let's say the test has a 1% probability of a false positive. For this example, let's also say that the probability of a patient with a negative family history having the mutation is 1 in 10,000. In contrast, the probability of a patient with a positive family history having the mutation is 1 in 10. This would mean that in low-risk patients, a positive test would be a false positive 99% of the time,[xxi] while in high-risk patients, a positive test would be a false positive only 10% of the time.[xxii]

There are real problems with doing screening tests in patients who are highly unlikely to have the condition for which they are being screened. False positive results mean high anxiety for the patient as well as expensive and potentially risky additional assessments. However, finding a mutation in a high-risk individual can lead to specific actions to reduce cancer risk. For example, in patients with BRCA mutations, prophylactic surgical removal of the breasts and ovaries can reduce the risk of developing breast or ovarian cancer.[214,215] This is not something you would want to consider if you had a false positive BRCA test.

There are also "categorical" variable risk factors. These are different from dichotomous "present-vs.-absent" factors. A categorical variable is one in which there can be a number of different possible outcomes (e.g., "a" vs. "b" vs. "c" vs. "d"). For example, most genes can have what are called "single nucleotide polymorphisms" or SNPs, involving a single nucleotide in the gene. Some members of the population will have one SNP, and others will have a different SNP. For many genes, different SNPs have little impact on cancer risk, but others do. For example, some SNPs associated with control of cell growth[216-218] or with less efficient elimination

---

[xxi] Since 1 in 10,000 low risk patients would have the mutation, but 100 in 10,000 would have a false positive.

[xxii] Since 1,000 out of every 10,000 high-risk patients would have the mutation, while 100 in 10,000 would have a false positive.

of cells with damaged DNA[219] may be associated with increased risk of development of lung cancer. However, any increase in risk may not be enough to classify the individual as being "high risk," and hence there are no currently recommended steps around screening or other interventions related to these SNPs.

Some popular genetic ancestry tests may tell you that you are at increased risk of a variety of conditions. In general, I feel that this information is not helpful. Being "low risk" does not indicate that it is safe for you to lead an unhealthy lifestyle. Conversely, we currently don't know many specific things you can do to substantially reduce "high risk." This might change as we learn more about the impact of specific SNPs. However, at this time, if you receive a solicitation to undergo one of these commercial ancestry tests, it will have a low probability of helping your health much.[xxiii]

Having said that, some SNPs can have huge implications for lung cancer. For example, some SNPs affect the body's ability to get rid of nicotine absorbed from tobacco smoke.[220] Nicotine is toxic enough that it is used as an insecticide. The amount present in several cigarette butts is enough to seriously harm a young child and larger amounts can potentially harm adults.[xxiv] Hence, being able to break down nicotine rapidly can help protect from its toxic effects. However, smokers addicted to nicotine need a steady supply of it to silence intense nicotine withdrawal symptoms. Consequently, smokers with SNPs that increase the rapid breakdown of nicotine tend to smoke much more than smokers with other SNPs.[220] In the process of smoking more cigarettes to lessen nicotine withdrawal symptoms, they inhale larger amounts of other cancer-causing tobacco chemicals. The ability to rapidly break down nicotine thereby increases their risk of developing lung cancer.

[xxiii] They can, however, stimulate interesting dinner conversations. My wife and I both underwent testing. We learned that the average Caucasian has DNA that is 2.1% Neanderthal. My wife and I both had only 1.3% Neanderthal DNA. We tell people that any problems we run into are due to our Neanderthal deficiency.
[xxiv] I grew up on a farm. When I was young, a neighbor became seriously ill when he used his bare hands to handle a nicotine-based insecticide.

While this information is interesting, to gauge lung cancer risk, you need only to look at the amount a person smokes.

The presence of "continuously variable" factors can also impact risk. A continuously variable factor is one that can vary between people and can be anywhere from none to a small amount to a large amount. To understand the impact of continuously variable factors on cancer risk, it is helpful to place patients in percentiles. If 99 out of every 100 other people had more of the substance than you have, then you would be at the 1st percentile. If 99 out of every 100 people had less of the factor than you did, then you would be at the 99th percentile. If 49% of people had less of it than you did, then you would be at the 49th percentile, while if 51% of people had less of it than you did, then you would be at the 51st percentile.

Continuously variable factors can impact your risk of developing cancer. However, as with categorical factors, it can be difficult to determine how much this affects your risk and what prevention approaches you should adopt. In studies assessing the impact of continuously variable factors, it is common to use "cut points." For example, let's assume the 50th percentile as the cut point and we find that patients above the cut point (in the 51st to 100th percentiles) are at higher risk of developing cancer than are patients below the cut point (in the 1st to 50th percentiles). This information is statistically interesting but is of little practical help. Why? In our example, it assumes that a patient at the 49th percentile is different from one at the 51st percentile- and the same as an individual at the 1st percentile. This is biologically irrational. On the other hand, comparing people in the 75th and 25th percentiles may be quite useful.

In medicine, there are several examples of continuously variable factors impacting medical decisions (e.g., blood pressure or serum cholesterol above or below a certain value as a basis to prescribe blood pressure or cholesterol medications) but it is not at all clear that this is the best way to do things. Knowing what your individual levels are for a continuously variable factor can help you and your physician make decisions about your care, but we need caution in interpreting cut points.

Other factors: There are also several other factors associated with cancer risk, including air pollution,[221] asbestos exposure,[222,223] cooking oil fumes,[224] use of solid fuels such as coal,[224,225] lung infections such as tuberculosis,[224] and exposure to herbicides/pesticides,[226] some chemicals encountered in factories,[224] and some lung irritants inhaled while mining,[227] welding,[228] etc. Space does not permit a full discussion of all of these.

Conclusions: The evidence that I have presented underscores the opening paragraph of this chapter. The first reason that cancer still sucks is that so many people develop it, and that so many different things in our day-to-day existence can increase our risk. As I stressed earlier, there is no way to completely eliminate the risk of cancer, but we can improve our odds of developing it by exercising moderation in all things associated with it, while continuing to enjoy our lives.

— 2 —

Lurking in the Depths: Cancer Screening

The second reason that cancer still sucks is that it can spread and become incurable long before it is ever detected, as it lurks hidden in the depths of the body.

Short Primer

Many patients have said to me, "I can't possibly have an incurable cancer! I feel so well!" I also frequently hear, "Why did my doctor initially treat me for pneumonia? He should have known that this was lung cancer!" A third frequent comment is, "They should have seen it on the x-ray I had two years ago! How could they have missed it?" And many patients ask, "How long have I had this?"

All good questions. It's frustrating that cancers often produce no symptoms until they are far advanced and incurable. This is especially true of lung, pancreas, ovary, stomach, colon, and brain cancers. We would be much better off if they made people sick when they were small and drew attention to themselves, but they usually don't do that. There are some exceptions. A person may see a skin cancer or find a breast lump when they are still small, and some cancers in the kidney, bladder, colon, rectum, and uterus may draw attention to themselves by producing visible bleeding.

When cancers do start producing symptoms, they tend to look just like something much less ominous. For example, when my lung cancer patients begin to cough, many of them are initially treated with an antibiotic for suspected pneumonia. Or their back pain from bone metastases may initially be treated as a disc problem. Too often, it is only when things keep getting worse that further tests are done that reveal a lung cancer.

Most patients seen by a family doctor for a new or worsening cough will have a respiratory infection[xxv] (or possibly an allergy). Most times that a person begins to feel unwell, it is something that will be reversible. There may be no way that a physician or patient can detect that "This time is different"—until symptoms keep worsening over time, despite the treatments that usually work.

Simply ordering more frequent scans isn't the solution. In Chapter 1, I discussed the fact that radiation from x-rays and scans carries a very small but real risk of causing a later cancer. We also discussed Bayes' theorem: if you do a test in someone at low risk of a problem, then there will be a high probability that an abnormal test will be a "false positive." False positives often lead to a lot more testing, anxiety, expense, and potential complications from further testing procedures. For both reasons, it is important to limit the testing to those who can be identified as being at higher risk based on the nature or evolution of their symptoms.

To complicate things even more, some small cancers may disappear on their own if a driving factor is withdrawn. For example, I mentioned in Chapter 1 that estrogen replacement therapy may increase the risk of breast cancer by driving cell growth, but the risk of breast cancer drops rapidly after estrogen replacement is discontinued. Some studies suggest that small breast cancers that develop during estrogen replacement therapy may disappear once the estrogen replacement therapy is stopped.[229] Hence, no intervention is required. Increasing screening to detect all these small cancers would increase radiation exposure, costs, and potentially harmful interventions to deal with very small tumors that were going to disappear on their own anyway.

[xxv] Such as a cold, pneumonia or a "COPD exacerbation". A COPD exacerbation is an acute infection superimposed on the chronic cough from smoking-related emphysema or chronic bronchitis. These are very common in current or former smokers and typically improve with an antibiotic and steroid.

Similarly, while some cancers are very aggressive, others are relatively indolent and may never cause a problem. For example, some prostate cancers are aggressive, and can rapidly metastasize, but others grow very slowly and never cause a problem even if they are not treated. It's doubtful that increased screening to detect indolent prostate cancers would translate into a substantial benefit.

How long has cancer been there before diagnosis: As I noted earlier, patients often ask me why their cancer was not seen on an earlier x-ray, and how long they have had their cancer. I tell them that when we look back at x-rays or scans done a couple of years earlier, we will sometimes see a subtle abnormality where a large cancer is now located. But it is often not possible to tell whether the previously unnoticed abnormality was an early cancer or whether it was an area of scarring that later developed into a cancer.[xxvi] In other cases, even recent tests may not have shown anything.

It's difficult to tell how long a person has had cancer since some cancers grow very slowly while others grow rapidly. Furthermore, a cancer may change from slow to rapid growth due to the development of a new mutation. I also often point out to patients that it can take ten or twenty different changes in a cell for it to become malignant, so some of the early changes may have been present (but undetectable) decades earlier.

I often explain that a cancer may look like it is changing from growing slowly to growing rapidly with no real change in its behavior. This can happen solely due to the "exponential growth" that most cancers exhibit. In exponential growth, the first cancer cell divides to form two, these two divide to form four, these four divide to form eight cells, etc. It works sort of like compound interest or pyramid scams work in the financial world. By the time one billion cancer cells divide to form two billion cancer cells, it may look like the cancer is growing much more rapidly than it was previously, but this is just because each time that the cells divide, there are twice as many of them to produce new cells.

I frequently use the analogy of a normal pregnancy to illustrate how rapidly cancers may evolve. A pregnancy starts as a single cell, but nine months later you can have a baby that weighs 6 pounds or more and is

---

[xxvi]   See Chapter 1 and the link between inflammation/scarring and cancer. Most scars do not develop into cancers, but an occasional one does.

17 inches long! Some cancers grow more slowly than normal pregnancies since the cancer cell is a sick cell and many of the cancer cells may die rather than divide to produce new daughter cells. However, I have also seen cancers grow as fast or faster than a normal pregnancy. So, the patient's first cancer cell may possibly have been there ten years earlier, but it also may just have been there a few months earlier: there is no way for me to tell this with certainty.

The value of early detection of a cancer: Some cancers develop incurable metastases very rapidly. Others do not. For some cancers, metastases can appear several years after removal of the original cancer. For other cancers, if there is no evidence of recurrence by two to three years, then there is only a low risk that it ever will recur.

The bigger a cancer is when first discovered, the higher the probability that it will already have spread. However, even some very large tumors never spread. Large cancers are also more likely to have cells in them with mutations that make them resistant to radiotherapy and chemotherapy.

One way or another, the longer a tumor is in place, the higher the risk of metastases that preclude cure. Early detection and early action are important. That makes screening of high-risk patients important. However, most screening tests are negative and of those that are positive, many will be false positives. Furthermore, the screening test may miss an existing cancer (a "false negative" test), or the cancer may not be detected since it was not present at the time of the screening test and only developed a few months to a year or more after the screening test.

Currently, most cancers are found because of symptoms and not by screening. Don't get me wrong. Screening is important and can reduce the risk of dying from some cancers, but there are patients who will die from a cancer that only developed after they were screened.

All said, the small but real risks involved with medical testing and the problems with false positives mean that screening is best not done indiscriminately. Most current screening programs have been implemented only when studies have shown that the overall benefits outweigh the risks, resulting in a reduction in cancer death rates.

"Relative" vs. "absolute" reduction in risk of death with screening: It is important to keep these concepts in mind when considering the impact of screening tests. They help to understand the benefits and limitations of screening.

It may be difficult for many people to understand the benefit from screening. Why? Proponents of screening usually talk about the "relative" reduction in risk of death. Here's an example of how relative risk would be calculated. Let's say we are conducting a test of a new cancer screening method. In a trial, 10,000 participants undergo screening and 10,000 do not. Forty of the patients who were screened died from the relevant cancer, but 50 who were not screened died. The relative risk for those undergoing screening would be 40/50 = 80%, and the relative reduction in risk of death would be 20%.

To some people, this would sound impressive. However, the "absolute" reduction in risk of death from the cancer would be much smaller. Of the 10,000 patients undergoing screening, only 10 fewer died with screening. So, the absolute reduction in death rate from the cancer would be 10/10,000 = 0.1%. In this example, we would need to screen 1,000 people to prevent one death and of every 1,000 people screened, 999 would derive no real benefit.

Various types of screening tests are now used for several cancers, but to illustrate the general principles, I will just discuss a few particularly important types (breast, prostate, colon, cervix, and lung).

Screening for breast cancer: Mammography is used for the detection of early breast cancer. As a breast cancer begins to develop, small calcium deposits can form in the cancer, and these calcium deposits may be visible on mammographic x-rays. The detection of suspicious-looking calcium deposits on the mammogram leads to further testing that might include ultrasound, magnetic resonance imaging (MRI), and biopsy.

Current recommendations are that mammography should be done every two to three years in women between the ages of 50 and 74. For women at average risk of developing breast cancer, mammography reduces the relative risk of death from breast cancer by 21% in those aged 50 to 69 and by 32% in those aged 70 to 74. In contrast, the absolute reduction in risk of death from breast cancer in these two age groups is 0.14% and 0.22%, respectively.

For women aged 40 to 49 who are at average risk of developing breast cancer, the relative reduction in risk of death from breast cancer is 15%. The absolute reduction in risk of death is 0.05%. About one-third of younger women have a false positive mammogram at one of their screenings. These require further testing or biopsies, with a small but real risk of complications. Overall, the general feeling is that the benefit vs. risk ratio is not high enough in this younger age group to justify mammography.

A general principle of screening is that individuals at higher risk benefit more than do individuals at low or average risk. This is the case with mammography. Women at high risk of developing breast cancer include those whose mother, sister, or daughter have developed breast cancer and those who are known to have inherited a mutation such as *BRCA1* or *BRCA2*. Overweight women with denser breasts are also at increased risk of developing breast cancer, as are those who have previously been treated for any reason with radiotherapy.

It's more likely that breast cancer will be detected in high-risk women than in low- to average risk women, and the probability of having a false positive is lower. In high-risk women and women with very dense breast tissue, adding MRI screening to mammography is more effective than mammography alone, although it does increase the risk of having a false positive test.

The bottom line is that screening for breast cancer unequivocally reduces the probability of death from breast cancer. However, the absolute reduction is small in women who are at low or average risk, and there are many false positive tests. For example, my wife Lesley has no first degree relatives[xxvii] with breast cancer, but she regularly undergoes mammography. She invariably laments that mammographic testing "must have been designed by a man," given how painful it can be. She has so far had two anxiety-provoking false positive mammograms that have meant further testing. Fortunately, the further testing has been negative. Dealing with false positive tests is not easy, but she will continue to undergo routine mammography.

Screening for prostate cancer: A blood test called prostate specific antigen, or PSA, can be used to screen for prostate cancer. The higher the PSA, the higher the risk that a man will have prostate cancer. PSA screening of men aged 50 to 74 results in a 21% relative reduction in risk

---

[xxvii] First degree relatives of an individual are their parents, siblings, and children.

of death from prostate cancer, and in a 0.07% absolute risk reduction. However, when death from any cause is assessed (all-cause mortality[xxviii]), PSA screening has no major impact. This is because only a minority of all deaths in the population are due to prostate cancer, and PSA testing has no impact on deaths from other causes.

Furthermore, some prostate cancers never cause a problem since they are very indolent, and about two-thirds of positive PSA tests are false positives. As with mammography in women, further tests in patients with false positives can sometimes cause serious complications. Consequently, only men at high risk of developing prostate cancer—like African Americans and men with a family history of prostate cancer—should undergo PSA testing.

But not all recommendations are followed. For example, I have elected to undergo PSA testing even though I have no family history of prostate cancer.[xxix] To date my PSA has been completely normal, and this is reassuring. However, I may face a dilemma if it begins to rise, knowing that it will have a high probability of being a false positive. Any further testing (such as a biopsy) will carry a risk of complications. I will be dealing with the uncomfortable consequences of having been screened despite being low risk.

Screening for colon cancers:[xxx] Benign growths called polyps may develop in the colon. These may turn into colon cancers over time. Both polyps and cancers can bleed. It is not normal to have blood in the stool. There may be a polyp or cancer if a test called fecal occult blood testing (FOBT) detects blood. This leads to other tests such as colonoscopy to investigate further. In average risk individuals, FOBT reduces the relative risk of dying from colon cancer by 14%. The absolute risk reduction is 0.1%. As with PSA testing for prostate cancer, FOBT has no major impact

---

[xxviii] All-cause mortality means death from any reason.

[xxix] In Ontario, I and other low-risk individuals pay for PSA testing out of pocket. The government pays for high-risk patients. Lesley insists that I have it done. Who am I to argue? At parties, I have heard several women say they had saved their husbands lives by insisting they have a PSA test. I am not necessarily convinced that they did.

[xxx] In this section, if I use the term "colon" I am referring to either the colon or the rectum.

on all-cause mortality since most deaths in people who might undergo screening are from other causes, and FOBT has no impact on deaths from causes other than colon cancer.

Another test—called flexible sigmoidoscopy—involves using a scope to look inside the bottom 25 centimeters of the bowel. Flexible sigmoidoscopy is associated with a 28% reduction in relative risk of dying from colon cancer and a 0.2% absolute risk reduction. Unlike FOBT, this 0.2% absolute risk reduction in death from colon cancer is high enough that flexible sigmoidoscopy has a detectable impact on all-cause mortality. It remains uncertain if colonoscopy (using a longer scope to look at the entire length of the colon) is any better than sigmoidoscopy in average risk individuals. Currently, either screening method may be used in average risk individuals.

As with other cancer types, screening for colorectal cancer has its greatest impact in high-risk individuals. High-risk factors for colorectal cancer include a history of inflammatory bowel disease[xxxi] or having a first degree relative with colorectal cancer, particularly if there is a family history of colorectal cancer developing at a young age. Males who develop iron deficiency should also be screened since the iron deficiency may be due to blood loss from a bowel polyp or cancer.[xxxii] Colonoscopy is usually used to screen high-risk individuals.

When her mother was diagnosed with colon cancer, Lesley was advised to undergo a screening colonoscopy. While having no risk factors, I also qualified for one since I was over the age of 50, and I volunteered to have one too, as moral support for her.

Having a colonoscopy can sound like a very unpleasant experience. Very occasionally, they can be associated with complications. However, Lesley and I and several of our friends who have had them agree that undergoing a colonoscopy is not nearly as bad as it sounds. The most unpleasant part is the preparation. You must take very potent laxatives to clean out the bowel so that the colonoscopist can see the bowel wall

---

[xxxi] Inflammatory bowel diseases include Crohn's disease and ulcerative colitis.
[xxxii] Iron deficiency is common in women due to blood loss during menstruation. Hence, it is less likely to be a sign of a colon cancer in women (particularly premenopausal women) than it is in men.

easily. This means spending several hours on the toilet purging prior to the procedure. However, the benefits outweigh the unpleasantness and the small but real risks.

Screening for cervical cancer: Screening for cancer of the cervix was one of the earliest victories in the cancer screening story. A Papanicolaou test, commonly known as a pap smear, can detect precancerous changes in the cervix. This permits treatment to reduce the risk of development of cervical cancer. Unlike other screening tests, we cannot tell with certainty the extent to which pap smears reduce the risk of dying from cervical cancer since there have been no randomized trials, with half the women undergoing pap smears and the other half being observed, without a pap smear. Hence, estimates of the benefit of pap smears are derived from changes in the incidence and death rate from cervical cancer over time, recognizing that other factors besides pap smears may have contributed to any improvement.

Pap smears became common in Canada in the 1970s. Since then, the incidence of cervical cancer and deaths from it have dropped by more than 60% in women over the age of 40.

Like mammograms and colonoscopies, pap smears can be unpleasant, but they save lives.

Screening for lung cancer: Low dose CT scans[xxxiii] are an effective way to screen high-risk individuals for lung cancer, but chest x-rays are not. CT screening reduced the relative risk of dying from lung cancer in current or former heavy smokers by 16%. The absolute risk reduction was 0.4%. Hence, CT screening is now recommended for heavy smokers, even if they quit smoking several years ago. Although we have no information on use of CT screening in individuals who have had less tobacco exposure, I would predict that it would have little impact.

[xxxiii] The term "low dose" CT scan means that a lower dose of radiation is used than with a standard CT scan. The radiologist can see somewhat less detail with a low dose CT scan than with a standard dose CT scan, but the scan does nevertheless permit detection of small nodules in the lungs. Using a lower dose of radiation reduces the very small but real risk of eventually developing another cancer from the radiation used during the scan.

Small cell lung cancers comprise about 12–15% of all lung cancers. It is the most rapidly growing type of lung cancer and is typically too far advanced at diagnosis for surgery. Interestingly, since the initiation of our lung cancer screening program, I have seen two patients with small cell lung cancers that were diagnosed when they were still very small. They were successfully removed by surgery.

Further Details and References

I will now go into a bit more detail on screening and provide references for those who are interested.

Why screening can make a difference: The bigger a cancer is when diagnosed, the higher the probability that it will already be widely metastatic.[230] If screening detects a cancer when it is still small, this increases the probability that it is still localized and can be cured.

Mutations are occurring with every cell division. The bigger the cancer, the more cells that divide, and the higher the total number of mutations in the cancer. Every time a cancer cell divides, it is likely that the daughter cells will have new mutations. These mutations may enable the cancer cells to divide even faster or to spread more efficiently to another part of the body. Furthermore, the bigger the cancer, the more cells that are available to spread.[xxxiv]

The development of new mutations as cells divide may not only enable cancers to grow and spread more rapidly. It also increases the probability of mutations that make a cancer resistant to therapy. Most of the cells in the initial cancer may be sensitive to a therapy. If, however, there is even

---

[xxxiv] It is estimated that it generally requires at least thirty tumor doublings for a cancer to grow from a single cell to a tumor that is 1 millimeter across, and that a 1-millimeter tumor would have about 2^{30} (i.e., 1 billion) cells. However, there are also many normal cells (called stromal cells) mixed with the tumor cells in a tumor. These normal cells include blood vessels, white blood cells, "structural" cells called fibroblasts, immune cells called macrophages, etc. Consequently, if a tumor has 2^{30} tumor cells it will actually be much bigger than 1 millimeter due to the presence of these normal cells that are contributing to the tumor's size, and a 1-millimeter tumor would typically have 1 billion cells of any type but far fewer than 1 billion of those cells would be tumor cells.

one cell in a million cells with a new resistance mutation, then that one cell will continue to divide while the sensitive cells are dying in response to the therapy. Hence, even though the cancer initially shrinks with treatment, it will start to grow again as the progeny of the one resistant cell increase and become the dominant population in the tumor.

It is important to know that new mutations do not guarantee the spread of a cancer. The probability of metastases varies substantially from one tumor to another, and is strongly influenced by the types of mutations that occur as the tumor cells divide.[230]

I often tell a patient that if we had looked carefully prior to their surgery or radiotherapy, we would probably have found millions of cancer cells circulating in their blood stream. A tumor sheds about 1 million tumor cells into the blood stream for every gram of tumor that is present.[231] Metastases will appear if these cells have undergone mutations that permit them to survive and grow in another part of the body. If there are metastases, the cancer probably would be incurable. However, it is also possible that none of the cancer cells circulating in the blood stream will be able to form metastases.[231] That is why more than 60% of patients with cancers (including some patients with very large cancers) can be cured by surgery or radiotherapy.

A rough estimate of the proportion of patients cured of cancer is the proportion alive and free of recurrence at five years.[xxxv] In fact, there is nothing magic about five years; it is simply a probability-based guess. For some cancers, circulating tumor cells may lodge in a distant part of the body and only decades later begin to divide, grow, and produce metastases.[232] For other cancers, absence of metastases two to three years after removal of the original cancer indicates a probable cure.

Screening for breast cancer: The US Preventive Services Task Force recommends mammography every two years for most women 50 to 74 years old.[233] The Canadian Task Force on Preventive Health Care recommends mammography every two to three years for this age group.[234]

The statistics and discussion that follow are all for women with an average risk of developing breast cancer. In women aged 40 to 49, the

---

[xxxv] Called the 5-year "relative survival" rate.

relative risk reduction of death from breast cancer is 15%, with an absolute risk reduction of 0.05% (4.7 fewer breast cancer deaths per 10,000 women screened). For women 50 to 69, the relative risk reduction of death from breast cancer is 21%, and the absolute risk reduction is 0.14% (14 fewer deaths per 10,000 women screened). For women 70 to 74, the relative risk reduction of death from breast cancer is 32%, and the absolute risk reduction is 0.22% (22 fewer deaths per 10,000 women screened).[234]

We do not have enough evidence in women older than age 75 to assess the pros and cons of mammography.

To prevent one death from breast cancer in women ages 50 to 69 would require mammograms in 720 women every two to three years for eleven years. We would see false positive mammograms in 204 (28%) and negative biopsies in 26 of the 720.[234] To prevent one breast cancer death in women aged 70 to 74 would require screening 450 women. There would be false positive mammograms in 96 (21%) of the 450 and negative biopsies in 11.[234]

It is obvious that the benefits from mammography are higher in older women than in younger women. Why is this? For one thing, breast cancer is less common in women aged 40 to 49 than it is in women over 50. The number of false positives and unnecessary biopsies is larger.[233,234] We would need to screen 2,100 women in the 40–49-year-old age group every two to three years for eleven years to prevent one death. However, 690 (33%) of these 2,100 women would have at least one false positive mammogram and 75 would undergo a biopsy that would prove negative.[234] At the same time, the small but real risk of eventually developing a cancer as a direct result of the radiation from a mammogram would be higher in younger women.[235] The general consensus is that the negatives outweigh the positives when it comes to mammography for women below the age of 50, unless they are high-risk individuals.

In addition, mammography may miss 10–30% of breast cancers,[236] particularly in dense breast tissue.[237] Breast density is higher in younger women than older women.[238]

Furthermore, the more times mammography is done, the more likely it is to find a slow-growing breast cancer that would have never threatened the life of the patient,[233] or to find a cancer that would have disappeared spontaneously on its own with menopause or with cessation of hormone replacement therapy.[229]

Detection of disease that is present but that would never have harmed the individual if it had been left untreated is called "overdiagnosis." Estimates are that for every life saved by mammography, there are 10 breast cancers detected and treated unnecessarily and 200 women who are subjected to substantial psychological distress by a false positive result.[239]

In average risk patients, there is insufficient evidence to recommend methods such as magnetic resonance imaging (MRI) as a screening tool. Teaching breast self-examination to women aged 31–64 did not reduce deaths from breast cancer, but it did significantly increase the probability of unnecessary biopsies.[234] Most abnormalities found by breast self-examination end up not being cancers.

The rest of this section deals with high-risk individuals. High-risk women include those with a first degree relative with breast cancer,[xxxvi] those known to have an inherited "germline" mutation[xxxvii] of the genes *BRCA1, BRCA2, TP53, PTEN* or *CDH1*, those with a parent, sibling, or child known to have one of these germline mutations, those with denser breasts (particularly if the individual is also overweight),[240] or those with prior treatment with radiotherapy.[233,241,242] For example, in women at high risk of having *BRCA* mutations, mammography detected breast cancer in 0.3–1% of individuals screened,[242,243] with false negatives in 4.8% of patients.[244]

MRI is better at detecting a breast cancer than mammography, with up to 3–8% of patients found to have breast cancer by MRI.[242-245] Combining mammography with MRI is even better. However, false positives are more frequent with MRI, whether used alone or combined with mammography.[242,246,247]

The proportion of patients with false positives in high-risk groups[242] is much lower than the risk of false positives in average risk individuals,[234] in keeping with Bayes' theorem. A high-risk patient will have about the same

---

[xxxvi] Risk is lower if a somewhat more distant relative such as a grandparent, aunt or cousin develops breast cancer.

[xxxvii] A germ line mutation is a mutation that you inherited from one of your parents and is present in all the cells in your body from conception.

probability as an average risk patient of having a false positive test but will have a much higher probability than the average risk person of having a "true positive" test. Therefore, of all positive tests in a high-risk patient, a lower <u>proportion</u> will be false positives.

Adding breast ultrasound to mammography may also increase cancer detection rates in some higher-risk groups with dense breast tissue.[248]

There are also other methods to reduce breast cancer risk. For example, in patients with *BRCA* mutations, surgical removal of the breasts in healthy individuals (called prophylactic mastectomy)[xxxviii] can reduce the risk of developing breast cancer by up to 95%. It is not yet clear whether it improves life expectancy,[249] but it would be reasonable to anticipate that it does.

Screening for prostate cancer: Prostate cancer can be very indolent and may never cause a problem in some men who have it. In males who die from other causes, autopsy studies have reported the presence of prostate cancer in about 5% of those under the age of 30, 15% of those ages 40 to 50, and 59% of those over the age of 79.[250] However, only 11–14% of males are diagnosed as having prostate cancer while alive, and prostate cancer accounts for only 2–3% of all male deaths in North America.[1,251]

Using PSA screening for prostate cancer in males ages 50 to 74 reduced the risk of death from prostate cancer from about 33 per 10,000 men without screening to about 26 per 10,000 with screening.[252] Hence the *relative* reduction in death rate from prostate cancer would be 7/33 = 21% (which sounds impressive). But the *absolute* reduction in death rate from prostate cancer was only 33-26 = 7 per 10,000 men screened, or 0.07%.

About 1,400 men would need to be screened to prevent one death from prostate cancer,[252] and death from <u>any cause</u> was the same in the screened group as in the control group that did not undergo screening.[253] At the same time, 8.2% of the patients in the screened group were diagnosed as having prostate cancer, compared to 4.8% in the control group. In other words, 8.2 - 4.8 = 3.6% of the control group (360 per 10,000) were living with a prostate cancer that had not been diagnosed and was not bothering them. Of those who actually had prostate cancer, 48 would

---

[xxxviii] As was the case with actor Angelina Jolie, as discussed in Chapter 1.

need to undergo treatment to prevent one death from prostate cancer. As noted above, while prostate cancer accounts for 2–3% of all male deaths in North America, a lot of individuals live with localized prostate cancer without it ever bothering them.[1,250,251]

Of men who are found to have a high PSA value, it is a false positive in about two-thirds (i.e., they do not have prostate cancer), and about 15% of men have false negative tests (i.e., prostate cancer is present despite a normal PSA).[254] Men undergoing biopsies based on the finding of a high PSA had a 32% probability of having a moderately severe complication from the biopsy, and a 1.4% risk of a serious complication such as a major infection requiring admission to hospital for intravenous antibiotics.[254]

Furthermore, of those cancers that <u>are</u> found by screening, it is often not possible to predict which ones will *vs.* will not eventually cause a problem. If therapy is offered, it may cause long-term complications such as incontinence of urine and inability to have an erection. It has been estimated that because of the potential negative impact of treatment, out of every 1,000 men screened, 3 more will end up with urinary incontinence and 25 more will end up with erectile dysfunction compared to those who are not screened.[254]

Based on the relative benefits vs. risks of PSA screening for prostate cancer, it has been recommended that most men <u>not</u> undergo it.[255] Similarly, digital rectal examination (i.e., a physician feeling the back of the prostate gland by inserting a gloved finger into the rectum) also has a very high rate of both false positives and false negatives. It may be of little value.[256]

As discussed earlier, the risk of developing prostate cancer is increased in African American men and in men with a family history of prostate cancer.[257] Men with *BRCA2* mutations who are diagnosed as having prostate cancer on average have more advanced disease and a lower probability of survival than do those without *BRCA2* mutations.[258]

There is too little information from PSA screening studies to permit conclusions on the relative benefit of screening in high-risk vs. average risk individuals, but in keeping with Bayes' theorem,[259] if screening is going to be of benefit anywhere, it is most likely to be of benefit in higher-risk groups. There is indirect evidence to support this assumption.[260,261] Considering other risk factors (such as family history, race, age, etc.) along with the PSA level can help in arriving at a rational decision with respect to who may not need a prostate biopsy despite having an elevated PSA level.[262]

Screening for colorectal cancers: In average risk patients, FOBT reduces the *relative* risk of dying from cancer of the colon or rectum by about 14%.[263] However, the *absolute* risk reduction is 0.1%, or 10 fewer colon cancer deaths per 10,000 individuals screened, and there is no detectable reduction in overall mortality (i.e., in the risk of dying from any cause).[263] About 1,000 average risk people would need to be screened by FOBT to prevent one death from colorectal cancer.

FOBT reduces the probability of *developing* colorectal cancer by 0.1%, or 10 people out 10,000. This is presumed to be due to detecting polyps that can be removed before they develop into cancers.[263]

A stool assessment called a fecal immunohistochemical test will detect about twice as many polyps or cancers as FOBT, but it also results in about twice as many false positives. It is not yet known if it reduces colon cancer death rates compared to FOBT.[264]

In average risk patients, flexible sigmoidoscopy reduced the *relative* risk of dying from colorectal cancer by about 28%, with an *absolute* reduction of about 0.2% (20 per 10,000 individuals screened).[263] In other words, it was about twice as effective as FOBT. Unlike FOBT, it also improved overall mortality rates.[263] About 500 average risk individuals would need to be screened by sigmoidoscopy to prevent one death from colorectal cancer.

Sigmoidoscopy also reduced the probability of developing colorectal cancer by 0.4% (40 per 10,000 people screened), presumably by permitting detection and removal of polyps before they developed into cancers.[263]

There is insufficient information available to be able to say with certainty that colonoscopy is any better than flexible sigmoidoscopy in an average risk individual. The information that is available suggests that it might be,[264] but colonoscopy is also more invasive than sigmoidoscopy— and it has a higher risk of complications.

Major complications of colon cancer screening can include bleeding, perforation, or death due to the screening procedure or to subsequent colonoscopy or surgery. Major complications occurred in 0.03% (3 in 10,000 screened) of those undergoing FOBT and 0.03–0.08% (3 to 8 per 10,000 screened) for those undergoing flexible sigmoidoscopy.[263] The risk of major complications from screening colonoscopy was 0.28% (28 per

10,000 screened).[265] Removing a polyp during a screening colonoscopy increased the risk of serious complications, with 85% of serious colonoscopy complications occurring in procedures involving removal of a polyp.[265]

Not surprisingly, screening high-risk groups is generally more useful than is screening in lower-risk people. Individuals at highest risk of developing colorectal cancer include those with inflammatory bowel diseases like ulcerative colitis and Crohn's disease and those with a first degree relative with colorectal cancer.[266]

High-risk individuals are offered screening by colonoscopy.[264] In patients with inflammatory bowel disease, 8% of patients screened by colonoscopy died of colorectal cancer compared to 22% of patients who did not undergo screening. Colorectal cancers that were diagnosed in the screened group were more likely to be early stage vs. late stage, a marked difference compared to those who did not undergo screening.[267] Patients with inflammatory bowel disease who are found during colonoscopy to have precancerous changes (dysplasia) may undergo complete removal of the colon because of the high risk of progression to full-blown cancer.[268]

Individuals who have any first degree relatives with colorectal cancer are themselves at increased risk. The risk increases if they have several first degree relatives with colorectal cancer or if any of their relatives developed it at a relatively young age. [269] If a person has a first degree relative who develops colorectal cancer younger than age 60 or if they have two first degree relatives who were diagnosed with colorectal cancer at any age, then this increases the probability that they have an inherited susceptibility to colorectal cancer. For these individuals, the US Multi-Society Task Force on Colorectal Cancer recommends, "colonoscopy every five years, beginning ten years before the age of the youngest affected relative or age 40, whichever is earlier."[269]

Risk of colorectal cancer is substantially increased in individuals with "Lynch's Syndrome" (also called familial non-polyposis colorectal cancer syndrome). Lynch's Syndrome is due to mutations of specific genes involved in repair of DNA damage. Guidelines recommend colonoscopy every 1–2 years in individuals with this genetic defect.[270] Surgery to remove the colon is recommended for people with another inherited genetic defect called familial adenomatous polyposis (or Gardner Syndrome) since these individuals have a very high risk of developing colon cancer.[271]

Screening for cervical cancer: The use of pap smears to screen for precancerous changes in the cervix was one of our earliest successful forays into cancer screening. Pap smears were first used in the late 1940s and 1950s, but their use became more prevalent in Canada in the 1970s after the introduction of Medicare.[272] From the early 1970s to the early 2000s, in *relative* terms, both the incidence of cancer of the cervix and the proportion of the population dying from it dropped by greater than 60% in women over the age of 40. In this period, the incidence rate dropped from about 31 per 100,000 to about 12 per 100,000 women in those older than 40, while the mortality rate dropped from about 14.3 per 100,000 to about 4.6 per 100,000.

Although treatments improved, the major reason for the drop in cervical cancer incidence was probably screening. Screening permitted detection and treatment of precancerous conditions before they developed into a cancer.[272] Cervical cancer is usually caused by a chronic infection with a virus called human papillomavirus (HPV) acquired during sexual activity. Changes in sexual practices might have led to a substantial increase in cervical cancers. But they didn't. It is reasonable to assume pap smears made a major difference. Impact of pap screening is about the same in the United States as in Canada.[272]

As with other cancers, the impressive 60% improvement in *relative* mortality rates translates into only a small *absolute* improvement of one less death for every 10,000 women over the age of 40. About 10,000 women would need to undergo a pap smear to prevent one death. Because this cancer becomes more common in older age groups, the changes in women under the age of 40 are much smaller. Mortality rates in younger women dropped from around 1.6 per 100,000 women to about 0.9 per 100,000, or one less death from cervical cancer per 143,000 younger women screened.[272]

The benefit from pap smears is probably actually greater than this. Not all women undergo screening. If we assume that only about half of women over the age of 40 have a pap smear, then out of every 10,000 women, 5,000 would have had a pap smear. If deaths from cervical cancer decreased by one in 10,000 for the overall population, this would mean that they decreased by one in 5,000 for those who were screened, with no reductions in the 5,000 who were not screened.

Another major development has been the availability of HPV vaccines to reduce the risk of developing cervical cancer. As discussed in Chapter 1, chronic HPV infection is associated with development of cancers of the cervix, vagina, vulva, head and neck, anus, penis, and others.[273] HPV vaccines reduce the risk of precancerous changes in the cervix.[274]

Screening for presence of HPV infections can improve the ability to detect and treat precancerous changes in the cervix.[275] Overall, women who do not have HPV infection detected on screening have a lower rate of later cervical precancerous changes than do women with HPV infection— and they may subsequently need less frequent pap smears.[276]

Screening for lung cancer: Screening people using chest x-rays has no impact on deaths from lung cancer.[277] However, an analysis combining results from nine different trials[xxxix] indicated that CT scans using low dose radiation make a difference in high-risk patients.[278] Participants in these trials had an average of a 40 pack-year smoking history, equivalent to having smoked one pack of cigarettes per day for 40 years (or two packs per day for 20 years).

Across the nine trials, the *relative* risk of dying from lung cancer was reduced by 16% in the group screened by CT scans, with the *absolute* reduction in lung cancer deaths being 0.4% (from about 237 lung cancer deaths per 10,000 unscreened patients to 197 deaths per 10,000 screened patients).[278] Screening of 265 high-risk individuals would prevent one lung cancer death.

Screening reduced the relative risk of dying from <u>any</u> cause by 4%. The reduction in risk of dying of any cause was lower than the reduction in risk of dying from lung cancer. This is because screening had minimal positive or negative impact on the risk of dying from causes other than lung cancer, and lung cancer only accounted for about one-quarter of all deaths among study participants.

CT scan screening resulted in false positives in about 8% of participating individuals. About 4 participants per 10,000 screened suffered a major complication as a result of conducting further tests to assess a false positive result.[278]

[xxxix] An analysis that combines results from several different studies is called "meta-analysis."

There is little information on use of low dose CT screening in average or low-risk individuals, but in keeping with Bayes' theorem, it is likely that any benefit would be small and that the false positive rate would be very high in lower-risk individuals.

Several jurisdictions, including the province of Ontario, are now offering low dose CT scan screening for individuals who are at high risk of lung cancer due to their smoking history.

Conclusions: The second reason that cancer still sucks is that many cancers are already at an advanced, incurable stage when they are first discovered. Screening high-risk patients does help, since finding cancers at an early stage increases the probability that they can be cured. Screening unequivocally saves lives. However, we have a long way to go. Reduction in *relative* death rates often sounds impressive but may be associated with only small *absolute* probability of benefit for most people.

The usefulness of screening is limited by several factors. First, there is at least a small risk of serious complications from some of the screening procedures—or from further investigations conducted because of false positive screening results. Second, most screening methods are not very sensitive and will miss some cancers. Third, they are not very specific. A high proportion of abnormal results will be false positives, particularly in individuals who are not at high risk. Fourth, many cancers are missed in patients undergoing regular screening since they develop in between screening episodes.

High-risk patients are particularly likely to benefit from current screening methods. However, individuals with only a low risk will have a low probability of a cancer being found by screening. In these low-risk people, any positive result will have a high probability of being a false positive. Some individuals at low or average risk will choose to undergo screening and some will elect not to. I tell patients that the right approach for them is what their gut tells them to do.

New approaches are being developed to supplement or replace current screening methods. For example, a lot of effort has gone into development of blood tests to detect circulating tumor cells or cancer-associated mutations. So far, none of these are sensitive or specific enough to be of much help, but I think there is a good possibility that this may change in the not-too-distant future.

— 3 —

Wasting Away: How Cancer Makes People Feel So Sick

The third reason that cancer still sucks is that it can cause several nasty symptoms.

Short Primer

In this chapter, I will give a brief overview of a few of the more common, distressing cancer symptoms. Treatment of the cancer is the most effective way to alleviate these symptoms. Shrinking a cancer even modestly may improve symptoms.

Pain: We are a lot better at controlling cancer pain than we used to be. Still, it is one of the most feared complications of cancer. There are three main types of pain.

Somatic pain comes from cancer damaging or compressing body parts like bone or skin that have lots of nerve fibers that carry pain signals. Examples of somatic pain include pain from a bone that breaks after being weakened by cancer or pain from a cancer invading the skin. With somatic pain, the patient can usually point right at the area that is painful.

Neuropathic or neuralgic pain results from the cancer invading or compressing nerves. Neuropathic pain is often more difficult to control than somatic or visceral pain.

Visceral pain is from growth of a cancer in an inner organ like the lung, liver, or pancreas. Internal organs have few nerve fibers designed to carry pain signals. Consequently, a cancer can reach a very large size without causing pain. When it does cause pain, the pain is vague and difficult to localize and describe. I often tell patients that we would be much better off if cancers caused symptoms like pain when they were very small and drew attention to themselves, but they usually do not do that.

No matter what the type of pain, radiotherapy, and systemic therapies[xl] that cause even modest tumor shrinkage can be more potent pain killers than morphine. These therapies reduce pain by reducing the tumor's pressure on surrounding structures and by enabling healing of damaged tissues. This is one of many mechanisms by which anticancer therapy can improve a cancer patient's quality of life.

One major development over the past few decades has been the emergence of the specialty of palliative care. It has improved our ability to help patients deal with pain. Consequently, this specialty has become the oncologist's key ally in the fight to alleviate the suffering of cancer patients.

Pain can still be a terrible consequence of cancer, but for a high proportion of patients, we are far better at controlling it now than we were just a few years ago.

Weight loss and loss of appetite: In most people's minds, weight loss (or cachexia) is strongly associated with advanced cancer. Patients with some cancers (like cancers of the stomach and pancreas) are more likely to develop cachexia than are patients with other types of cancer.

The major contributing factors to cancer-related cachexia are cancer-induced wastage of energy sources, destruction of muscle, fat, and other tissues and reduced food intake. With respect to energy wastage, cancer cells may burn up far more glucose (i.e., sugar) compared to normal cells. Many people have difficulty losing weight by going on a diet but if they have an undiagnosed cancer, they may find that they are suddenly losing weight quite easily because of this calorie wastage. I tell my friends, "Beware the diet that finally works."

The reliance of cancers on high amounts of glucose is the reason that they may light up on a PET scan. Many cancers will take up large amounts of the radioactive glucose that is injected intravenously for a PET scan.

[xl] Systemic therapies include chemotherapy, hormone therapy, targeted therapies, and immunotherapy.

The PET scan camera detects this radioactive glucose and helps reveal areas of cancer spread.

As I discuss further in Chapter 9, some people have argued that eating a low sugar diet should help fight the cancer. The reason that this is unlikely to work is that the cancer can produce a ready supply of glucose for itself just by breaking down a patient's normal tissues. This happens since the body reacts to the cancer by producing immune system cells that infiltrate into the cancer. These immune system cells release substances called cytokines (such as tumor necrosis factor-alpha, or TNA-α) that break down muscles, fat, and other tissues to produce the glucose that can then be used by the cancer. In Chapter 7, I discuss further how the immune system may fight cancer, but it is a double-edged sword that can also cause problems, such as this tissue breakdown which leads to weight loss and loss of muscle mass. Effective treatment of the cancer can stop this abnormal tissue breakdown.

Another way that cancer leads to weight loss is by causing loss of appetite or "anorexia."[xli] The cytokines that cause tissue breakdown and other factors released by the cancer can suppress the brain's appetite center. These substances may also alter the sense of taste, and they may interfere with the ability of stomach muscles to contract. If the stomach is not contracting properly, it can take a long time for the stomach to empty. Consequently, after eating breakfast it may be several hours before the stomach has the capacity to accept any more food. Depression and anxiety can also decrease appetite.

As with cancer-related pain, the most effective way to counteract weight loss is by treating the cancer. Even if the treatment itself causes nausea, the patient's appetite may improve substantially if the cancer shrinks.

Corticosteroids like prednisone or dexamethasone may also markedly stimulate the appetite. I often tell patients, "It's hard to keep the fridge full if you are on steroids." However, corticosteroids can also cause problems, particularly if they are continued for a long duration of time. For example, they may cause wasting of the muscles of the hips, and this may cause

---

[xli] Many people associate the word "anorexia" with the eating disorder "anorexia nervosa." But the term anorexia applies to loss of appetite from any cause.

marked weakness and impaired mobility. They can also cause insomnia, irritability, and other side effects.

Marijuana and cannabis derivatives have been assessed for their ability to improve appetite, but there is little evidence that they help much. Many of my older patients do not tolerate them very well while many of my younger patients feel that they are helpful. I often tell patients that this suggests to me that there may well be a biological basis to the "generation gap" in recreational drug use. I personally do not prescribe cannabinoids but tell my patients that I do not have any problem if they want to purchase them legally.

Fatigue/weakness: While fatigue can be caused by many things and is very common across the general, healthy population, increased fatigue is often one of the early symptoms of cancer. The fatigue and weakness associated with cancer is also called "asthenia." A cancer can cause this fatigue by wasting energy, as discussed above, and by causing muscle breakdown. While both systemic therapies and radiotherapy often directly cause fatigue, many patients notice a substantial improvement in their energy level as their cancer shrinks with therapy.

Another very important factor in combating fatigue is to stay as active as possible. We used to tell fatigued patients to rest a lot, but this was bad advice. We now know that inactivity leads to increased muscle loss, and this worsens not only fatigue but also increases the risk of other problems. These include confusion/delirium, blood clots, pneumonia, and bed sores. I often tell patients that they should really push themselves to stay active. This helps maintain muscle mass and health.

Cancer and cancer treatment can also cause low hemoglobin (anemia) that can contribute to fatigue. Some cancers also release substances that make the blood calcium level go up or that make sodium levels go down. High calcium levels and low sodium levels can cause drowsiness, weakness, and confusion. Treatment of the cancer, using transfusions to correct low hemoglobin levels, and using other approaches to correct high calcium or low sodium can all help with management of cancer-related fatigue.

Shortness of breath: Cancers involving the lung can also cause shortness of breath (dyspnea) by blocking an airway, by compressing lung tissue, or by interfering with transfer of oxygen into the lung's blood vessels. Not surprisingly, the best way to improve the dyspnea is to treat the cancer. For example, in one of my patients who presented initially with severe dyspnea, a scan showed a cancer that had spread very extensively to both lungs. There was very little normal lung left to take up oxygen. By the time I saw him, he was requiring the highest amount of supplemental oxygen[xlii] that could be delivered without putting him on a ventilator in the Intensive Care Unit. This high flow supplemental oxygen was barely keeping him alive.

We started him on urgent chemotherapy, and he improved rapidly. A few weeks after starting chemotherapy, a repeat CT scan showed that his cancer had regressed markedly. He was eventually, able to get by without the use of any supplemental oxygen.

While he has had two further episodes of tumor growth accompanied by increased dyspnea, changes in therapy have managed to again shrink his cancer. He is now more than three years from presentation and functioning well, with no current need for any further supplemental oxygen. While some patients do not improve with therapy, many do.

A cancer can also cause anemia that will reduce the ability of blood to transport oxygen from the lung to other tissues. Here, a transfusion can help the dyspnea.

A lung cancer may increase the risk of pneumonia by obstructing the flow of mucous out of the lung. This reduces the ability of the lung to get rid of bacteria. Antibiotics and corticosteroids often help in this situation.

Cancers may also cause a pleural effusion (an accumulation of fluid in the pleural space that separates the lung from chest wall). Pleural effusions cause dyspnea by compressing adjacent lung tissue. Similarly, cancers may also cause pericardial effusions (an accumulation of fluid in the sac surrounding the heart). By compressing the heart, pericardial effusions impede the flow of blood back into the heart. Consequently, there is less blood available to pump into the lungs to pick up oxygen. Draining pleural and pericardial effusions using a needle inserted through the chest wall can yield immediate improvement in dyspnea.

Many cancers increase the risk of blood clots. These clots may develop in veins near the skin (and these are called superficial phlebitis) or they

---

[xlii] Supplemental oxygen means using a tube and either nasal prongs or a mask to deliver high levels of oxygen from a tank.

may develop in larger, deeper veins such as in the legs or pelvis (referred to as deep venous thrombosis, or DVT). Part of a DVT clot may break off and float up into the heart. From there, it may be pumped into the lung (referred to as a pulmonary embolus) where it blocks lung blood vessels. A pulmonary embolus may cause the sudden onset of very severe dyspnea. When blood clots are discovered, anticoagulants are administered, and these often effectively control the problem.

Cough: Lung cancer patients often experience a persistent cough as one of their first symptoms. In most people, a new cough is caused by something simple like an infection or allergy. It is only when it keeps worsening that it leads to the tests that reveal the lung cancer. When new lung cancer patients tell me that they are upset that their family doctor initially missed their lung cancer, I point out to them that what happened to them is very typical for lung cancer. The overwhelming majority of patients who see their family doctor for a cough will not have lung cancer. It is only when the cough keeps getting worse despite initial management that a CT scan should be considered.

However, coughing up blood (hemoptysis) increases the suspicion of lung cancer, and the onset of hemoptysis will typically result in rapid further testing.

In some patients, an incessant cough will be bad enough that it causes chest wall pain, and it may seriously interfere with sleep. The cough may make lying in bed intolerable and the only way that some patients can get any sleep is in a recliner chair, with their head propped up.

If a lung cancer shrinks with treatment, then the cough will generally improve. Narcotics like codeine, morphine, or hydrocodone may also help reduce cough.

Infection: Infections are very common in cancer patients. Cancers may cause infections by eroding through barriers, such as the skin or the lining of the gastrointestinal tract. This erosion permits bacteria to invade into deep tissue that are normally protected by these barriers.

Another way that cancers can increase the risk of infection is by blocking the normal flow of body fluids. Usually, body fluids flush out bacteria that enter the urinary tract, lung, lymphatic system, and biliary system.[xliii] If fluid flow is blocked by a growing tumor, the fluid can stagnate, like water in a swamp, permitting the bacteria to grow and cause an infection.[xliv]

Bone marrow that is packed by malignant cells may be rendered incapable of making the white blood cells required for protection from infection. As cancers grow, they may also suppress the immune system. This can increase the risk of reactivation of some viruses, such as the herpes zoster (chicken pox) virus that causes shingles.

Bleeding: Bleeding is another common complication of cancers. Bleeding can occur because an ulcerating cancer invades through the skin or through the wall of the intestine, airway, or any other organ. Cancers can cause sudden massive bleeding if they invade into a major blood vessel.

Neurological complications: Tumors that spread to the brain can cause fluid leakage into the surrounding brain, creating cerebral edema. This increases the pressure inside the skull. This increased pressure can cause headaches and can interfere with function of the surrounding brain.

Tumor deposits in the brain can also lead to seizures (uncontrolled shaking) by causing the equivalent of a short circuit in the brain. The seizure may be "generalized," involving the entire body, and leading to loss of consciousness. Or it may be "focal" and involve just one part of the body such as an arm, leg, a single finger, or facial muscle.

Since tumor blood vessels are more fragile than normal blood vessels, they can cause bleeding into tumor deposits in the brain. If the bleed is small, it may cause no symptoms. Conversely, if it is large, it can cause coma or death.

Metastases to the backbone can press against the spinal cord, causing spinal cord compression. This can block blood flow to the spinal cord, seriously damaging it and potentially causing permanent paralysis.

Antibodies the immune system produces against a cancer can also sometimes attack the nervous system. This is called a "paraneoplastic

---

[xliii] In the case of the lung, mucous is swept up out of the lung, carrying bacteria with it.

[xliv] Urinary tract infection if the ureter or urethra are blocked, "obstructive pneumonitis" if an airway is blocked, "cellulitis" if lymphatic channels are blocked and "cholangitis" if the biliary system is blocked.

syndrome." Neurologic paraneoplastic syndromes most commonly occur in neuroendocrine cancers that have some proteins in common with the nervous system. Different paraneoplastic syndromes can damage different parts of the brain, causing problems such as dementia, weakness, loss of balance, numbness, or tremors. The paraneoplastic syndrome can develop months or years before the cancer is detected. Treatment of the cancer will sometimes help neurological paraneoplastic syndrome symptoms, but often it does not. For example, one patient I saw recently had suffered from the sudden onset of dementia and unsteadiness on his feet. He had a neuroendocrine cancer called small cell carcinoma of the lung. Treatment with chemotherapy and radiotherapy resulted in disappearance of all evidence of the cancer, but his neurological symptoms never improved.

Another recent patient developed uncontrolled random jerking of his eyes and legs. He could neither see properly nor walk. He was diagnosed as having the opsoclonus/myoclonus ("dancing eyes/dancing feet") syndrome. Like the previous patient, he had small cell carcinoma of the lung. His cancer disappeared rapidly with chemotherapy and radiotherapy. He was also treated with intravenous immunoglobulin treatment—and unlike the other patient discussed above, had marked neurological improvement.

A third recent patient developed a rash and severe weakness in mid-2020. He was diagnosed as having dermatomyositis. Extensive scans were negative. However, in July 2021, CT scans showed evidence of a cancer involving his liver and the lymph nodes in his chest. A biopsy showed that he had large cell neuroendocrine carcinoma of the lung. His neurological symptoms have not improved despite regression of his cancer with chemotherapy.

Swelling: Swelling (also called edema) is very common in patients with advanced cancer. It can arise several ways. For example, increased pressure in veins (due to blockage by a blood clot[xlv] or tumor) or lymphatic channels (due to blockage by tumor or scar tissue) will increase pressure within the veins or lymphatic channels. This can lead to leakage of fluid into surrounding tissues.

[xlv]   The blood is abnormally "sticky" in cancer patients, so blood clots are common. Swelling of one leg but not the other raises suspicion of a blood clot in the swollen leg.

Also, if the main vein draining the upper part of the body (the superior vena cava, or SVC) is compressed by a tumor, this may cause edema of the face and arms. One of the earliest symptoms of SVC obstruction is often edema around the eyes on awakening in the morning, with reduction in the edema (due to impact of gravity on drainage of the fluid) after the patient gets up and around in the morning. There can also be generalized edema if a malignant pericardial effusion increases venous pressure by impeding return of blood to the heart.

Edema can also be caused by increasing leakiness of capillaries and by decreased blood albumin levels in patients with cancer. Inactivity also increases edema. Muscles contracting while walking help force tissue fluids back into the blood stream, and inactivity decreases these muscle contractions. If a person spends most of the day sitting, gravity will also contribute to development of leg edema. A paralyzed limb will also frequently be edematous since it is usually both inactive and hanging down rather than being elevated. Patients who cannot lie down to sleep due to pain or shortness of breath will also develop leg edema.[xlvi]

Hoarseness: Anyone can develop laryngitis due to inflammation of the vocal cords. However, hoarseness may also occur if a vocal cord is invaded or paralyzed by cancer. Lung cancer may cause vocal cord paralysis by compressing the nerve that makes the vocal cord move.

Further Details and References

I will now go into further detail on how cancers can cause symptoms.

Pain: As discussed earlier, pain is one of the most feared symptoms of cancer, but about one-third of patients with advanced cancers have no cancer-related pain.[279] For those with pain, we are now much better at controlling it than we were a few decades ago, but we still have a long way to go.[280]

Pain is your body's warning signal that you are at risk of major damage. If a child puts their hand in a flame (for example), they will feel immediate

---

[xlvi]   I frequently see this in patients who must sleep in a recliner chair due to pain, dyspnea or cough.

pain and will reduce damage by pulling their hand out of the fire. Pain helps us identify and mitigate serious risks.

Somatic pain: Pain researchers describe three different types of pain: somatic, visceral, and neuropathic. Cancer-induced somatic pain happens when cancer invades bone, joint, muscle, or skin. This pain is caused by cancer-induced local inflammation, by minor or major bone fractures, by breakdown of bone in response to chemicals released by cancer cells, by compression of nerves by the tumor, and by spasm of muscles overlying tissue damage, etc.[280,281] With somatic pain, a patient can accurately pinpoint the site of the problem.

Neuropathic pain: Cancer may also cause neuropathic or neuralgic pain by directly damaging nerves.[281] Neuropathic pain can be particularly difficult to control, although narcotics, anticonvulsant drugs like pregabalin, some antidepressants like amitriptyline and some topical therapies can help.[282,283]

Visceral pain: On the other hand, pain related to internal organs (called visceral pain) is often generalized and difficult to localize and describe.[281,284] Furthermore, there may be no pain at all from internal organs despite extensive invasion or destruction. Pain may not occur until the tumor invades the organ's outside surface capsule, until it invades adjacent tissues, or until it almost completely blocks a tubular organ such as the bowel.[281] Consequently, a tumor may reach a large size inside these organs before it produces any symptoms that lead to its discovery. As an example, pancreatic cancer is often extremely painful, but it does not start causing this pain until it is large enough that it is invading adjacent tissues.

Headache as a variant of visceral pain: The brain is one example of a site generally characterized by visceral pain, with no pain or symptoms until the tumor is quite large. Tumors in the brain cause pain primarily by causing increased pressure. The inside of the skull is a fixed, confined space, and growth of tumor in the brain crams increasing bulk into this confined space. Furthermore, the blood vessels in and around the tumor become leakier so that fluid (edema) leaks out into the brain tissue around the tumor. This edema further increases the bulk in this confined space. The pressure caused by this increased bulk is experienced as a headache.

When a patient with a brain tumor lies down at night, gravity causes an increase in the fluid leaking out into the brain around the tumor, and when the patient gets up in the morning, gravity accompanying the upright position results in fluid draining back into the blood stream again. Consequently, headaches from brain tumors are often at their worst on awakening in the morning, and then improve when the patient gets out of bed. Many healthy people may have headaches, but oncologists are concerned if their cancer patient experiences headaches that are at their worst upon awakening in the morning. This is particularly true for cancers like lung cancer which have a relatively high risk of metastasizing to the brain.

Treatment of pain: As discussed earlier, shrinking a cancer with treatment is one of the most effective painkillers we have. Overall, our management of cancer pain is much better now than it was a few decades ago, and the evolution of the specialty of palliative care has played an important part in this.

There is a long-standing interest in marijuana derivatives (cannabinoids) for treatment of cancer-related pain. Clinical trials to date have given conflicting results. There is no definite evidence of benefit for most types of cancer pain,[285,286] and cannabinoids can have substantial side effects.[48,285] However, there is some evidence that cannabinoids may be useful for neuropathic cancer pain.[48,285] This could be valuable in light of the difficulty often faced in treating neuropathic pain. Needed additional cannabinoid clinical trials are ongoing.

Physician-assisted suicide: In several jurisdictions in the world, physician-assisted suicide (also called "medical assistance in dying" or MAID) has become available for patients with intolerable suffering. When it was legalized in Canada in 2016, I had serious concerns that it might be abused. The dark truth is that for an overburdened healthcare system, the cheapest patient is a dead patient. Caring for a suffering patient can be expensive for a healthcare system and burdensome for a patient's family. Economic analyses indicate that MAID saves money.[287] I was concerned that there might be subtle or overt pressure put on patients to coerce them into requesting MAID to reduce this burden on society and families. I am relieved that to date I have seen little indication of this happening, but the experience from other countries tell us that this clearly is a risk.[288]

Many of us were also concerned that MAID eventually would result in decreased funding for the effective cancer treatments and palliative care that are essential to keep patients both comfortable and alive. As the situation evolves, it will be essential that we continue to monitor this.

Weight loss and loss of appetite: It is common for cancer patients to experience weight loss or cachexia. Cachexia in cancer patients is associated with higher hospitalization rates,[289] increased risk of complications from anticancer therapies,[289] and shorter life expectancy.[289-292] Cachexia is also seen with other chronic illnesses such as uncontrolled AIDS and tuberculosis.[293]

The probability of developing cachexia varies with tumor type.[292] It's seen in about 80% of patients with advanced cancers of the stomach or pancreas, about 50% of patients with advanced cancers of the lung, prostate, or colon, and in about 40% of patients with other types of advanced malignancy.[293]

As discussed earlier, cancer cachexia is caused primarily by inefficient use of nutrients by the cancer, breakdown of muscle, fat and other tissues, and impaired appetite.

Wastage of energy sources: Cancers waste energy.[289,292] There can be increased energy consumption by the tumor itself[289,292] (an estimated 100–1,400 kilocalories per day).[289] The cancer can also produce substances that cause normal tissues to consume excessive amounts of energy.[292] Cachectic cancer patients have abnormalities of metabolism of protein, fat, and carbohydrate.[292]

Normal cells break down glucose to produce energy in the form of adenosine triphosphate or ATP. Carbon dioxide is also produced. The cell normally uses what is called oxidative metabolism[xlvii] to do this. Oxidative metabolism requires the presence of oxygen.

If oxygen is in short supply, the cell will use a process called anaerobic glycolysis instead. In anaerobic glycolysis, there is substantially less ATP produced than there is with oxidative metabolism. With glycolysis, glucose breakdown produces lactate instead of carbon dioxide.

Instead of using either oxidative metabolism or anaerobic glycolysis, cancer cells may use a process referred to as aerobic glycolysis. Aerobic glycolysis is like anaerobic glycolysis except that it occurs despite the

---

[xlvii] Also called the citric acid cycle or oxidative phosphorylation.

presence of oxygen. This paradoxical reliance by cancer cells on aerobic glycolysis rather than on oxidative metabolism is referred to as the Warburg effect, after the scientist who first described it.[294] With the Warburg effect, only about 5% as much energy is produced per molecule of glucose than would be produced by oxidative metabolism. For every glucose molecule broken down, only two ATP molecules are produced by aerobic glycolysis compared to thirty-eight ATP molecules produced by oxidative metabolism.[295]

However, aerobic glycolysis produces those two ATP molecules much faster than oxidative metabolism produces <u>any</u> ATP molecules. If there is a sufficient source of sugar, this rapid access to ATP gives the cancer cell a survival advantage over competing normal cells. For example, the cancer cell using up large amounts of glucose may mean that there is not enough glucose to fuel immune cells that come to kill the cancer.[295] This fast supply of ATP also helps the cancer cell immediately adapt to a rapidly changing, hostile environment.[296]

Cancer cells may use both oxidative metabolism and the aerobic glycolysis of the Warburg effect,[295,297] but on average, two-thirds of a cancer cell's glucose molecules are metabolized through aerobic glycolysis.[297] The Warburg effect is essential for development of a cancer. The oncogene mutations that lead to development of a cancer also induce these new cancer cells to convert their metabolism to aerobic glycolysis.[295]

Aerobic glycolysis also produces a large amount of lactate (unlike the carbon dioxide produced by oxidative metabolism). This lactate makes cancer tissue quite acidic,[295] which may help cancer cells invade adjacent tissues.[295] The acidity can be toxic to surrounding normal cells that compete with the cancer cell for nutrients[298] and to immune system cells that try to kill cancer cells. But this acidity may also make cancer cells more sensitive than normal cells to some chemotherapy drugs such as cisplatin that can reach high concentrations in acidic tissues.[299]

Using glycolysis rather than oxidative metabolism also means that a cancer cell is less dependent than a normal cell on a ready supply of oxygen from local blood vessels. This, in turn, means that a cancer cell may survive and reproduce even if the growth of new blood vessels has not kept up with the growth of the cancer.[297]

The large number of carbon molecules and other elements produced by aerobic glycolysis are available to the tumor as building blocks needed for rapid growth.[295] Aerobic glycolysis may also support production of other molecules that are needed for "cell signaling" that can promote tumor cell survival and growth.[295]

Hence, the Warburg effect provides several advantages to cancer cells, but it also means that in the process of ATP production, cancer cells waste a substantial amount of glucose. This wastage can contribute to a cancer patient's weight loss. The energy wastage can be further increased because the immune system and the tumor produce proteins that induce fat cells to generate heat instead of energy.[293]

Breakdown of muscle, fat, and other tissues: Immune cells responding to the cancer can create substances called cytokines that can break down normal tissues. This "auto-cannibalism" is a major contributing factor to cachexia. It ensures that the cancer has a steady supply of glucose to satisfy its inefficient energy needs, no matter how little the patient eats.[289,292]

Cachexia is much different from the weight loss associated with starvation.[293] In starvation, the body burns stored fat and preserves muscle, but in cachexia, the body burns muscle in addition to fat.[293] Cachexia impacts many organs in the body, including muscle, fat, heart, liver, bone, gastrointestinal tract, and brain.[300] For example, up to one-third of patients may have detectable impairment of memory and mental function when they are first diagnosed with cancer, and this may worsen with initiation of treatment.[301]

Cancers commonly induce inflammation (including the reaction of the immune system to a cancer). This is a key contributor to auto-cannibalism and cachexia.[300] Immune cells called macrophages and lymphocytes produce pro-inflammatory proteins called cytokines. Cytokines such as tumor necrosis factor-alpha (TNF-α, also known as cachectin), interleukin-1, interleukin-6, interleukin-8, and interferon-gamma all contribute to cachexia.[289,292,293] TNF-α promotes increased production of glucose through breakdown of muscle and fat, and inhibits the synthesis of the proteins needed for building muscle.[292,293]

These TNF-α effects may be magnified by interferon-gamma.[293] TNF-α, interleukin-6, and various other factors also induce breakdown of fat and inhibit fat formation.[292,293] They also may play a role in the impairment of memory and thinking commonly associated with malignancies.[301]

Various other factors released by cancers can directly or indirectly lead to metabolic derangement[289] and breakdown of muscle and fat.[300] For example, the liver may be induced by the cancer to produce substances that break down muscle protein into amino acids. The liver then uses these amino acids to produce glucose using inefficient mechanisms that waste energy.[293] At the same time, the tumor will rapidly take up nitrogen and amino acids produced by the liver, making them unavailable to muscle for protein synthesis.[292,293]

Fever from any cause also can increase the muscle breakdown caused by the cancer,[292] and cancers may either cause fever or increase the risk of fever-inducing infections.

Treatment of the cancer is the key to preventing and reversing this tissue breakdown. Exercise is also very important in preserving muscle mass in cancer patients.[289] I tell my patients that it may be difficult when they feel tired and sick, but staying active and exercising is very important to maintain their strength and quality of life. Various other approaches have been tested to treat cachexia. There are a few new medications that look promising but are not yet available for general use.[289]

Appetite loss ("Anorexia"): Another key cause of cachexia in cancer patients is loss of appetite, or anorexia. Anorexia is very common, occurring in 40% or more of patients with advanced cancers.[290,291] Anorexia is caused in part by the impact of several cancer-associated substances on the areas of the brain that control appetite.[290] For example, pro-inflammatory cytokines that are produced by immune system lymphocytes and macrophages in response to the cancer and that can cause muscle breakdown also may suppress appetite.[290] Cytokines such as TNF-α,[292,293,302] interleukin-1,[293,302] interleukin-6,[290,293] interleukin-8,[293] and interferon[302] may be particularly important causes of cancer-associated anorexia.

Even if a patient is able to eat, cancers may decrease absorption of nutrients from the gastrointestinal tract in a variety of ways.[292]

Depression, psychological distress and severely impaired kidney and liver function can also contribute to anorexia in cancer patients.[292] So can high lactate levels associated with the Warburg effect.[290,292,302] A cancer also can result in increased levels of the neurotransmitter serotonin,[290,292,302] of various peptides,[290,302] and of the inflammation marker C-reactive protein,[290] all of which can contribute to anorexia.

Anorexia may also in some cases be a "learned food aversion." Typically, this is an association of a particular food or taste with an unpleasant internal symptom like nausea and vomiting.[292]

An altered sense of taste (called dysgeusia) or altered sense of smell in cancer patients is another key factor.[290,292] Foods no longer taste good, and there may in particular be an aversion to meat.[290,292] Cancers can also cause a decrease in the body's zinc levels through a number of mechanisms, and low zinc levels may impair the sense of smell[292] and the function of the taste buds.[292,302]

Some patients will feel hungry, but when they try to eat, they chew their food, but cannot get it to go down. This may be related to cancer-related increases in brain serotonin levels.[292] Cancer patients also frequently develop a symptom referred to as "early satiety":[290] they may initially feel hungry but then begin to feel full after eating only a small amount of food.[292] An impact on brain appetite centers of the high lactate levels produced by the Warburg effect may contribute to early satiety,[293] as may a substance called leptin that is increased in response to the cytokine interleukin-1.[293]

Decreased contractions of the stomach muscles may cause the stomach to empty more slowly than normal.[290,293] This "decreased gastric motility" may be caused by an effect on the brain of the same substances that cause anorexia.[290] Since the stomach only empties slowly after a meal, patients often notice that they can eat a reasonably good breakfast (since the stomach had gradually emptied overnight), but then they are not able to eat much for lunch or dinner.[292] Some patients find that eating five or six small meals a day works better than eating three normal size meals. Alternatively, drinking high calorie liquid dietary supplements can be easier than eating solid food. In some patients, the drug metoclopramide can help by stimulating the stomach to empty faster.

Other factors causing decreased food intake: Head or neck cancer may interfere with chewing or swallowing, either by causing pain or else by directly blocking food from going down. In some patients, the muscles of the throat or esophagus may not contract properly. In this situation, the individual may aspirate liquids into the lungs, causing potentially fatal choking or "aspiration pneumonia." Nevertheless, they may be able to swallow thick liquids. Specialized professionals called speech and language pathologists may be very helpful in the management of this by teaching the patient different tricks to help with swallowing.

If a patient is unable to swallow, high calorie liquids can be given by a tube inserted through the nose (a "nasogastric feeding tube") or directly through the abdominal wall into the stomach (a "percutaneous endoscopic gastrostomy" or "PEG" tube). However, some patients find these tubes uncomfortable. Many patients also do not tolerate tube feedings because decreased gastric motility slows stomach emptying, or worse, because the fluids flow back up the esophagus and into their lungs, causing potentially deadly choking.

Metastases may directly block the small bowel.[292] Surgery does not usually help alleviate this blockage since if a surgeon operates, they often find multiple sites of obstruction, making it impossible to bypass them all.[292] A blocked small bowel may open and start working again for a period of time if pressure is taken off the bowel by temporarily putting it at rest. To put the bowel at rest, a nasogastric tube may be placed through the nose into the stomach to suck fluids out of the gastrointestinal tract for a few days. Administration of drugs like dexamethasone and octreotide may also help reopen a blocked bowel. This may get the system working well enough to get the patient eating and home again for a few days or a few weeks. But it often does not work for long unless chemotherapy succeeds in shrinking the cancer.

The bowel may stop contracting sufficiently to push digested food through. This is referred to as "paralytic ileus" and can be caused by narcotic pain killers, by severe electrolyte imbalances (such as a low blood potassium level), or sometimes by chemotherapy or other medications. Resting the bowel by nasogastric suction, correcting electrolyte imbalances by intravenous replacement of deficient factors, and drugs such as metoclopramide may help with this.

For some illnesses in which the bowels are not functioning properly, patients may be given high calorie liquids intravenously, in a process called "total parenteral nutrition" or TPN. However, TPN does not prolong average life expectancy in patients with advanced cancers, and is associated with increased risk of infections and other complications.[291] Consequently, TPN is felt to be a bad idea for most patients with advanced cancers,[291] although under some circumstances an occasional patient may benefit.

Counteracting appetite loss and energy wastage: Anticancer therapies like chemotherapy can be a double-edged sword in dealing with appetite loss and energy wastage. They may cause nausea and vomiting. But, if they shrink the cancer, many patients then experience an improved appetite and sense of taste.[292] This is one example of how systemic therapies improve quality of life in the average patient despite the very real side effects they can cause.

Some hormonal agents can also stimulate appetite. Despite this, they do not have any major impact on life expectancy or overall quality of life.[291] The hormonal agent megestrol acetate can increase both appetite and weight gain in cancer patients, but can also be associated with increased risk of serious blood clots.[291]

Corticosteroids like dexamethasone and prednisone may stimulate appetite.[291] Corticosteroids can improve patient quality of life, but they can cause several complications including insomnia, mood swings (sometimes euphoria, sometimes marked irritability, and rarely psychoses), worsened diabetic control, cataracts, bone fractures, life-threatening infections, and stomach bleeding.[291]

Of most importance, corticosteroids often cause a "proximal myopathy." They weaken the muscles of the shoulders and hips, leading to difficulty lifting the arms above the head or difficulty getting up out of a chair or climbing stairs. This myopathy can be disabling and may not improve if the corticosteroids are discontinued. Hence, corticosteroids can be helpful, but they must be managed carefully. We typically try to give them for as short a time as possible and at the lowest dose that does the job. Because of the risk of complications, some experts feel that they should only be used for appetite stimulation in patients with a life expectancy that is no more than a few weeks.[291]

Marijuana and cannabis derivatives have been assessed for their ability to improve appetite, calorie intake, and weight loss in cancer patients. There is not yet good evidence that they are of much help for most patients.[48,286]

Fatigue/weakness: Cancers frequently cause fatigue. Below, I will discuss a few mechanisms by which they do this.

Fatigue caused by cachexia: Cancer patients feel tired and weak because of the energy wastage and breakdown of muscle tissue caused by their cancer.[293,300] This fatigue may manifest not just as tiredness and weakness but also as impaired concentration, poor memory, and emotional lability.[293] Insomnia due to worry and pain can contribute to the fatigue.[303]

As noted before, it is important to stay active.[293,300] Exercise programs have clear benefit in cancer patients.[300] In addition to helping maintain and build muscle mass, exercise can help reduce the inflammation that drives cachexia.[300]

Fatigue caused by anemia: A cancer may also increase fatigue by causing anemia. With anemia, there are fewer red blood cells to carry oxygen to tissues. Cancers may ulcerate, leaving an exposed surface that oozes blood. This is common in cancers of the bowel. This chronic, slow blood loss can result in iron deficiency, and this in turn can contribute to the development of anemia. The available red blood cells may also not carry oxygen efficiently if the patient is deficient in phosphorus due to anorexia.[304]

Cancer-induced inflammatory changes similar to the ones that lead to cachexia and anorexia also can inhibit formation of new red blood cells by the bone marrow.[305] This can lead to "anemia of chronic disease"—a condition seen with cancers, various infections, and autoimmune diseases.[305]

Cancer metastases to bone may sometimes cause anemia by crowding out or replacing the bone marrow. This is seen most frequently with leukemias and lymphomas but also can be seen with small cell lung cancer and cancers of the prostate, breast, and occasionally with other malignancies.

Occasionally, red blood cells may be broken down abnormally rapidly. Termed "autoimmune hemolysis," this happens because the immune system targets red blood cells. Autoimmune hemolysis can be seen in some types of leukemias and lymphomas, but rarely in other malignancies.[306] Red blood cells also may be torn apart by tumor cells or by strands of blood clotting proteins in small blood vessels, causing what is termed "microangiopathic hemolytic anemia."[307]

Fatigue caused by high calcium levels: Some cancers release substances that leach calcium from the bones. This may cause high calcium levels in the blood.[308,309] Other factors that can contribute to high calcium levels include dehydration, immobility, breakdown of bone by metastases, or consumption of calcium-rich foods and vitamin D.

High calcium levels may cause fatigue, drowsiness or confusion, and may prove fatal if not treated.[309] In addition to tiredness, other early symptoms of high blood calcium levels are nausea and increased frequency of urination (called polyuria) combined with increased thirst (called polydipsia).[309] This polyuria and polydipsia is similar to what is seen in diabetics with high blood sugar levels. The oncologist's top suspicion when a patient complains of increased urine output and thirst is that they may have abnormally high blood levels of either calcium or sugar. High calcium levels may improve with cancer treatment, intravenous saline administration, and medications such as bisphosphonates.[309,310]

Fatigue caused by low sodium levels: Fatigue also may be caused by abnormally low blood sodium levels.[311] The pituitary gland secretes anti-diuretic hormone, or ADH, which acts on the kidneys to reduce the amount of fluid they release from the blood into the urinary tract. If the pituitary senses that a person is becoming more dehydrated, it secretes more ADH so that more fluid is retained in the blood stream. Conversely, if the pituitary senses that the person has too much fluid, it secretes less ADH, and more fluid is excreted as urine.

Small cell lung cancers (and occasionally other malignancies) may produce increased amounts of ADH. The result is the "syndrome of inappropriate ADH" or SIADH. With SIADH, too much fluid is retained in the blood stream.[312] This dilutes sodium, resulting in very low serum sodium concentrations.

If mild, low sodium levels may cause no symptoms. But if the sodium is very low, it can cause anything from fatigue to life-threatening seizures.[312] In most cases, the major treatment for SIADH is to restrict fluid intake to less than 1.5 liters per day, while at the same time treating the cancer. More aggressive measures are needed in some patients.[313]

Both cancer-related elevation of calcium levels and SIADH are referred to as endocrine "paraneoplastic syndromes."

Shortness of breath: Shortness of breath, or dyspnea, is very common in lung cancer and can have a major negative impact on quality of life.[314] Dyspnea can be caused by a lung cancer blocking an airway, leading to collapse (atelectasis) of the part of the lung that is upstream from the obstruction. Hence, there is less lung available to take up oxygen. Numerous metastases in the lung (or even a few large tumor masses) also can compress lung tissues and decrease the volume of lung available for oxygen uptake. Additionally, there can be extensive growth of a cancer through the lymphatic tissues in the lung, called lymphangitic carcinomatosis. Lymphangitic carcinomatosis impedes oxygen uptake into the blood. It does this by increasing the thickness of tissue that oxygen must pass through to get from the lung air pockets into the blood stream.

Cancer-related anemia also can contribute to dyspnea. With low hemoglobin levels, the blood has decreased capacity to transport oxygen from the lungs to the tissues. This increases dyspnea, but the dyspnea can be improved with red blood cell transfusions.

Lung cancers also increase the risk of developing pneumonia by blocking airways and thereby making it harder to clear mucous from the lungs. Bacteria may grow in this stagnant mucous. Some lung cancers also produce large amounts of mucous that fill the airways, limiting the ability of oxygen to get into the lungs. If a lung cancer is bleeding, the blood may fill the airways and block oxygen uptake. Cancer-related cachexia also can contribute to dyspnea by causing weakness of respiratory muscles.[315]

Dyspnea caused by cancer may be particularly problematic in patients who had damaged lungs before they developed the cancer. Patients who have developed emphysema due to smoking and individuals with chronic lung scarring (referred to as "pulmonary fibrosis") are at increased risk of developing lung cancers. The cancer may increase the dyspnea that was already present from the emphysema or scarring.

Therefore, the most effective way to treat dyspnea is to treat the cancer with either systemic therapies or radiation therapy. Supplemental oxygen, narcotics like morphine (to reduce the sensation of dyspnea), corticosteroids, and transfusion for cancer-associated anemia also can help.[316]

Malignant pleural effusion: The lung is surrounded by a sac called the pleura that forms a barrier between the lung tissue and the ribs. The area within the sac separating the lung from the ribs is called the pleural cavity. Usually this sac is empty (like a collapsed balloon) so that the lung is right up against the ribs, separated only by a thin layer of pleura. However, tumor deposits on the pleura may make it leaky. Fluid can then accumulate in the pleural cavity, causing what is called a "pleural effusion." A large collection of this fluid can compress the lung, resulting in less lung volume available to take up oxygen.

Pleural effusions may also form if there is extensive cancer involvement of lymph nodes in the center of the chest. The presence of tumor in the lymph nodes may impede the flow of fluid from lung tissues through lymphatic channels and back into the blood stream.[xlviii] A tumor bleeding into the pleural space also can cause sudden onset of pleural effusion.

The fluid can be drained by inserting a needle through the chest wall into the pleural cavity, potentially permitting re-expansion of the lung and rapid improvement in breathing. If this fluid drainage helps the patient's symptoms, then a tube such as a PleurX catheter can be inserted into the pleural cavity and left there long term.[317] The PleurX catheter has a clamp that can be opened to drain fluid two or three times per week. This can be an effective way to manage effusion-related dyspnea. The amount of fluid frequently will decrease over time, particularly if the patient is also receiving an effective systemic therapy for their cancer.

Draining the pleural fluid helps many patients, but not all. In some patients, scar tissue forms on the outside surface of the lung and prevents lung re-expansion despite drainage of the fluid. This is called a "trapped lung."[318] In other patients, fibrinous material may be deposited within the pleural cavity. This can result in the fluid being separated into several discrete pockets, with no flow of fluid from one pocket to the other. This is termed a "loculated pleural effusion."[317] A pocket in which a tube or needle is inserted may drain, but this will not help much as there will be many other pockets that cannot be drained. In some patients, injection of medications such as tissue plasminogen activator (tPA) into the pleural cavity will break down the fibrinous material enough to enable drainage of the entire pleural cavity and re-expansion of the lung.[317]

Malignant pericardial effusion: Just as the lung sits in a sac formed by the pleura, the heart sits in a sac called the pericardium. As with the pleura,

---

[xlviii] If the pleural effusion is due to metastases of the cancer to the pleura, then the cancer would not be curable. However, if it is due to involvement of central lymph nodes in the chest, it may potentially be curable. Chronic inflammation of the pleura due to radiotherapy may also sometimes cause a pleural effusion.

metastases to the pericardium can make it leaky.[xlix] [319] A buildup of fluid inside the pericardium (a "pericardial effusion") can compress the heart,[319] reducing the ability of blood to get back into the heart from the rest of the body. This means less blood entering the heart for it to pump to the lungs to take up oxygen and less oxygenated blood getting back into the heart from the lungs. Hence, a pericardial effusion can also cause dyspnea.[320]

A large volume of pericardial fluid can cause "cardiac tamponade," with bulging of the neck veins (since blood is impeded from draining into the heart) and dangerously low blood pressure (since the heart does not have enough blood to pump out into the circulation). An ultrasound of the heart called an echocardiogram will show diagnostic changes as the situation evolves from inconsequential pericardial effusion to cardiac tamponade. Cardiac tamponade is a medical emergency. Draining the pericardial effusion by inserting a needle through the chest wall into the pericardial sac can give marked and rapid improvement in a patient's symptoms.[319]

Blood clots: Cancers produce several substances that make the blood more sticky than normal. This increases the risk of blood clots in both veins and arteries.[321] The highest risk of blood clots is associated with cancers of the pancreas, uterus, lung, stomach, kidney, and brain.[321] These clots will often form in the large veins in the legs and pelvis, and are called a deep venous thrombosis or DVT. If part of such a clot breaks off, it may rise and lodge in the blood vessels in the lungs as a "pulmonary embolus."

A small pulmonary embolus may cause no symptoms, but a larger one can cause sudden, severe dyspnea or even sudden death. When a cancer patient reports that they have developed sudden onset of dyspnea, the oncologist's first thought is pulmonary embolus, particularly if is associated with chest pain that increases with deep inspiration (referred to as "pleuritic chest pain").[l] In a high proportion of patients, such blood

---

[xlix] As with a pleural effusion, pericardial effusions can also be caused by inflammation from unrelated causes.

[l] A cracked or broken rib or a chest wall muscle strain may also cause chest pain that increases with a deep breath. If these are the cause, the patient will often have tenderness when pressure is applied to the chest wall, while the pain from a pulmonary embolus is not usually associated with chest wall tenderness.

clots can be treated effectively with injectable or oral blood thinning medications called anticoagulants.[322]

Pneumothorax: A leak may develop between an airway and the pleural space. This can cause sudden onset of dyspnea and pleuritic pain from lung collapse. This can be rapidly alleviated by inserting either a chest tube or a needle attached to a Heimlich valve[323] to drain the air.

Cough: More than 60% of patients with lung cancer have a cough at the time of diagnosis.[324] In fact, a progressive, worsening cough is a common first symptom. However, the overwhelming majority of people who develop a chronic cough do not have lung cancer,[324] so cancer-related cough is often initially treated as pneumonia, bronchitis, allergy, etc. The scans that reveal the presence of a lung cancer are often not done unless the cough continues to worsen over time or if there is development of other symptoms like weight loss or pain.

Cough is more likely to occur with a cancer that started in the lung than with metastases to the lung from another site. This is because primary lung cancers start in the airway where they can irritate "cough receptors," while metastases are typically found deeper in the lung, away from the cough receptors of the major airways.[324] Primary lung cancers that start in the airway are also more likely than metastases to cause the obstruction of mucous flow that may cause cough-inducing airway infections.[324] Despite being far from the major airways, malignant pleural effusions may also cause a dry cough.[325]

As discussed earlier, a severe cough can cause chest wall pain and may seriously interfere with sleep. Overall, cough can have a major negative impact on patient quality of life.[326]

Narcotics like hydrocodone syrup will help the cough in some patients, but it often does not help much.[324] Some inhaled medications can also help cough,[324] as may the antiseizure/pain medication gabapentin.[327] However, by far the best cough treatment is treatment of the cancer with systemic therapies or radiotherapy.[324] This will often improve both the cough and the patient's quality of life.

Coughing blood (Hemoptysis): Initially, a cough alone may not raise suspicion that there is a lung cancer, but hemoptysis will.[328] It is a frequent early symptom of lung cancer.[329] In about 40% of patients developing hemoptysis, the underlying cause is lung cancer, although it also may occur with pneumonia, COPD exacerbation, tuberculosis, or pulmonary embolism.[330] Approximately 20% of lung cancer patients will experience hemoptysis at some point during the course of their illness,[329] and "massive hemoptysis" (i.e., hemoptysis that is severe or life-threatening from erosion of a cancer into a major blood vessel) occurs in about 3% of lung cancer patients.[329] Radiotherapy or systemic therapies can rapidly improve hemoptysis. In my experience, antibiotic treatment also helps, presumably by reducing infection in ulcerating tumors.

For massive hemoptysis, a radiologist may inject dye to identify the blood vessel that is the source of bleeding. The bleeding can be stopped by injecting materials that block the blood vessel—a process called "bronchial artery embolization." Bronchial artery embolization will stop the bleeding in 70–99% of patients, at least temporarily.[331]

Infection: Infections are among the most common cancer complications.[332] In leukemias and lymphomas there may be insufficient production of white blood cells or there may be other immune system abnormalities. In most cancers, however, the increased risk of infection is due either to break down of mechanical barriers to bacteria or blockage of flow of fluids.[332] For example, an ulcerating tumor involving the skin (e.g., with advanced breast cancer) may allow bacteria that would have been blocked by the skin to get into the deeper tissues where it can cause serious infection.[332] The same can happen if there is an ulcerating tumor involving the mouth, throat, bowel, etc. The damaged epithelium no longer provides an effective barrier to bacteria.[332] At their worst, cancers in one organ can invade an adjacent organ, forming an opening called a fistula. This permits bacteria to travel from an organ in which they are tolerated (e.g., the bowel) into an organ where they can cause serious infection (e.g., the bladder).[332]

Furthermore, any bacteria that enter the urinary tract (for example) are ordinarily eliminated when the individual urinates. If a partial blockage impedes emptying of the urinary tract, then the retained urine can stagnate, permitting growth of the bacteria and a potentially serious infection.[332] Similarly, the airways produce mucous that is moved up and out of the airways, clearing bacteria from the lungs in the process. However, in a blocked airway, the secretions stagnate, increasing the risk of an "obstructive pneumonitis."[332] Blocked lymphatic channels slow the flow of lymphatic fluids and increases the risk of skin infections such as cellulitis and erysipelas.[333] Blocked bile ducts may result in "ascending cholangitis" with infection of the biliary system.[334] The same goes for impairment of fluid drainage from any part of the body.

Any mechanical device that is inserted into the body to help manage the cancer may also become infected. This includes stents to relieve areas of obstruction, metal pins and rods used to stabilize broken bones, and catheters to permit ongoing intravenous fluids or to drain areas such as a pleural effusion or a blocked urinary tract or bile duct.[332]

Cancer can reactivate viral infections like hepatitis B and C.[332] If someone had chicken pox as a child, the virus lives for the rest of their life in a part of the nervous system called the dorsal root ganglia. Various stresses such as cancer can potentially reactivate the virus as shingles, causing painful skin blisters which follow the course of the infected nerve from the spinal cord outwards. While several types of cancer are associated with an increased risk of shingles, the risk is highest with lymphomas.[335]

Bleeding: Bleeding is another common complication of cancer.[336] Cancers cause bleeding since the blood vessels that form as a cancer grows can be abnormally fragile. In addition, cancers may ooze blood after ulcerating through the skin or the wall of any organ, such as an airway, the urinary tract, or the bowel. In some cases, a cancer may erode into large blood vessels such as a carotid or pulmonary artery, causing sudden, massive, life-threatening bleeding.

As discussed earlier, treating the cancer may stop bleeding, but embolization may be needed urgently if a large vessel is bleeding.[336] Applying pressure, dressings and packing can help control gradual blood loss from a superficial site.[336] The drug tranexamic acid also may reduce bleeding by promoting clotting.[336] Treating with antibiotics may help if infection of the tumor is contributing to bleeding.

As I said earlier, many cancers produce substances that increase the formation of blood clots. When anticoagulants are used to treat these clots, they may result in bleeding or in extensive bruising (called ecchymoses). Patients who require anticoagulants to treat blood clots are at a particularly high risk of bleeding if they also have tumor ulcerating through the skin or through the wall of a hollow organ, like an airway or bowel.

Occasionally, a cancer may induce so much abnormal blood clotting that it exhausts the body's supply of clotting factors.[336] Uncontrolled clotting is then followed by uncontrolled bleeding in a process called consumptive coagulopathy, or disseminated intravascular coagulation.

Increased intracranial pressure, headaches, and impaired function: Cancers can cause various neurological complications. Malignancies that involve the brain can cause headaches by generating increased pressure inside the skull. Tumor cells produce "vascular endothelial growth factor" (VEGF) that stimulates production of abnormal, leaky blood vessels. These abnormal vessels permit flow of fluid and protein from blood vessels into brain tissues, creating cerebral edema.[337] Increased protein in the interstitial spaces[li] of the brain creates an osmotic force that pulls even more fluid into these spaces. The size of the space inside the skull cannot increase, so pressure inside the skull goes up because of growing tumor and cerebral edema being added to the brain tissues that are already there. This increased pressure causes headaches, and it also compresses local blood vessels. This compression of local blood vessels decreases blood flow, and impedes function of the tissues supplied by these vessels.[337]

Corticosteroids like dexamethasone may reduce edema around a tumor by decreasing VEGF expression.[337] This can give temporary but rapid improvement in intracranial pressure and neurological dysfunction.[338]

As discussed earlier, headaches from brain tumors are typically worst on first awakening in the morning. Once upright, gravity helps drain some of the extra edema fluid out of the brain. The effect of gravity goes away when lying down, so the fluid that had drained during the day reaccumulates.

Seizures: Brain tumors may cause seizures by increasing the concentrations of neurotransmitters like glutamate, by altering cellular chloride transport, or through other mechanisms.[339] Seizures can be either generalized (involving the entire body) or focal (involving just one body area). The probability of developing seizures varies with tumor type. Among primary brain tumors (i.e., tumors that initially start in the brain), seizures occur in 60–75% of low grade gliomas, in 25–60% of high grade gliomas such as glioblastomas, and in 20–50% of meningiomas.[339] In one series of patients with brain metastases, seizures occurred in 16% of patients with a

---

[li]   Interstitial spaces are the spaces in between cells.

breast cancer primary site, 21% with a gastrointestinal primary, 29% with a lung primary, and 67% with a melanoma.[340] The probability of seizure development also varies with the part of the brain involved by the tumor. The highest risk of seizures is when parts of the brain called the frontal or temporal lobes or the insula are involved.[339]

Treatment of the brain tumor with surgery, radiotherapy, or chemotherapy can reduce the risk of recurrence of seizures, as can antiseizure medications such as Keppra (levetiracetam).[339] However, even with therapy, it may not be possible to guarantee that seizures will not recur. A focal seizure can rapidly spread and involve the entire body, with loss of consciousness. Hence, patients who have experienced a brain tumor-related seizure should always exercise caution. They should not get involved in activities with a high risk of serious consequences if they were to suddenly lose consciousness. In many jurisdictions, physicians are legally required to report their patient to the local motor vehicle licensing bureau if they think the patient is at an increased risk of having a seizure.

Bleeding into brain tumors: There is also a risk of bleeding into a tumor deposit in the brain. As noted previously, new blood vessels formed in brain metastases are abnormal, and can be fragile. For patients with brain metastases, the risk of bleeding varies with tumor type, with the highest risk in malignant melanoma and cancer of the kidney.[341] It is relatively common for scans to reveal evidence of small, asymptomatic areas of bleeding that do not require specific interventions. However, there is also a small risk of sudden, major bleeding that requires surgical treatment, and that can cause serious incapacitation or death.[341]

As discussed above, blood clots are very common in cancer patients, and the presence of blood clots generally requires that a patient take blood thinners (anticoagulants). Anticoagulants are associated with a risk of bleeding complications. Despite this, anticoagulants appear to have minimal impact on the risk of bleeding into most types of brain tumors.[341,342] However, they are associated with an increased risk of bleeding into primary brain tumors and brain metastases from malignant melanoma and kidney cancers.[342]

Spinal cord compression: The bones in the back (vertebrae) are each composed of a large vertebral body (somewhat like the body of a padlock) to which is attached the ring-like vertebral arch (somewhat like the closed shackle of a padlock). The circular space between the vertebral body and the vertebral arch (analogous to the space between the body and closed shackle of a padlock) constitutes the "spinal canal." The spinal cord runs down through the spinal canal, protected by the bony structures of the vertebral bodies and vertebral arches. However, metastatic tumors growing in the body or arch of a vertebrae can invade the spinal canal and push the spinal cord against the bony structures on the other side of the spinal canal.[343] This compresses the blood vessels in the spinal cord, thereby reducing spinal cord blood flow. The reduced spinal cord blood flow causes swelling (edema) in the spinal cord, and this edema puts further pressure on the blood vessels. Eventually, it reaches the point that there is inadequate blood flow to keep the spinal cord alive, destroying that area of spinal cord and leading to permanent paralysis below the level of the blockage.[343]

In a small proportion of cases, spinal cord compression can result from a metastasis inside the spinal cord itself (particularly with small cell lung cancer) pushing the surrounding spinal cord against the bony ring of the spinal canal.[344] Spinal cord compression also can result from tumor invasion along a nerve root into the spinal canal. This is most likely to happen with lymphomas.[343]

While any cancer may cause spinal cord compression, the most frequent causes are cancers of the prostate, breast, and lung. These are common cancers and they often metastasize to the bones of the spine.[343]

Most patients with spinal cord compression have back pain due to involvement of the vertebrae by the tumor. A high proportion will also have weakness and numbness of the legs, and difficulty urinating.[343] Some will also have bowel symptoms, including incontinence due to loss of sphincter control.[343]

Treatment of spinal cord compression is a medical emergency. A patient may go from the first symptoms of spinal cord compression to permanent paralysis within just a few hours.[343] The probability of major improvement with treatment is low if a patient already has paralysis of their legs when treatment is started.[343]

Generally, the corticosteroid dexamethasone is administered as soon as there is clinical suspicion of spinal cord compression.[343] The dexamethasone buys time by reducing edema, thereby temporarily improving spinal cord blood flow. An MRI scan is performed as rapidly as possible to confirm the presence and site of spinal cord compression.[343] Radiotherapy is then started urgently in the hope that it will shrink the tumor and prevent progression to permanent paralysis.[343] With rapid action, spinal cord function and ability to walk may be preserved, and this has a major impact on patient quality of life.[343]

Surgery is also used in some patients. It may result in a better outcome than radiotherapy alone,[343] but it is a major operation. Many patients with advanced cancers would not tolerate the surgery. Surgery also could delay the initiation of chemotherapy that is needed to control the cancer in other parts of the body. Surgery would be of limited benefit or feasibility if the cancer involved several different vertebrae in the spinal column.

Neurological paraneoplastic syndromes: About 0.01% of all cancer patients will develop a neurologic "paraneoplastic" syndrome.[345] In neurologic paraneoplastic syndromes, a patient's immune system generates anticancer antibodies that attack and damage the brain, nerves, or muscles.[345]

Neurologic paraneoplastic syndromes are most likely to develop in patients with "neuroendocrine carcinomas". The tumor cells in neuroendocrine carcinomas such as small cell lung cancer have some characteristics similar to neurological tissues, and antibodies created by the immune system to attack the cancer may also attack components of the nervous system that share characteristics with the cancer.

The neurological symptoms may develop months or years before the cancer itself becomes obvious. Of patients who develop what looks like a potential neurological paraneoplastic syndrome, a cancer is eventually diagnosed in as few as 5% of patients with some specific syndromes, but in up to 60% of patients presenting with other syndromes.[345] When neurologists diagnose one of these disorders, they often then recommend scans to look for a previously undetected cancer.

While neurologic paraneoplastic syndromes are quite uncommon, they can be devastating, causing rapidly progressive, irreversible damage.[345] For example, a type of paraneoplastic syndrome called limbic encephalitis can cause rapid onset of irreversible dementia (often with seizures). Another, called subacute cerebellar ataxia, causes incoordination and unsteadiness.[345] A paraneoplastic syndrome called sensory neuronopathy can cause painful numbness of the feet, and dermatomyositis causes a characteristic skin rash and muscle weakness, particularly involving the shoulders and hips.[345]

While the nervous system damage may be irreversible, patients may at least stabilize if the cancer is treated successfully. A few patients improve with corticosteroids, with intravenous immunoglobulins or with a procedure called plasmapheresis in which the patient's plasma (containing the abnormal antibodies) is removed and replaced by plasma from a normal donor.[345]

Swelling (edema): Edema may be caused by cancer compressing or blocking veins or lymphatic channels. This increases back pressure that can cause fluid leakage into surrounding tissues.[346]

Edema can also be caused by increased permeability or leakiness of capillary walls caused by inflammatory cytokines such as TNF-α and interleukin-1. These cytokines are produced by the immune cells (such as macrophages and lymphocytes) invading a cancer.[346]

In addition, "colloid osmotic pressure" is required inside blood vessels to hold fluid within the blood vessels. The protein albumin (produced by the liver) is the major contributor to the blood's colloid osmotic pressure.[347] If serum albumin levels are low, then fluid will leak out of blood vessels into surrounding tissues.

Serum albumin levels are often low in patients with advanced cancers. The same inflammatory factors that make vessel walls permeable to fluids also make them permeable to albumin, so albumin leaks out of the blood stream.[348] Albumin levels may also be low since TNF-α and related inflammatory cytokines inhibit production of albumin by a cancer patient's liver[349] and because cancer cells break down albumin as a source of building blocks for tumor growth.[348]

Inactivity can cause edema of the lower legs. Sitting with the legs down increases hydrostatic pressure, which results in more fluid leaking out of the blood and lymphatic vessels into surrounding tissues. However, with exercise, there is increased venous blood and lymphatic flow, resulting in decreased pressure. At the same time, the increased tissue pressure of contracting calf muscles acts as a pump to force fluids back into vessels.[350] Leg edema tends to occur in cancer patients who walk less due to weakness and fatigue. Sitting with the legs elevated can help. A partially paralyzed limb also may be associated with edema.

I noted above that some patients sleep sitting in a recliner because of the severity of their cancer-associated cough, or because of dyspnea or pain. In this situation, neither gravity nor muscle action is helping promote overnight movement of tissue fluids from the legs back into the blood stream. These patients will often develop increasingly severe leg edema.

Just as tumor involving the pleura or pericardium can cause pleural and pericardial effusions,[lii] tumor involving the inner lining of the abdomen (the peritoneum) may result in the abdomen becoming uncomfortably distended as it fills with fluid (referred to as ascites).[351] Tumor blockage of the lymphatic channels in the abdomen can also cause ascites, as may blockage by tumor, blood clot, or scar tissue of the portal vein (the major vein that carries blood and nutrients from the intestine to the liver).[351]

The same type of PleurX catheter that is inserted into the pleural cavity to permit repeated drainage of a pleural effusion also can be inserted into the abdominal cavity to permit repeated drainage of ascites.[352]

In addition to edema being uncomfortable, it also can increase the risk of infection. I noted previously that impaired flow of fluid though lymphatic channels increases the risk of skin infections such as cellulitis and erysipelas.[333] No matter what part of the body, stagnation of fluids increases the risk of infection.

Hoarseness: Hoarseness is often an early symptom of cancer of the vocal cords.[353] Hoarseness also may be a symptom of lung cancer. The nerve to operate the right vocal cord goes from the brain to the vocal cord following a route that takes it near the uppermost part of the right lung.[354] The nerve for the left vocal cord goes all the way from the brain down into the lung, under the arch of the aorta, then through the left side of the "mediastinum",[liii] and back up to the vocal cord.[354] A tumor in the left lung that is invading into the mediastinum can damage this nerve, causing hoarseness by paralyzing the left vocal cord.[354] Hence, paralysis of the left vocal cord increases the probability that a cancer of the left lung is invading the mediastinum and will not be removable by surgery. Even if a surgeon tries to remove the cancer, there is always some left behind. The cancer may possibly still be curable by radiotherapy, but not by surgery.

---

lii   As discussed in earlier sections.
liii   The mediastinum is the tissue in between the two lungs.

Hoarseness can impede patient quality of life by making talking very tiring.[355] The nerve function and hoarseness will sometimes improve with treatment of the cancer, but in many patients it will not. Otolaryngologists can sometimes help by injecting the paralyzed vocal cord to make it stiffer.[356]

Conclusions: This chapter is just a brief overview of some of the many symptoms that can be caused by different cancers. The third reason that cancer still sucks is that it can cause immense suffering. The good news is that we are making progress. We have therapies that will improve quality of life by alleviating many of these symptoms. While we still have a very long way to go, our cancer therapies are steadily becoming more effective. Together, these facts—that cancer can cause so much suffering and new therapies can have a major impact on alleviating this suffering—drive home how important it is to move rapidly to develop effective new therapies.

— 4 —

The Master Swordsmen:
Cancer Surgery

The fourth reason that cancer still sucks is that it will recur in some patients even if it appears to have been removed completely by surgery. Surgery is not even an option for many patients.

Short Primer

Patient "resectability" and "operability": The decision about whether to offer a patient surgery is always a balancing act. The probability of success is weighed against the chances of a cancer recurring or of the surgery leaving the patient severely impaired. It is important to involve patients in this decision-making process by informing them of the risks and rewards. They need to know as much as possible to make informed decisions.

Surgeons will generally try to remove a cancer if they think that they have a reasonably good chance of curing the patient. Other things they consider would be the risk of the patient dying from the surgery, the disability and disfigurement the surgery could cause, and the availability of other options, such as radiotherapy. For example, radiotherapy can be effective against a head and neck cancer, although not quite as good as surgery. All things being equal, the surgeon might recommend radiotherapy if surgery might leave the patient permanently disfigured.

Similarly, lung cancer might be treated with radiotherapy if surgery would leave the patient a "respiratory cripple," unable to function because of permanent severe shortness of breath.

Two factors are important in deciding whether to operate. First, a cancer must be "resectable." That means it can be completely removed by surgery without an unacceptable level of risk. Second, the individual must be "medically operable." An inoperable patient might have high-risk major medical problems (such as bad heart disease) or might be otherwise too weak to tolerate surgery.

In general, surgery is not appropriate if the cancer has spread to other sites. In these situations, the surgeon would probably not be able to cure the patient and would feel that the minimal possible benefit would not justify the risks and discomfort of surgery.

Patients with widespread lung cancer often ask me why the surgeons will not remove the lung where the cancer started. They know this is possible because they have heard of people who have successfully undergone removal of a lung. I tell them that taking out a tumor in one area will not cure a patient if there are several other tumors that cannot be removed. Nothing would be gained by major surgery, and it could cause serious complications.

If a cancer is "locally extensive" it may not be resectable. For example, surgery usually doesn't help if a lung cancer is invading the region between the lungs called the mediastinum. A lot of major blood vessels and nerves travel through the mediastinum, and surgery cannot completely remove a tumor in this area without causing major irreparable damage.

Here's a simple way to understand why a cancer may not be resectable. Think of being at the beach and dropping your hamburger in the sand. You pick it up and remove all the sand you can see, but when you bite into it you inevitably hit more sand. The same thing happens if a cancer has spread widely into adjacent tissues. No matter how extensive the surgery, there will be some tumor left behind.

If a lung cancer has just invaded into the adjacent pleura (i.e., the outside lining of the lung), it may be possible to remove it by surgery. But if there is extensive tumor involvement of the pleura, it would not be considered resectable. Even if the lung is completely removed there will be residual unresectable tumor deposits in the adjacent chest wall.

A potentially resectable tumor also might be considered inoperable if there is a very high probability of cancer recurrence, despite complete removal of the cancer. For example, cancers often spread through "lymphatic channels" to form metastases in local lymph nodes. Tumor cells that survive in lymph nodes increase the probability that tumor cells circulating in the blood stream would also form metastases. The more extensive the lymph node involvement, the higher the probability that blood-born metastases eventually will appear elsewhere. Hence, even if the lymph node metastases are resectable, surgery may not be worth the risk.

Typically, the larger the number of lymph nodes involved by the cancer and the further the involved nodes are from the primary tumor, the higher the risk of eventual appearance of incurable distant metastases. For example, with lung cancer, surgery may be worth it if there are cancer-infected lymph nodes within the lung itself. On the other hand, surgery is generally not worth it if there is extensive involvement of lymph nodes in the mediastinum. In this case, radiotherapy is often offered instead. As noted above, radiotherapy may be somewhat less effective than surgery, but it is also less likely to cause major complications—and debilitating loss of quality of life. If the treatment is unsuccessful the patient has at least lost less with radiotherapy than with unsuccessful surgery.

Surgical myth: According to an old wives' tale, surgery makes a cancer grow faster "by exposing it to air." This is not correct. I address this further in Chapter 10, Oncology Myths, and Legends.

Multidisciplinary decision-making: In difficult cases, surgeons usually have help deciding about who should or should not have surgery. When the answers aren't clear cut, many centers have weekly "multidisciplinary tumor rounds" where several specialties meet to discuss the best option. For example, in Ottawa, our weekly Multidisciplinary Thoracic Oncology Rounds on lung cancer patients involve thoracic surgeons, medical oncologists, radiation oncologists, pulmonologists, radiologists, nuclear medicine specialists, pathologists, nurses, social workers, and others. Each group contributes to the discussion and a consensus is reached on the best option to be offered to a patient.

Similar multidisciplinary rounds are held regularly for most tumor types. In each case, the group decisions are informed by evidence-based and best-practice guidelines that have been developed by international or national experts.

Why surgery can fail: Sometimes during surgery cancers that initially appear to be resectable are found not to be—usually because they have invaded surrounding structures that cannot be removed. In addition, distant metastases may show up months or years after surgery. I noted in Chapter 2 that there may be millions of tumor cells circulating in the blood stream that have escaped from the primary tumor. However, in some patients, none of these cells circulating in the blood stream are capable of establishing metastases. This is the reason that some cancers may be cured by surgery despite these circulating tumor cells.

However, if one circulating cancer cell can establish itself as a metastasis in a distant site, then there are probably many others that will do the same. Metastases smaller than about the size of the nail on my small finger may be too small to be seen on a scan. A patient may have none of these or may have millions of them that only show up long after surgery.

Assessing the risk of postoperative cancer recurrence to guide use of adjuvant therapies: Once a patient undergoes surgery, a pathologist carefully examines the resected tissue. They confirm not just the type of the cancer, but also look for factors that will affect the risk of recurrence. For example, they will report microscopic evidence—if any—of the cancer extending to the "surgical margins" (i.e., to the very edge of the tissue that the surgeon removed). If the pathologist detects microscopic evidence of tumor cells extending right to the edge of the tissue that the surgeon removed, this will indicate a very high probability that there were tumor cells left behind at the time of surgery. There will be a high risk that the cancer will recur unless something more is done. In some cases, the surgeon might then do further surgery to try to remove the residual cancer, or radiotherapy might be given to try to eradicate surviving tumor cells.

Again, a surgeon always weighs multiple factors. The more tissue removed, the better the chance that all the cancer is gone. However, the larger the amount of normal tissue removed, the higher the risk of a major negative impact on the patient's quality of life. A surgeon might consider taking the patient back to surgery for a "wide excision" if the pathologist sees tumor cells at the surgical margins. A larger amount of tissue around the original tumor site would then be removed.

For example, basal cell carcinomas of the skin due to sunlight exposure are extremely common, and many of us have had them. They are also highly curable by surgery. However, my wife Lesley had an uncommon type of basal cell carcinoma removed from her back. It had an increased risk of local recurrence due to invasion into surrounding tissues. Consequently, she underwent a "wide excision" of the area. No residual cancer was found, and she has had no evidence of recurrence. When I had a basal cell carcinoma on my nose, it was removed by a 4 mm "punch biopsy." When the pathologists examined it under a microscope, the margins were negative for cancer and it was a standard "low-risk" type of basal cell carcinoma, so no further surgery was needed.

If residual tumor is found at the margins of surgery, radiotherapy might be given to the area around the original tumor to try to kill off any remaining tumor deposits in the area. In breast cancer, it is usual to do a "lumpectomy" for selected tumors (just removing the tumor that can be seen) without doing disfiguring extensive surgery. This limited surgery is then routinely followed by radiotherapy to deal with any local tumor(s) that may be left behind.

The pathologist also assesses other characteristics associated with an increased risk for developing distant metastases. This includes information about the size of the cancer and on "tumor differentiation." In a "poorly differentiated" cancer, the tumor cells look markedly different from normal cells. In a "well differentiated" cancer, the tumor cells look similar to normal cells. Poorly differentiated tumors are more likely to recur and behave in a highly malignant fashion than are well differentiated tumors.

The pathologist will also look for things like involvement of lymph nodes in the resected tissue, other small "satellite" tumor deposits separated from the main tumor mass, invasion of the tumor into blood vessels and lymphatic channels, penetration of the tumor to involve the outside surface of the organ, spread of lung cancer cells along an airway, etc. Presence of any of these factors increases the risk of eventual recurrence of the cancer.

Taken together, this information is used to establish the "stage" of a patient's cancer[liv] and to make a best guess as to the probability of recurrence. This information is then used to advise the patient on possible next steps.

After removal of a cancer,[lv] systemic "adjuvant therapy" may be offered to patients at moderate or high risk[lvi] of recurrence to try to reduce the probability of recurrence. Systemic therapies like chemotherapy, hormone therapy, and immunotherapy are given intravenously or by mouth and are distributed through the blood stream to all parts of the body. For most cancers, these systemic therapies cannot cure patients with advanced metastatic disease. They may, however, increase cure rates after surgery or radiotherapy by killing "micrometastases" (i.e., metastases that are initially too small to show up on a scan but that would eventually grow and take the patient's life).

Adjuvant therapies can help for most types of cancer that are at moderate or high risk of recurrence. For most malignancies, the size of the gain with these adjuvant therapies is small (e.g., increasing long-term survival rates by 5–15%). However, many patients feel that this is worth it. I tell my patients that if I give them adjuvant chemotherapy after removal of their lung cancer, I might increase their chance of cure by about 5%. However, if the probability of cure was only 10% with surgery alone, then adding another 5% can be quite meaningful. On the other hand, if the chance of cure with surgery alone was 80%, then adding another 5% may

---

[liv] Stage definitions vary somewhat across tumor types, but for most cancers, stage I cancers are typically relatively small with no lymph node involvement, stage II cancers are larger or have involvement of lymph nodes close to the cancer, stage III is larger still or with invasion into adjacent structures or with involvement of more distant lymph nodes, and stage IV indicates that there is distant spread of the cancer. The higher the stage, the lower the probability of cure.

[lv] Or after eradication by high dose radiotherapy.

[lvi] Higher stage, larger tumor size, poorer tumor differentiation, presence of lymph node metastases, invasion of tumor into blood vessels or lymphatic channels, presence of "satellite" tumor nodules separate from the main tumor mass and some other features that vary between tumor types are all characteristics associated with a higher risk of eventual development of distant metastases.

not seem to be that major. Just as screening is most valuable in high-risk patients,[lvii] the same holds true for postoperative adjuvant therapies.

I also tell my patients that if they are alive and well twenty years after adjuvant chemotherapy, there is no way to know with certainty that the adjuvant therapy was responsible. The same result might have happened with surgery alone. Similarly, a recurrence in, say, six months after declining adjuvant therapy does not mean that the adjuvant therapy would have prevented the cancer coming back. It could have happened anyway.

All we know is that the therapy modestly improves their chances but comes with side effects and inconvenience. To some patients this is worth it, and to others it is not. I tell them that the right answer for them is what their gut tells them to do.

It would be different if the adjuvant therapy guaranteed a cure or if it had no benefit at all. It can be a difficult decision when there is a real, proven benefit but it is small.

To reduce recurrence, adjuvant radiotherapy may also be given to areas from which involved lymph nodes had been removed surgically (e.g., the armpit or "axilla" with breast cancer, or the mediastinum with lung cancer) to reduce the risk of eventual recurrence in these areas.

Further Details and References

Below, I discuss surgery in a bit more detail, with references.

Why surgery can fail: I discussed earlier that surgery may fail to cure a patient. Here's a recap of the reasons. The cancer may be invading into adjacent tissues that cannot be removed. Alternatively, cancer cells circulating in the blood stream may lodge in different areas and eventually become metastases. For every gram of tumor, an average of one million tumor cells will enter the blood stream.[357] Many of these escaping cells will be sick or will not have the characteristics required for the establishment of metastases.[358] Consequently, a cancer may be cured by surgery or radiotherapy despite the presence of circulating tumor cells.

[lvii]   As discussed in Chapter 2.

Factors associated with an increased risk of cancer recurrence after surgery: I have already discussed the tumor features pathologists currently use to assess the risk of cancer recurrence. The persistence of tumor DNA in the blood stream is likely to soon become an important piece of information for determining prognosis. A low probability of relapse is associated with the disappearance of tumor DNA in the blood stream after a cancer is removed. Persistence of tumor DNA in the blood stream is associated with a high risk of eventual relapse.[359] Assessment of residual circulating tumor DNA is currently regarded as "experimental," but I think within the next few years, it will become one of the most important factors that we look at.

Postoperative adjuvant therapies: Giving adjuvant chemotherapy after surgery may reduce the risk of later tumor recurrence by eradicating micrometastases. For some cancers (e.g., cancers of the breast,[360,361] colon,[362] and lung[363]), clinical trials have clearly demonstrated that adjuvant chemotherapy after surgery can increase the probability of long-term survival in moderate and high-risk patients. Adjuvant hormonal therapies[364] or therapies targeting Her2[365] can also improve long-term survival rates in breast cancers driven by estrogen or by a genetic alteration called a "Her2/neu amplification." The targeted therapy osimertinib as a postoperative adjuvant therapy markedly reduces the risk of cancer recurrence in lung cancers with an *EGFR* mutation.[lviii] [366] Preliminary data also suggest that immunotherapy given either before or after surgery will prove useful as an adjuvant therapy for a variety of malignancies.

Surgery for "oligometastatic" disease: Surgery is not usually done in patients who have distant metastases. An important exception is "oligometastatic" disease. This is when there are a very small number of metastases, and they may truly be the only ones that exist.[367] A small

---

[lviii] It is not yet clear how much impact adjuvant osimertinib will have on long-term survival in this group of patients. Since osimertinib is also effective at controlling metastatic *EGFR*-mutant lung cancers, it remains possible (but still uncertain) that just giving osimertinib to those patients who do suffer a recurrence will have almost as much impact on survival as giving the drug to all *EGFR*-mutant patients who have undergone surgery. Some of these *EGFR*-mutant patients will have been cured by surgery and would never require osimertinib.

proportion of patients with oligometastatic disease can be cured by surgery or radiotherapy. The odds of a cure are greater if there are only one or two metastases, and if these metastases only show up months or years after treatment of the primary tumor.[lix] [367,368]

Approximately 23% of all patients with oligometastases survive more than eight years if all areas of tumor can be treated by surgery or high dose radiation,[367] and at least some of these long-term survivors will have been cured. While most patients with distant metastases are not candidates for surgery or high dose radiotherapy because there are too many metastases or because the primary tumor is not controlled,[358] a small proportion of patients may be cured if there are only a very few metastases and the primary tumor can be controlled.

The brain and lung are the most common sites of oligometastases, but oligometastases may also be seen in the adrenal gland, bone, liver, and other sites.[367] One might imagine that surgery to remove a solitary brain metastasis is a highly risky undertaking. However, it can in fact be very well tolerated, provided that the tumor is in a part of the brain that permits surgery with minimal damage to surrounding brain structures that are important for speech, movement, etc.

In fact, surgical removal of a brain metastasis can be one of the best tolerated types of cancer surgery—although occasionally there are severe complications. Patients may be up walking the same day as the surgery. I vividly remember one patient who was so debilitated by emphysema that she could not lie down for surgery or receive a general anesthetic. Nevertheless, she was able to tolerate having a brain metastasis removed under local anesthetic while she sat awake and upright in a chair.

There are only a few situations where removal of multiple metastases is likely to be beneficial. One example is removal of part of the liver in patients with colorectal cancer.[358] As blood flows through the walls of the bowel, it enters the "portal venous system" which passes through the liver on its way back to the heart. This is how nourishment from food is brought from the bowel to the liver for processing. However, tumor cells also travel

---

[lix]   Metastases that show up months or years after the primary tumor is discovered are called "metachronous metastases," while those that are already apparent at the time the primary tumor is discovered are referred to as "synchronous metastases."

from a colon cancer into the portal vein. In this case, the liver can act as a filter, with the circulating tumor cells settling in the liver. Consequently, while there may be many liver metastases, the liver may be the only site of involvement.[358]

The liver has two major lobes (a right lobe and a left lobe). If all metastases are in just one lobe of the liver, a "partial hepatectomy" can remove that lobe. A surgeon can remove up to 80% of the liver, and the liver can regenerate from the remaining 20%,[358] and the patient may be cured. Partial hepatectomy is a major operation and was once dangerous. Advances in surgical techniques over the past few decades have made it much safer. Only a minority of patients with colon cancer metastatic to the liver are candidates for partial hepatectomy, but it can be effective. More than 20% of patients able to undergo the surgery may become long-term survivors, with the probability of cure being higher if there is a single metastasis than if there are multiple metastases.[358]

Another situation where removal of several metastases may prove useful is with lung metastases, particularly with some types of sarcomas. Again, probability of survival decreases with increasing numbers of metastases, but long-term survivors have been seen with removal of up to fifteen lung metastases.[369]

After chemotherapy treatment, widely metastatic testicular cancers may transform into "mature teratoma."[lx] There is a high risk that these teratomas will become malignant tumors if they are left in place. Luckily, even very extensive surgery to remove multiple teratoma deposits results in a high probability of long-term survival. [370] A testicular cancer patient may have both extensive lung metastases and large lymph nodes successfully removed from the abdomen— with an incision from the pelvis to the

---

[lx] Testicular cancer cells divide very rapidly, and under a microscope, they look very different from normal cells. But exposure to chemotherapy may change the characteristics of these cells, even if it does not kill them. In a mature teratoma caused by exposing testicular cancer cells to chemotherapy, the cells divide much more slowly than in the untreated cancer, and they look a lot more like normal cells. However, if they are not removed by surgery, they may eventually change back again into highly malignant cells.

neck.[371] There is no other type of malignancy where such extensive surgery to attempt to remove all metastases is helpful.

"Palliative" surgery: In the above examples, the objective is to try to cure the patient despite the presence of metastases. If there is no possibility of cure, however, there are situations where "palliative" surgery can help symptom control and quality of life. One example of this is removal of a primary colon cancer despite presence of distant metastases. This can reduce the risk of a bowel obstruction that would prevent the patient from eating or drinking anything. Another example is removal of a large brain metastasis that is interfering with brain function or that is increasing pressure in the brain. Such surgery is particularly likely to be helpful for a large metastasis in the lower back part of the brain called the "posterior fossa." Large metastases in this area can cause "obstructive hydrocephalus." Obstructive hydrocephalus increases the risk of rapid deterioration and death because fluid flowing out of the brain's ventricles is blocked. It can also be helpful to insert a metal rod into the bone of an arm or a leg to reduce the risk that a bone weakened by metastases will break.

Conclusions: The good news is that many cancers can be cured by surgery. Surgery is an extremely important part of cancer treatment, but like all therapy modalities, it has limitations.

The bad news and the fourth reason that cancer still sucks is that many cancers are too far advanced for surgery. Furthermore, many patients who undergo an apparently complete removal of their cancer will develop a recurrence due to tumor cells that were released into circulation before the cancer was removed. The recurrence can happen months or years later.

— 5 —

Rays of Hope: Radiation Therapy
in the Treatment of Cancer

The fifth reason that cancer still sucks is that although radiation therapy[lxi] can be an effective cancer treatment, it may fail as a cure. It may fail to kill some of the tumor cells in its path, or it may miss some tumor cells completely. In addition, while "palliative" radiotherapy can dramatically improve symptoms in some incurable patients, it may have minimal impact or provide only very brief relief to others.

Short Primer

Benefits of radiotherapy: Radiotherapy (treatment of cancers using radiation) can cure some localized cancers that are not amenable to surgery. In general, it has a lower probability than surgery of curing a patient. With surgery, the entire tumor mass may be removed, while with radiotherapy, a small number of tumor cells within the tumor mass may survive the treatment. However, radiotherapy is an option when surgery is not feasible, and it unequivocally can cure many patients with localized cancers.

If administered following surgery, radiotherapy also may reduce the risk of a cancer recurring. In patients in whom the cancer is incurable since it is widespread, radiation can provide effective palliation (i.e., improvement in some types of cancer symptoms). In selected patients

---

[lxi]   Also called radiotherapy.

with incurable cancers, it can also be highly effective in improving cancer symptoms like pain, shortness of breath, and bleeding.[lxii]

How radiotherapy works: Radiotherapy acts like a bolt of lightning that breaks DNA in tumor cells. This DNA damage can either kill the cell or can cause changes that prevent the cell from dividing again.

Some of the DNA damage caused by radiotherapy can be repaired. Normal cells are somewhat more effective at repairing the damage than tumor cells, particularly when relatively low doses of radiation are given. This is why "curative intent" radiation is often given once or twice per day, Monday to Friday for several weeks in a row. The day-to-day cumulative DNA damage in tumor cells may be greater than in normal cells since the tumor cells do not do as good a job in repairing DNA damage as the normal cells. The result can be eradication of the tumor without excessive toxicity to normal tissues.

Some, but not all the DNA damage in normal cells will be repaired. If a cancer grows back in an area that was previously treated with high dose radiotherapy, a patient might tolerate a low dose of additional radiotherapy to try to temporarily reduce symptoms. However, they could not be treated again with a high dose of radiotherapy since it could cause permanent, severe, disabling, or life-threatening damage to skin, major blood vessels, or other organs.

Because of this permanent tissue damage, radiation oncologists keep very careful track of the areas they have treated and of the radiation doses these areas have received.

Surgery is also often not an option for a tumor that has recurred in a previously radiated area. Tissues that have received high dose radiotherapy do not heal well after surgery. Radiation-induced scarring also makes the tissue very hard and difficult to remove. It also is difficult to see and maneuver around major blood vessels during surgery in regions of the body that have been radiated. Doing surgery in these areas carries an increased risk of major, life-threatening bleeding. If a team starts out with a plan to give high-dose radiotherapy followed by surgery, they will try to ensure

---

[lxii]   Whether the radiotherapy will help will depend on a variety of factors, such as the mechanism by which the cancer is causing the symptom, the type of cancer, etc.

that the surgery is done within just a few weeks after completion of the radiotherapy before substantial scar tissue has formed.

Radiotherapy methods: More than 100 years ago, radiotherapy involved the direct application of radioactive materials like radium to superficial tumors. It then evolved to simple x-ray machines, then to cobalt beam radiotherapy, to linear accelerators, and then to proton beam therapy. The result of this evolution is a progressively more DNA damage in the tumor and less damage in surrounding normal tissues. Very precise methods are now used to plan the radiation's route from radiotherapy machine to the tumor, again reducing damage to normal tissues. This permits higher, more effective radiotherapy doses.

The larger a tumor treated with radiation, the larger the amount of normal tissue that is also hit by the radiotherapy. Conversely, the amount of normal tissue that is damaged when a small tumor is treated may be limited enough that it is safe to give very high radiation doses to the tumor over just a few days. In these cases, techniques like CyberKnife, Stereotactic Radiotherapy, or brachytherapy are used[lxiii].

I regard the physicists who oversee the highly technical, computerized planning required to guide precision radiotherapy as being equivalent to

---

[lxiii] With standard radiotherapy approaches, the radiation goes in one side of the body, and out the other side. It can damage normal tissues on its way into the tumor and on its way out the other side. The normal tissues in the path of the radiotherapy beam receive high doses of radiotherapy and sustain at least some degree of damage from it. With CyberKnife and Stereotactic Radiotherapy, numerous small radiotherapy beams come in from several different angles (and hence also go out at several different angles). Hence, normal tissues are hit by the radiation, but each area of normal tissue is hit by only a small, relatively nontoxic radiation dose. However, all the radiation beams are aimed so that they pass through the tumor. Consequently, the tumor itself is hit by a high dose of radiation despite any one area of normal tissues around it only being hit by low doses. Brachytherapy involves injecting a radioactive seed into a tumor. The radiation goes outward into the tumor from the seed, so not much of it goes out into surrounding normal tissues. Each of these approaches are useful if used for treatment of relatively small tumors less than 3 cm in diameter. To sufficiently cover a larger tumors with radiation, too much normal tissue ends up being damaged for these approaches to offer any advantage. Extra caution is also required if the tumor is close to a large blood vessel or airway. There can be serious consequences if a large blood vessel or airway is damaged by the high dose radiation.

the Star Wars Jedi Knights[372] of cancer therapy! They carefully mold the radiotherapy fields to try to maximize destruction of tumor cells while limiting damage to normal tissues.

Why radiotherapy may fail: Radiotherapy may fail because some stubborn tumor cells unfortunately can repair the DNA damage caused by radiotherapy. In addition, "hypoxic" tissues (i.e., tissues that are oxygen-deficient) have increased resistance to radiotherapy. Large tumors are more likely than small tumors to have hypoxic areas. Thus, they are less likely to be cured by radiotherapy.

In addition, a large tumor has a higher number of tumor cells than a small tumor. Consequently, it is more likely to have cells with resistance-causing mutations or cells that escape into the blood stream to cause incurable distant metastases.

Only a very small fraction of tumor cells may have developed resistance-inducing mutations. But if just one out of a billion cancer cells survive the radiation, the tumor eventually regrows. These regrowing cells may be resistant to further radiotherapy since they inherited these resistance-inducing mutations.

Rapidly growing tumor cells are more sensitive to radiotherapy than slowly growing cells. But if the cells grow very rapidly, all the cells that were killed by a radiotherapy treatment may already have been replaced by new cells by the time the next treatment is given the following day.

Antioxidants like vitamins C and E can increase the resistance of tumor cells to radiotherapy. Hence, I advise my patients not to take high dose antioxidants when they are undergoing therapy.

Tumor location can also play a role in the success of radiotherapy. Take as an example a patient with a cancer at the bottom of the lung but with spread to lymph nodes high up in the chest. To treat all the involved areas, damaging radiotherapy might need to hit too large an amount of normal lung tissue. Also, the liver is very sensitive to radiation, so it can be difficult to safely use radiotherapy to treat cancer deposits in the liver. The spinal cord, intestine, and heart are also sensitive to radiation, and care must be taken to avoid high radiation doses to these areas.

Drugs that increase sensitivity of cancers to radiotherapy: Combining some chemotherapy agents with radiotherapy can make it work better. For many cancers, it is now the "standard of care" to combine selected chemotherapy agents with "curative intent" high dose radiotherapy. Not only may the chemotherapy make radiotherapy more effective in the areas that the radiation hits, but it also may kill very small tumor deposits (too small to show up on scans) outside of the area treated by the radiotherapy. If these small tumor deposits were not eliminated by the chemotherapy, then they would grow and show up months after completion of the radiation, rendering the patient incurable.

Factors increasing sensitivity of normal tissues: The normal tissues in some patients have increased sensitivity to radiation. This makes it difficult to treat these patients with high-dose radiotherapy.

Other medical conditions may also limit the ability to give radiotherapy. For example, patients must lie flat to receive radiotherapy, and some patients are unable to do this because of severe shortness of breath or pain. Also, the tables that patients must lie on to receive radiotherapy may have a maximum weight that they can support (e.g., 300 pounds), so it may not be possible to treat very obese patients.

Other local approaches that are sometimes useful: Some other approaches occasionally prove helpful in localized cancers that can't be treated with radiotherapy or surgery. For example, an ultrasound probe may be inserted into the tumor to heat it up to the point that the tumor cells are killed. This is called radiofrequency ablation. This might be considered in a patient who cannot be given radiotherapy. Alternatively, a cryoprobe that is cooled by a method such as liquid nitrogen may also be inserted into a tumor to kill it by freezing it. Both these approaches have technical limitations that mean that they can only be used in a very small proportion of patients. But in the right patient, they can be very helpful.

Further Details and References

References and further details follow.[lxiv]

How radiotherapy works: Radiotherapy kills tumor cells by causing DNA breaks. The radiation may break DNA directly[373] or it may knock an electron off a water molecule.[373] A remnant of the water molecule becomes

---

[lxiv]   In addition, please see Chapter 8 for a more detailed discussion of toxicity of radiotherapy and other types of therapy.

what is called a "hydroxyl radical." Hydroxyl radicals are "highly reactive", meaning that they react rapidly with other molecules in their vicinity. Like a ravenous animal seeking prey, the hydroxyl radical aggressively seeks molecules with a high hydrogen content, and "absracts" a hydrogen from the molecule. This causes major damage to the molecule losing the hydrogen. DNA is rich in hydrogens, and the hydroxyl radicals can break the DNA strand as a result of this attack.[373-376]

The cell's DNA repair machinery recognizes this break and attempts to repair it. If repair is not possible, it induces the cell to become senescent (i.e., alive but incapable of trying to divide again) or it turns on the machinery that leads to cell death via a suicide process called apoptosis.[373,374] The higher the radiation dose, the more DNA damage that is caused and the lower the probability of repair.[374]

Unfortunately, the radiation does not just damage the DNA in tumor cells: it also damages the DNA in nearby normal cells.[374] The radiation passes through normal tissues on its way in to the tumor and goes through more normal tissue after passing through the tumor. It also hits normal tissue immediately around the tumor.

A high dose of radiation will severely damage the DNA in both tumor cells and normal cells. At lower radiation doses, the cells may repair some of the DNA damage. Normal cells are more effective than tumor cells at repairing the DNA damage done by low doses of radiotherapy.[374] Therefore, a low dose of radiation is often given daily Monday to Friday for six to seven weeks since this approach may be successful in killing tumor cells while sparing normal cells. [374]

In the process of treating a cancer, a large volume of normal tissue will be hit by the radiation if a tumor is large. Only a small volume of normal tissue is hit if the tumor is small. Both tumor tissues and normal tissues are less able to repair damage from very high individual radiation doses. Consequently, very high individual radiation doses may be more effective than lower doses, but they will only be tolerated if they are delivered to relatively small areas of the body.[377] Therefore, high dose approaches such as stereotactic radiotherapy and CyberKnife may be used against relatively small cancers,[lxv] but cannot safely be used for larger cancers.[378]

[lxv] Typically, up to 3 cm in diameter.

Over the past 100 years, numerous technical advances have improved radiotherapy, increasing its effectiveness against tumor cells while reducing its toxicity to normal tissues.[379]

Why radiotherapy may fail: As discussed earlier, cells may repair the DNA damage done by the radiation. Some malignancies may be resistant to radiotherapy since they have a high capacity to repair this damage.[380,381]

There are also other factors that may reduce radiation efficacy. For example, an electron that is knocked off the water molecule by radiation can snap back on to it. If this happens, the hydroxyl radical that is needed to damage DNA is converted back to a water molecule that is no longer toxic to the tumor.[382] If there is something in the area that can trap the electron, then the electron is no longer available to recombine with the hydroxyl radical, making it more likely that the hydroxyl radical will reach its DNA target. The radiation, therefore, is more effective.

Oxygen can trap free electrons and also has other effects that can increase the DNA damage done by radiation.[375] Oxygen increases the effectiveness of radiotherapy. Tumor cells that are "hypoxic" (i.e., deficient in oxygen) are relatively resistant to radiation.[383]

Tumor cells that are a relatively long distance from the nearest blood vessel may be hypoxic. Furthermore, blood flow through the tumor's blood vessels may be impaired by a variety of factors,[384,385] and may vary substantially from one part of a tumor to another.[386]

Several other factors besides hypoxia can make cancers resistant to radiotherapy. Larger cancers are less likely to be cured than are smaller cancers. Tumors become more hypoxic as they increase in size.[386] Furthermore, large tumors have more cells than small tumors. More cells mean an increased risk that at least one tumor cell will have a mutation that makes it resistant to radiation.[368] More cells in the tumor also mean more cells available to enter the blood stream to potentially cause distant metastases.[368]

In general, slowly growing tumors are more resistant to radiotherapy than are those that grow more rapidly. Here's the reason. As cells divide, they pass through five distinct phases of the "cell cycle." Namely from the G0 phase (during which the cell rests between episodes of cell division) to G1 to S to G2 to M and then back to G0 or G1 again.[374]

Cells in M phase are more sensitive to radiation than cells in other cell cycle phases. This happens since DNA is tightly packed together during M phase. Radiation causes more damage to DNA if it is tightly packed. In addition, during the M phase there are no active DNA repair mechanisms.

It takes a cell a relatively short time to go through M phase, so only a small proportion of the tumor cells will be in the sensitive M phase at any one time.[374] If a cell is growing slowly, many of the cells will be in the resting G0 phase of the cell cycle. If many are dividing, more will be passing through the sensitive M phase when the radiation hits.

Tumor growth rate will often slow as the tumor becomes larger. This is referred to as "Gompertzian growth."[387] Gompertzian growth happens because there are fewer nutrients and less space as the tumor grows. Slower growth contributes to greater radiation resistance in larger tumors because there are fewer tumor cells in the sensitive M phase of the cell cycle.

Rapidly growing tumors might be more sensitive to radiation, but rapid tumor cell division is a double-edged sword. This is the case since the rapidly growing cells may be more sensitive to the killing effects of radiation, but any rapidly dividing tumor cells that survive may simply replace the tumor cells that have been killed.

As a tumor shrinks in response to radiation, various factors accelerate cell division among the tumor cells that have survived.[388] For example, blood flow to the surviving tumor cells may improve, providing more nutrients to surviving tumor cells. The net result of this "accelerated repopulation" may be that tumors shrink early during the treatment but grow back more and more rapidly as the treatment progresses. This accelerated repopulation increases the probability that at least a few tumor cells will survive. This could lead to recurrence of the cancer months after completion of radiation.

Some malignancies may also be resistant to radiation since they have increased amounts of antioxidants like glutathione,[389-393] beta carotene,[394] or vitamins C[392,393] or E.[390,392] These antioxidants may neutralize the radiation-induced hydroxyl radicals that are key to radiation's ability to damage tumor cell DNA. In fact, it has been demonstrated that administration of some antioxidants during radiotherapy may reduce its effectiveness.[395]

These antioxidants may be important in protecting us from the DNA damage that can lead to cancer-causing mutations in the first place, but once a cancer develops, they may worsen the cancer by protecting the tumor cells from the harsh, hypoxic, acidic environment created by unbridled tumor cell growth.[396-400] This may be part of the reason why therapy with antioxidants such as high dose vitamin C has failed to demonstrate much benefit in patients with advanced cancers,[lxvi 401] and why dietary supplementation with antioxidants does not reduce the risk of developing cancer.[lxvii]

The growth factors and mutated receptors (discussed in Chapter 1) that promote tumor cell growth also can help protect the cancer from the effects of radiation. They do this by promoting rapid division of tumor cells that survive the radiotherapy. Some of these growth factors, receptors, and related molecular pathways also can directly protect cancer cells by enhancing repair of radiation-induced DNA damage.[402]

Drugs that increase sensitivity of cancers to radiotherapy: Radiotherapy cannot cure cancers that have spread to areas of the body that the radiation does not hit.[lxviii] However, if chemotherapy is added to the radiation, it is sometimes able to eradicate very small, invisible tumor deposits that the radiotherapy misses. Furthermore, as noted previously, some chemotherapy drugs can potentiate the effect of radiotherapy. For example, when oxygen is in short supply, there are some drugs that can take the place of oxygen and can improve radiation effectiveness by prolonging the availability of radiation-induced hydroxyl radicals. They do this by trapping the electrons that radiation knocks off water molecules. For example, the platinum portion of the chemotherapy agents cisplatin and carboplatin can trap electrons, and also can potentiate the effect of radiotherapy by inhibiting cell repair of radiation-induced DNA damage.[382,403]

Since they are weak acids, cisplatin and carboplatin may diffuse preferentially into oxygen-deficient tissues. Drugs that are weak acids

---

lxvi Discussed in more detail in Chapter 9 on alternative and complementary therapies.

lxvii See Chapter 1.

lxviii And radiotherapy is too toxic to give to very large areas of the body.

accumulate preferentially in tissues that are acidic.[404] Oxygen-deficient tissues tend to be more acidic than are oxygen-rich tissues,[384,386] so platinum agents might reach high concentrations in these tissues.

Cisplatin reaches high concentrations in areas of human tumors that are "necrotic" (i.e., tumor areas with poor blood flow and with a few living cells hidden in the middle of large numbers of dead cells).[405] In fact, the cisplatin concentrations found in necrotic areas of tumor are about the same as they are in viable areas of tumor. Furthermore, distribution of cisplatin in normal human tissues does not corelate with the usual rate of blood flow to those tissues. It reaches similar concentrations in areas of high and low blood flow.[406] In laboratory experiments, cisplatin and carboplatin substantially increase the effectiveness of radiotherapy, particularly in hypoxic conditions.[403,407]

In the late 1970s and early 1980s, we[408,409] and others[382] first began to test ways to safely combine cisplatin with radiation clinically. It has been confirmed that giving platinum-based chemotherapy regimens concurrently with radiation safely improves the effectiveness of radiation against a variety of common cancers.[410-417]

Several other chemotherapy agents also can make radiation work more effectively by mechanisms that include interfering with DNA repair and blocking cells in the radiation-sensitive M phase of the cell cycle.[418]

Radiotherapy for oligometastases and areas of oligoprogression: Generally, radiotherapy does not cure cancers that have metastasized to areas other than lymph nodes. But sometimes it does. In Chapter 4, I discussed surgery for "oligometastases," where only a small number of metastases (typically from one to five) can be seen. As an alternative to surgery, high dose radiotherapy to all involved sites can cure at least some patients with oligometastases.[367]

In some patients, systemic therapy achieves ongoing control of a metastatic cancer but with progression at just a single site (termed "oligoprogression"). In this case, radiotherapy to the site of progression may facilitate more prolonged control of the patient's cancer.[367]

Factors increasing sensitivity of normal tissues to radiotherapy: As noted previously, some patients are more prone to develop radiotherapy toxicity since their tissues have an increased sensitivity to radiation. For example, people with scarring diseases like scleroderma,[419,420] related "collagen vascular diseases,"[419,420] and pulmonary fibrosis[lxix] from other causes[421] may develop life-threatening or fatal lung inflammation from relatively low doses of radiation.

People also may inherit genes that may make tissues more sensitive to radiation.[lxx] For example, some variants of genes related to inflammation[422] or DNA repair capability[423] may increase the probability of radiation toxicity. As things currently stand, we cannot predict who will and who will not be sensitive to radiation, so radiation doses are chosen that can be tolerated by most patients.

There are two problems with this. The first problem is that the radiation dose will still be too high for some individuals. They may develop severe short-term or long-term toxicities. The second problem is that the individual may have inherited genes that make their tissues relatively resistant to radiation. Their cancer cells also will have inherited these same resistance genes. So, a dose of radiation that might have been effective in the average patient may be ineffective against their cancer.

Within the next few decades, it is likely that most people will undergo genotyping (i.e., assessment of their gene variants) early in life and this will be used to guide not only the doses of radiotherapy with which they are treated but also doses and types of medications that are optimal for them for a wide variety of conditions.

Conclusions: In summary, radiotherapy can be a potent weapon in the fight against cancer and can cure or effectively palliate some patients. However, the fifth reason that cancer still sucks is that radiotherapy will fail in some patients and cause major toxicity in others.

[lxix] Pulmonary fibrosis is diffuse scarring of the lungs.

[lxx] See Chapter 1 for a discussion of "single nucleotide polymorphisms" (SNPs).

— 6 —

Magic Potions: Systemic Therapies and the Treatment of Cancer

The term "systemic therapy" refers to the treatment of cancers using drugs. These drugs are called systemic therapies because they go through the blood stream to all parts of the body. They are used against widespread cancers as they may get to cancer cells no matter where they are lurking.

In some cases, they kill tumor cells and shrink cancers. However, the sixth reason that cancer still sucks is that some patients derive no benefit from these therapies, and we cannot cure most patients with widely metastatic cancers.

Short Primer

Types of systemic therapies: Systemic therapies include chemotherapy, targeted therapy, hormonal therapy, and immunotherapy. I will discuss immunotherapy separately in Chapter 7. Generally, medical oncologists like me oversee the use of systemic therapies for most types of "solid tumors" such as cancers of the lung, breast, colon, and prostate, although gynecologic oncologists often oversee systemic therapies for gynecologic malignancies. Hematologists oversee systemic therapies for leukemias, lymphomas, and myeloma. Some physicians have expertise in both oncology and hematology.

Chemotherapy: Chemotherapy works against cancers by damaging cells and killing them. The most important targets hit by chemotherapy are the DNA[lxxi] or a protein called tubulin that is important in maintaining cellular structure and function. Chemotherapy drugs can also damage normal cells[lxxii] and are generally only useful if the cancer is more sensitive to their effects than most normal cells.

Hormonal therapy: In contrast, hormonal therapies decrease the ability of the body's hormones to drive cancer growth. Breast cancer cells may be driven by the female hormone estrogen and prostate cancer is driven by male hormones such as testosterone. Hormonal therapies either decrease the levels of a hormone in the patient's body, or else block tumor cell "receptors" for the hormone. If the receptors are blocked, the hormone is unable to access the machinery in tumor cells by which it drives tumor growth.

Targeted therapy: Targeted therapies bind to cellular molecules such as "growth factor receptors" that drive tumor cell growth.[lxxiii] When targeted therapies were first being developed, we hoped that they might slow down cancer growth. In the early clinical trials, some patients unexpectedly experienced marked tumor shrinkage. Researchers subsequently learned that in those patients, the cancers generally had mutations or multiple extra copies of the gene for the target. The tumor cells with the mutation were "addicted" to the mutant molecule, and when the drug blocked the target in a mutated cell, it could cause rapid cell death.

What drives oncologists: I graduated from Queen's University medical school in 1974. Every five years since then, members of my class of Meds '74 have had an October reunion. A few years ago, one of my classmates— an obstetrician—asked me why anyone would ever want to be an oncologist. He asked, "Isn't it incredibly depressing, with all those patients dying, no matter what you do?"

I told him that most oncologists I know love what they do. I explained that many of the patients we see with advanced metastatic disease are in pain and feel extremely sick. Many will die within weeks unless we are able to do something to change things for them.

[lxxi] See Chapter 1 for a discussion on DNA

[lxxii] I discuss therapy toxicity in more detail in chapter 8.

[lxxiii] See Chapter 1 for a discussion of oncogenes and the proteins that they code for.

Every week, I see some patients I cannot help. I see others who I will make feel even worse than they already feel due to side effects of the therapies I prescribe. But I see many others who begin to feel markedly better as their cancers shrink with our therapies, and instead of dying within weeks, they live to see another Christmas, or their daughter's wedding, or the birth of their first grandchild. Life's goals change radically when you suddenly realize that the remaining time before you may be very short.

I explained how inspiring it is to me to see the courage and strength of my patients facing certain death, and how gratifying it is to receive heartfelt thanks from my patients and their families. I also explained that, despite long odds, a few patients do remarkably better than the average- and are alive and well years after they would have died in pain if we had done nothing.

"Ah," said my classmate. "Now I understand. If you as an oncologist get one unexpectedly good result a year, it makes your year. If I, as an obstetrician, get one bad result a year, it ruins my year!" The good news is that every day, we celebrate numerous small victories. In addition, and against all odds, we experience some surprising spectacular victories. These victories, large and small, give me and my patients hope.

A few years after this conversation, my classmate shared with me that his wife had recently died after a long, difficult battle with leukemia. He and their family mourned her loss. However, they were extremely grateful for the added time that therapeutic interventions had bought for them. In particular, they were able to spend her last special Christmas with family gathered together at their beloved country retreat.

Death of a loved one by cancer is tremendously difficult for families. Perhaps the only positive about it is the knowledge that death is rapidly approaching. This can provide the opportunity to say goodbye. In 1986, my parents, Archie and Iris Stewart, were killed by a drunk driver with a suspended driver's license and three prior impaired driving convictions. Like so many families where a death is sudden and unexpected, we never got to say goodbye.

What systemic therapies can offer: By shrinking a cancer even modestly, systemic therapies can be a more potent pain killer than morphine. Systemic therapies can—but don't always—alleviate other cancer-related symptoms such as loss of appetite, profound cancer-related fatigue, and the unrelenting cough and distressing, smothering shortness of breath caused by extensive lung metastases.

Systemic therapies can also prolong life expectancy for many patients. While some survive only a few extra months, others may survive years.

The potential alleviation of cancer symptoms and the potential meaningful prolongation of life expectancy are the reasons why patients may elect to take systemic therapies despite their side effects. It is also the reason that healthcare systems pay for them.

We cannot reliably predict life expectancy in an individual cancer patient, and the potential impact of therapy and average survival time vary widely across cancer types. If we use metastatic non-small cell lung cancer as an example, average life expectancy is only about four or five months if no systemic therapy is given, and about 4% of remaining patients die each week that initiation of therapy is delayed.[424] However, with standard chemotherapy, average life expectancy is now more than one year. Combining new immunotherapy approaches[lxxiv] with chemotherapy increases average life expectancy to more than 1.5 years. If a patient has a relevant tumor mutation and can be treated with one of the new targeted therapies, average life expectancy can be three years or more.

An average life expectancy of three years in patients with metastatic lung cancer suggests that about 25% of patients could still be alive at six years and about 6% might still be alive at twelve years.[lxxv] [425] We still have a long way to go, but this is much better than the 4–5 month average life expectancy for a patient who does not receive any systemic therapy.

For almost all types of cancer there are systemic therapies that can be effective—at least in some patients. These therapies cannot cure most types of cancer when they are widespread, but they can cure a few. In particular, they can cure at least some patients with some types of leukemias and lymphomas, some childhood malignancies, testicular

---

[lxxiv] I more fully discuss immunotherapy in Chapter 7.

[lxxv] We can do this type of calculation since, across a population, survival usually approximates "first order kinetics."

cancer, and gestational choriocarcinoma.[lxxvi] As discussed in Chapters 4 and 5, adjuvant systemic therapies may also increase cure rates if given after surgical removal of a localized cancer or concurrently with radiotherapy.

Another important consideration is the rapid emergence of new therapies. I have many patients whose life expectancy was prolonged by 1–2 years with modestly effective therapy. During that time, a new therapy came along for some that was even more effective and added significantly to their lives.

An example is a young man I cared for in 1977 when I was doing my oncology training at MD Anderson Cancer Center. He had an uncommon type of leukemia called "hairy cell leukemia." At that time, the only therapy for hairy cell leukemia was to remove the spleen. He underwent this procedure, but several months later, his condition worsened. His life expectancy was at most a few months. The leukemic cells had crowded out the normal white blood cells that he needed to protect him from infection.

With no other viable alternatives, we treated him with an experimental chemotherapy agent called rubidazone. He was dangerously ill for several weeks, but he recovered the ability to make normal white blood cells even though there were still leukemic cells in his bone marrow.

The residual leukemia progressed slowly. It eventually began to crowd out his normal bone marrow cells again. However, by that time, another new drug had come along that again knocked his leukemia back. A few years later, it began to worsen again. By then, still another effective new drug had come along. He repeatedly managed to "ride the wave," with one new drug after another controlling his leukemia until an even better new drug came along. The last I heard in 2003, he was still doing well more than twenty-five years after first starting therapy for an "untreatable" malignancy. He is very much the exception rather than the rule, but he illustrates what is at least possible.

Limitations of systemic therapy: I want to emphasize that we are still unable to cure most types of metastatic cancers. In addition, the

---

[lxxvi] Gestational choriocarcinoma is a cancer that arises directly from pregnancy-associated tissues such as the placenta. Before the availability of chemotherapy, it was a highly lethal cancer, but with modern chemotherapy, it can now be cured in a high proportion of patients, even when widely metastatic.

improvements in cure rates of localized cancers by using adjuvant systemic therapies after surgery are generally only very modest. Indeed, some patients with widespread cancers derive little or no benefit from systemic therapies. Most widespread cancers that respond to systemic therapies eventually become resistant to them and tumors regrow, sometimes rapidly.

Many patients may never even make it to treatment.[lxxvii] As discussed above, 4% of remaining patients with metastatic non-small cell lung cancer die each week that initiation of therapy is delayed.[424] Many others deteriorate rapidly to the point that they would no longer tolerate treatment.[426] In the province of Ontario, Canada, the government pays for systemic therapies for patients with metastatic non-small cell lung cancer, but fewer than 25% of patients ever receive one of these therapies because many deteriorate and die too rapidly.[427]

Individual patient decisions about taking systemic therapy: Some patients to whom I offer a systemic therapy decide not to take it. They feel that the potential benefits are not enough to justify the potential side effects. When I discuss the potential benefits and side effects of systemic therapies with patients, some know right away that they want to take it, some know right away that they do not want to take it, and some need time to think about it. I often tell patients that if I could tell them with certainty how rapidly their cancer was going to worsen without treatment and how much benefit and toxicity they would have if they took treatment, then it would be much easier for them to decide. However, these are things that I cannot accurately predict. I tell them that there is no right or wrong answer as to whether they should proceed with a therapy. The right answer for them is what their gut tells them to do.

Therapy may produce marked, prolonged shrinkage of cancer in one patient. Frustratingly, there may be little or no benefit for another patient with the same type of cancer. Because of this, I make it a point to tell my patients to be prepared for the possibility that they may deteriorate and die quickly, no matter what we do. On the other hand, there is also a possibility that they may be alive and doing well four or five years or more from now. I tell them the average life expectancy with the therapy. I also

---

[lxxvii] See Chapter 14 for a discussion of why some patients don't make it to treatment.

remind them that half the patients will be above that average, and half will be below it. I stress to them that we cannot predict what will happen to them as an individual.

Some families of patients have told me that they regret that their loved ones took the therapy. They experienced no improvement in their symptoms, had severe side effects, and died within a very few weeks. However, I have many others who would have deteriorated rapidly and died within weeks of me first seeing them if they had not taken treatment. They had a dramatic response to therapy, are alive and well several years after therapy initiation, and are very thankful that they decided to take treatment. Many families have also thanked me profusely for at least giving their loved one the option to try, even though the therapy did not work for them.

People who are feeling very sick from their cancer and are facing certain death often look at options much differently than do healthy people. As I discuss further in Chapter 11, Slevin and colleagues found that healthy individuals would want a 50% chance of cure, a five year prolongation of life expectancy or a 75% probability that their symptoms would improve in return for taking a highly toxic therapy. On average, cancer patients reported that they would take the highly toxic therapy in return for a 1% chance of cure, a twelve month improvement in life expectancy, or a 10% chance their symptoms would improve.[428] All it would require for an average cancer patient to take a somewhat less toxic therapy was a three month prolongation of life expectancy or a 1% chance that their symptoms would improve.[428]

Perspectives can change radically when you are looking down the barrel of a gun, and the change can be sudden. Many people go to bed at night thinking they are healthy but wake up in the morning with the first symptom of an advanced, incurable cancer.

Other things I often discuss with my patients: Over the decades that I have been an oncologist, my message to patients and their families has gradually evolved as I have found what seems to be helpful and what does not work. Currently, when I see a new patient, I discuss their prognosis and therapy options. I also generally try to cover some other important things as well. I advise them to make sure that their will is in order and that they have designated the person they would want to act as power of attorney, to make decisions for them in case they are unconscious and unable to make decisions. That way, if they deteriorate rapidly despite our interventions, they don't have to worry about these details—and it can make things a lot easier for their loved ones.

I may also advise my patient to speak with their families about what they would want done if they were to stop breathing or have a cardiac arrest. Would they want full resuscitation and to be put on a ventilator in the Intensive Care Unit? I advise that this not be done if their deterioration is due to worsening of their cancer. But it might be reasonable to do it if the deterioration is due to something potentially reversible, like an infection.

I also frequently tell patients that while optimism does not make a treatment work any better, it can definitely make their quality of life better, so they might as well be optimistic.

Further Details and References

I will now discuss some additional details.

Chemotherapy: Chemotherapy generally kills tumor cells in one of three ways. It binds to and damages DNA, or it blocks the production of new DNA that is required for repair of DNA damage, or it interferes with the function of a key structural part of the cell called tubulin.

Sensitivity to chemotherapy may be different in different parts of the cell cycle.[lxxviii] As with radiotherapy, rapidly growing tumor cells may be more sensitive to chemotherapy than slowly growing cells. However, the situation is complex. The phases of the cell cycle that are sensitive to chemotherapy vary from drug to drug. The same is true for radiotherapy. Combining chemotherapy with radiotherapy can be more effective than radiotherapy alone. One reason is that cells in a particular stage of the cell cycle can be resistant to radiation but sensitive to chemotherapy—and vice versa.

[lxxviii] See chapter 5 for a short discussion of the cell cycle.

With chemotherapy, very specific drugs work against very specific cancers.[lxxix] That is why the chemotherapy for a patient with colon cancer will be different from the chemotherapy used to treat lung cancer.

We are not yet able to predict reliably who will and who will not respond to a given chemotherapy. We could spare patients from the toxicity of a therapy if we knew what prevented that therapy from being effective—but we don't. Numerous factors can make one patient's cancer resistant to a therapy that works well in many other patients.[299,429]

Oxygen concentration can affect the efficacy of some chemotherapy agents. As with radiotherapy, some are most effective in the presence of high oxygen concentrations. However, some other chemotherapy drugs are most effective against "hypoxic" cells with low oxygen levels, and oxygen content has little positive or negative impact on efficacy of still other drugs.[429]

Chemical characteristics of a drug can make it more acidic[lxxx] or more alkaline.[lxxxi] This impacts how it will work in different situations. Some more alkaline drugs reach their highest concentrations in well oxygenated areas very close to blood vessels where tissues are less acidic. Other more acidic drugs can diffuse easily into hypoxic, acidic tissues that are far from the nearest blood vessel.[429]

For most cancers, combinations of two or more chemotherapy drugs together will work better than a single drug by itself.[430] In part, this is because a tumor cell that is resistant to one drug may be quite sensitive to another drug. Also, one drug may enhance the effectiveness of another. For example, cisplatin (a weak acid) can penetrate deeply into acidic tissues. It may increase the ability of a weak alkaline drug like doxorubicin to penetrate deeply into areas that it would ordinarily not be able to reach.[431]

Resistance that is present from the initiation of a therapy is referred to as "intrinsic" or "de novo" resistance.[429] Factors that cause intrinsic

---

[lxxix] Different chemotherapy agents are also toxic to different types of normal cells. See Chapter 8 for a discussion on some specifics of chemotherapy toxicity.

[lxxx] More acidic drugs behave like a weak version of acetic acid (vinegar) and are referred to as having a "low pKa."

[lxxxi] More alkaline (or "basic") drugs behave like a weak version of sodium bicarbonate and are referred to as having a "high pKa."

resistance to one drug may not affect how well another drug works. This probably explains why a chemotherapy drug may work against one type of cancer but not against a different type of cancer.[429] Conversely, "acquired resistance" refers to the development of resistance to a drug that was initially effective.[429]

In most cancers, there are probably far more factors that could potentially contribute to acquired resistance than to intrinsic resistance.[429] When a cancer that was initially sensitive develops acquired resistance, it will often be resistant to a broad spectrum of other drugs as well.[429] Hence, while a given intrinsic resistance factor may impact the efficacy of just a few drug types, a given acquired resistance factor may reduce efficacy of a broad range of agents.[429] Consequently, a therapy that is given "second-line" after failure on a different therapy generally will be less effective than if it was given as first-line treatment before the cancer had been exposed to any other type of chemotherapy.[432]

Acquired resistance may develop because a few cells with preexisting resistance mutations survive chemotherapy. When sensitive cells are killed by the therapy, these resistant cells would then have less competition for nutrients. An analogy would be the cutting down of a tree. When the tree is cut, other plants thrive since they now have improved access to sunlight and no longer must compete with the tree for nutrients. If the sensitive tumor cells are killed, the resistant cells can then grow and divide to the point that they became the predominant cell type.[433]

The larger a tumor or the longer it has been present, the higher the probability that there would be at least a few mutant resistant cells present.[368] Initially, the cancer might shrink with treatment. Eventually, however, the tumor would begin to grow again as residual resistant cells divided and passed on their resistance mutations to their progeny.

Tumor cells can also undergo "epigenetic" changes that make them resistant.[429] Epigenetic changes are not due to gene mutations. Instead, they are due to different genes being turned on or off. For example, some cancer cells may use epigenetic changes to reduce molecules on their surface that are responsible for taking up into the cell both nutrients and chemotherapy.[434,435] It is as if the cells "pull up the drawbridge." Since far fewer nutrients are getting into the tumor cell, the cell grows more slowly,[435] and this slow growth decreases sensitivity to chemotherapy. The same cell membrane changes can decrease uptake of both nutrients and chemotherapy.

The good news is that if a patient's cancer initially shrinks with chemotherapy, they might feel a lot better than before the therapy started, even if there is minimal further ongoing killing of tumor cells. It might be quite a while before the cells in their cancer began to take up nutrients and grow again. Some of my patients who take a treatment break after having responded to chemotherapy do not begin to experience new tumor growth for one to two years.[lxxxii] The bad news is that the cancer cells are still alive and will eventually begin to grow again.

In the early days of chemotherapy, it became apparent that patients with some malignancies like acute leukemias would respond better and live longer if treated with higher doses of chemotherapy. Similarly, high concentrations of drugs were more effective in the test tube. Consequently, in my early days as an oncologist, I believed that high dose chemotherapy would be the key to conquering all cancers.

However, this has not happened. For most malignancies, some chemotherapy is better than none, but very high doses are not more effective than somewhat lower doses.[436,437] The term "dose-response curve" (or DRC) refers to a graph plotting number of surviving tumor cells (on a log scale) against drug dose. DRCs may initially be steep at lower drug concentrations, with increasing effectiveness as the drug dose increases. However, as the drug dose or concentration increases further, the DRC begins to flatten, with little further effect as the drug doses are raised.[438]

Most common malignancies behave as if the DRC is flattening at higher chemotherapy doses.[436,437] This is why high dose chemotherapy with bone marrow transplant (which may be of value in some leukemias, lymphomas, germ cell tumors, and multiple myeloma) is of little value in most of the more common cancers.[436]

It is not known why DRCs flatten at higher drug doses. The cancers are behaving as if they are "running out of" a target or process that is essential for drug effect.[436-438] For example, the machinery that is needed to transport a drug into the tumor cell or to convert the chemotherapy to

---

[lxxxii] Although in other patients, cancer growth occurs very rapidly after stopping treatment.

an active form might be saturated. It also might be something simple: once sensitive, rapidly dividing cells have been eradicated, the remaining cells may be highly resistant since they are resting and not dividing.

Patient-related factors may also limit chemotherapy efficacy. Potential side effects can make it difficult to treat very elderly patients with chemotherapy (although I do have some patients in their 90s who do well with it and are grateful for the improvement in their quality of life). It can also be difficult to treat patients with kidney failure, severe heart disease, or other major illnesses. In addition, patients who are so weak that they are spending much of the day in a chair or bed[lxxxiii] may have a high probability of major complications from the chemotherapy. They also will have a relatively low probability of major benefit.[439] Despite this, even patients with poor performance status may derive at least some benefit from chemotherapy.[440,441]

We all inherit a wide range of genes that impact our ability to metabolize drugs, repair damaged DNA, etc.[429] These genes come in different forms in different people. Normal variations in these genes are referred to as "single nucleotide polymorphisms," or SNPs.[lxxxiv] Patients with some SNPs may suffer greater toxicity from the chemotherapy than others. But their cancer will also carry these same SNPs, so it may be more sensitive to the chemotherapy than a cancer developing in a patient with resistance-associated SNPs.[429]

"Pharmacogenetic" studies assess the association between a patient's SNPs and a therapy outcome. In the next few decades, everyone will probably have a broad range of their SNPs characterized early in life. Physicians will use this information to decide what drugs or drug doses to use for them.

Hormonal therapy: Growth of tumor cells in many women with breast cancer is driven by estrogen[lxxxv] and growth of most prostate cancers is driven by testosterone and related hormones. These hormones interact with cell molecules called estrogen receptors and androgen receptors,

---

[lxxxiii] These debilitated patients are referred to as "poor performance status" patients.
[lxxxiv] I have discussed SNPs in more detail in Chapter 1.
[lxxxv] This is discussed in more detail in Chapter 1.

respectively, to set off a chain reaction in the cell that drives cell division and tumor growth.[442,443]

The cancer cells may stop dividing or die when we reduce production of the relevant hormone or if we can block its interaction with its receptor. When combined with surgery or radiotherapy for localized cancers, hormonal therapies may improve long-term survival rates for breast[442] and prostate[444,445] cancers.

In patients with advanced, metastatic disease, these hormonal therapies can also shrink tumors, reduce cancer symptoms, improve quality of life, and prolong life expectancy. However, they cannot permanently cure patients.[442,446] As with chemotherapy, hormonal therapy may not work at all in some patients but can control the cancer and prolong life expectancy by years in others.

One major advantage of hormonal therapy over chemotherapy is that it usually causes fewer side effects. It may be well tolerated and highly effective even in very elderly or sick patients.

In the early days of hormonal therapy for breast cancer, estrogen production was reduced by surgically removing the ovaries.[442] When the cancer began to grow again, removing the adrenals or the pituitary gland could control the cancer in some patients.[442] The male hormone testosterone[442] or drugs related to the hormone progesterone were also of benefit in some breast cancer patients.[442]

Currently the major hormonal therapies for breast cancer are agents like tamoxifen (which acts predominantly by blocking the estrogen receptor),[442] fulvestrant (which decreases the amount of estrogen receptor in the cell, rather than by blocking its action),[447] and "LHRH agonists"[448] and "aromatase inhibitors"[lxxxvi][449] which can reduce estrogen production. When combined with breast cancer hormonal therapies, a new class of agents called CDK4/6 inhibitors may improve therapy effectiveness by helping block tumor cell division.[450,451]

In 1941, Charles Huggins first reported that surgical removal of the testicles could control metastatic prostate cancer. It removed the body's major source of testosterone. In 1966, he received the Nobel Prize in

---

[lxxxvi] The drugs anastrozole and letrozole are examples of aromatase inhibitors.

Physiology and Medicine for this work.[452] Luckily, there are now several different hormonal therapies that can turn off testosterone production by the testes (and adrenal glands) or that will block the androgen receptor.[446]

However, as with chemotherapy, resistant cells eventually emerge. Typically, these resistant cells will either have a mutation of the receptor (so that it is no longer sensitive to the hormonal therapy) or altered expression of the receptor. In some cases, the tumor cells will have developed "bypass pathways" that drive tumor cell growth despite blockage of the hormone pathway.[442,443]

In both breast cancer and prostate cancer, treatment with hormone receptor inhibitors may also select for emergence of tumor cells with a mutated receptor that results in the hormonal therapy actually stimulating tumor cell growth instead of inhibiting it.[442,443] In these situations, if one then stops the hormonal therapy, one may get a "withdrawal response," with tumor shrinkage after the drug is discontinued.[442,443]

Targeted therapies: As discussed in Chapter 1, oncogenes are mutated genes that are responsible for tumor cell growth. The normal, non-mutated version of these oncogenes produce proteins that direct cells dividing in a controlled fashion when your body needs them to divide. However, the abnormal proteins produced by mutant oncogenes may drive uncontrolled cell growth. The protein doesn't turn on only when required but remains turned on all the time. This drives cancer growth.

"Targeted therapies" inhibit tumor cell growth by blocking the function of the mutated oncogene proteins. As I discussed earlier, when the early versions of these targeted therapies were first developed, we did not know that there were important genetic mutations affecting cell growth proteins. We knew that cancer cells often had abnormally high levels of some cell growth proteins. We thought that if we blocked these proteins, we might slow down tumor growth and perhaps prolong patient life expectancy.[453]

The drugs gefitinib and erlotinib were among the first targeted therapies. They targeted the epidermal growth factor receptor (EGFR) protein. The first clinical trials of these drugs were held at the end of the 1990s. The results were surprising. A few patients experienced unexpected, dramatic tumor shrinkage, with rapid improvement in symptoms.[454,455]

Then in 2004, researchers in Boston discovered that these patients had a previously unappreciated mutation of the *EGFR* gene.[456,457]

We now know that 10–15% of patients with lung adenocarcinomas have *EGFR* mutations.[458] We also know that tumor cells with relevant *EGFR* mutations generally have "oncogene addiction" and are very sensitive to blockage of the EGFR protein by gefitinib or erlotinib.[459] Mutated tumor cells are much more sensitive to these inhibitors than are normal cells or tumor cells that do not have an *EGFR* mutation. These addicted tumor cells may rapidly die if EGFR is inhibited. Effective targeted therapies may also bind more tightly to the mutant form of the protein than to the normal form of the protein.[460] Consequently, they can block the mutant protein and kill the tumor cells at concentrations that may cause only minor problems in normal cells.

It was also in the late 1990s that the targeted therapy imatinib was first reported to be of great benefit in patients with chronic myelogenous leukemia (CML). CML cell growth is driven by a gene alteration called a *Bcr/Abl* fusion, and imatinib blocks this fusion gene.[461]

About this same time, clinical trials started using the "monoclonal antibody" trastuzumab. Trastuzumab targets the growth factor HER2/neu and proved to be beneficial in patients with breast cancer if the expression of the HER2/neu protein on the tumor cells was greatly increased as a result of amplification[lxxxvii] of the *HER2/neu* gene.[462,463]

These observations set the stage for the development of "precision oncology" (also sometimes called "personalized medicine") in which a specific therapy targets a matching specific molecular abnormality in a patient's tumor.[152] There are now numerous examples of effective precision therapies. They target mutations or gene fusions involving the *EGFR*,[464]

[lxxxvii] Usually there is only a single copy of a gene on a given chromosome. If there are several extra copies of the gene on a chromosome, this is referred to as "gene amplification." Amplification of a growth factor gene may be an important driving factor for some malignancies. Cancer cells may also have far more than the normal number of chromosomes. If a cell has several extra copies of a gene due to having many extra chromosomes, but only one copy of the gene on each extra chromosome, this is referred to as "polysomy." Unlike gene amplification, gene polysomy does not usually drive tumor cell growth.

ALK,[465] *ROS1,*[466] *KRAS,*[467] *RET,*[468] *MET,*[469] *NTRK,*[470] and *HER2*[471] genes in adenocarcinomas of the lung, *BRAF* mutations in malignant melanoma,[472] lung cancer[473] and some other malignancies,[474] *c-KIT* mutations in gastrointestinal stromal tumors,[475] *BRCA* mutations in carcinomas of the breast,[476,477] ovary,[477] pancreas,[477] and prostate,[477] *Hedgehog* pathway mutations in metastatic basal cell carcinoma of the skin,[478] and *HER2* amplification in cancers of the breast[462,463] and stomach.[479]

Some of these have been proven to be more effective and less toxic than chemotherapy.[465,480,481] They are now the preferred first-line therapy in patients with incurable metastatic disease. Some of these agents produce at least some tumor regression in more than 95% of patients with the relevant mutation. The EGFR inhibitor osimertinib also results in a markedly reduced probability of tumor recurrence in patients with localized *EGFR*-mutant lung cancers who undergo surgical removal of their cancer.[366]

KRAS mutations are common in several tumor types. Until recently, KRAS has been very difficult to inhibit with a drug.[482] However, there are a variety of subtypes of *KRAS* mutations, and drugs have now been developed that can work well against the *KRAS G12C* mutation subtype that is most common in lung cancer.[467] The *KRAS* mutations that are seen most frequently in other tumor types like cancers of the pancreas and colon cannot yet be treated effectively, but the recent success against *KRAS G12C* in lung cancer is encouraging.

While these targeted therapies can cause marked tumor shrinkage with symptom improvement and prolongation of life expectancy in patients with a relevant mutation, they generally do not cure patients with metastatic disease. In the billions of cancer cell divisions occurring prior to the discovery of the cancer, at least a few mutations occur with each cell division. In other words, billions of mutations would have occurred before the cancer was discovered, and at least some cells within the tumor develop a mutation that make it resistant to the targeted therapy. Killing of the sensitive tumor cells by the targeted therapy decreases competition for nutrients and space, permitting the resistant cells to grow, ultimately leading to tumor progression, recurrence of symptoms, and ultimately the patient's death.[433]

We are rapidly learning more about the mutations that are responsible for this acquired resistance. New drugs are being developed to target these resistance mutations. For example, about half of tumors with *EGFR* mutations that become resistant to gefitinib and erlotinib have a new type of *EGFR* mutation called *T790M*. Osimertinib is highly effective against cancers with *T790M* mutations,[483] and it is also more effective than gefitinib or erlotinib when used as front-line therapy, prior to the emergence of the *T790M* clone as the dominant tumor cell population.[464] However, despite osimertinib's greater effectiveness than older agents, other mutations eventually emerge that are resistant to osimertinib.[484] Early clinical trials have now identified new drugs that may work in some patients whose tumors have become resistant to osimertinib.

No matter what type of targeted agent might work against a particular type of metastatic cancer, resistance eventually develops, and the patient's tumor begins to grow. There is now a huge effort underway to try to tackle this problem.

While targeting specific mutations has proven to be a useful strategy in treating patients with some malignancies, there are other cancers where specific molecular targets have not yet been identified. However, in at least some of these instances, other therapies have proven useful that are "targeted" (since they hit a specific type of molecule in the cell) but are not "precision oncology" (since we do not know how to identify patients who would be most likely to benefit). These therapies typically do not work as dramatically as the drugs targeting a specific mutated oncogene, and we do not have good ways of predicting who will derive the most benefit. Nevertheless, these agents are helpful for some patients.

Examples include drugs called "mTOR inhibitors" (e.g., everolimus) in breast cancer[485,486] and kidney cancer,[487] "CDK4/6 inhibitors" in patients with some types of breast cancer,[450,451] the drugs sorafenib[488] and sunitinib[489] in kidney cancer, monoclonal antibodies like cetuximab in cancers of the colon[490] and head and neck,[491] the monoclonal antibody rituximab in some types of lymphomas and leukemias,[492] "antiangiogenic" monoclonal antibodies like bevacizumab that reduce formation of the new blood vessels that are required for tumor growth,[493] and others.

"Drug-conjugated monoclonal antibodies" are another class of targeted therapy that is gaining traction. With these agents, a monoclonal antibody is developed that can link to a specific target on a tumor cell. A toxin is attached to the antibody. In this way, the monoclonal antibody can be used to specifically deliver a toxin to tumor cells. An example that has proven useful is trastuzumab emtansine that targets HER2 on breast cancer cells with *HER2* amplification.[494]

Monoclonal antibodies have also been developed for which one arm of the antibody attaches to one target and the other arm attaches to a different target. For example, amivantamab targets both MET and EGFR. There are early indications that this agent may be effective in patients with an *EGFR* mutation called an exon 20 insertion. This uncommon mutation does not respond well to standard EGFR inhibitors.[495]

Targeted therapies have emerged rapidly as a very important part of our anticancer arsenal. A huge effort is underway to expand and enhance our options.

Conclusions: Overall, systemic therapies are highly important in the management of patients with incurable, metastatic malignancies. They can improve patient symptoms, quality of life, and life expectancy. In patients with localized cancers that are amenable to potentially curative therapy with surgery or radiation, adjuvant systemic therapies can improve cure rates for some cancers. For most malignancies, the systemic therapies that we have now are much better than what we had just a few decades ago. However, the sixth reason that cancer still sucks is that gains with adjuvant systemic therapies are only modest. Most metastatic cancers eventually become resistant to these therapies and most patients with metastatic malignancies will eventually die from uncontrolled cancer as the resistant cells become the dominant tumor cell population.

— 7 —

T-cell Tom vs. The Masked Mutants: The Immune System and Immunotherapy of Cancer

In Chapter 1, I discussed how we all develop numerous mutations in our cells as they divide to replace damaged cells. One reason that we do not all develop cancer when young is that most of these mutations are of little consequence. It takes several "bad" mutations in one cell to convert it into a cancer cell. However, we are also protected by our immune system. It can identify and eradicate some early tumors.

For people who develop widespread cancers, new "immunotherapy" approaches can harness the immune system to tackle the cancer. In some cancer patients, immunotherapy can result in marked tumor shrinkage and improvement in cancer symptoms, quality of life, and life expectancy.

However, the seventh reason that cancer still sucks is that some malignant cells can escape the immune system to develop into full-blown cancers. Once a cancer does develop, immunotherapy is only effective in some patients, and in most patients who initially respond to immunotherapy, the cancer eventually becomes resistant.

Short Primer

Brief history of immunotherapy: In 1971, my summer job between the second and third years of medical school was in the laboratory of Dr. David Eidinger at Queen's University, Kingston, Ontario. My task was to test different immunotherapy approaches in mice for their ability to control implanted tumors. As so frequently happens with research, I was completely unsuccessful in my attempts.

At the time I was doing this research, "immunosurveillance" by the immune system was believed to kill tumor cells routinely and effectively as they emerged, with a cancer developing only if the immune system failed. There was widespread hope that cancer researchers were close to harnessing the immune system to eradicate any malignancies that had slipped by it. Animal studies identified effective new immunotherapy approaches, and early clinical trials suggested that stimulating the immune system with factors such as the bacterial product BCG could be beneficial.

By the early 1980s, initial studies with other new immunotherapy approaches using agents such as interferon and interleukin-2 were also producing promising results. By the late 1990s "monoclonal antibodies"[lxxxviii] were proving to be effective in at least some patients with certain types of cancers. It was also recognized that an immunological "graft—vs.—host" response[lxxxix] after bone marrow transplantation was associated with

---

[lxxxviii] Also see Chapter 6. Antibodies are large proteins produced by the immune system that can attach to specific targets called antigens. They can kill bacteria, viruses, and some types of cells. "Monoclonal" antibodies are highly purified antibodies that are produced in the laboratory and are designed to attack very specific targets. The anti-cancer monoclonals that have proven useful clinically target either a protein on the surface of a cancer cell or else target a circulating protein that can promote cancer growth. Monoclonal antibodies have also proven useful in treating a range of other diseases like Crohn's disease, rheumatoid arthritis, psoriasis, etc. Of historical interest, celebrity Martha Stewart went to jail for five months in 2004 for lying to federal agents about insider trading in the company ImClone Systems that owned the monoclonal antibody Erbitux (cetuximab).

[lxxxix] In the "graft—vs.—host" response, bone marrow cells transplanted from a donor mount an immune response against the tissues of the leukemia patient into whom they are transplanted. This can cause very serious problems in the bone marrow recipient, so drugs are used to suppress the immune system. However, patients who develop a graft—vs.—host reaction tend to have better control of their leukemia than those who do not develop this reaction. This suggests that the transplanted cells are establishing an immune response that helps control the leukemia cells.

improved control of some leukemias. Clinical trials also were underway of several anticancer vaccines.

The hype: The popular press eagerly seized on the romantic notion that the "answer to cancer" lay in a patient's own immune system. One of our lecturers reported to my medical school class his highly inaccurate observation that the immune system was so important that he had never seen a patient with allergies develop a malignancy. Story after story talked about exciting and promising immunotherapy research "breakthroughs." The solution appeared to lie within our grasp. However, as the decades passed, it appeared to me that the benefits of most of the new immunotherapy treatments were overstated and limited. Unequivocally, some of these therapies did help some patients, but for most patients, the reality did not match the hype. My message to trainees was that "Immunotherapy has been <u>about</u> to cure cancer for the past fifty years!" I was very skeptical that immunotherapy would ever amount to much.

My conversion to a believer, "immune checkpoint inhibitors": My faith in immunotherapy changed radically beginning in 2014–2015. Clinical trials of new inhibitors of the PD-1/PD-L1 "immune checkpoint" began to shrink a wide range of different cancer types.

Immune checkpoints are systems that prevent white blood cells called T-cells from destroying your normal tissues. When a T-cell approaches a normal cell, a protein on the normal cell called a ligand binds to a specific protein on the T-cell called a receptor. When the ligand binds to the receptor, this turns the T-cell off. This protects your normal cells from your T-cells.

Tumor cells may also be protected from the immune system by immune checkpoints. It's as if the tumor cells are wearing an immune checkpoint mask to protect themselves from the immune system.

Immune checkpoint inhibitors are monoclonal antibodies[xc] that bind to a ligand or receptor for one of the immune checkpoint systems. These antibodies prevent a ligand from interacting with its receptor. Consequently, the T-cell is not inactivated, and it can potentially kill the cell that it is approaching. The inhibitors effectively "unmask" the tumor cells hiding behind the immune checkpoint. Since some of the immune checkpoints appear to be most important in protecting normal cells from the immune system early in life, immune checkpoint inhibitors may preferentially activate the immune system against tumor cells, with relatively little impact on normal tissues.[xci]

Immune checkpoint inhibitors such as nivolumab and pembrolizumab target the PD-1/PD-L1 immune checkpoint system, while ipilimumab targets the "CTLA-4" immune checkpoint.

Power lines and switches: For many decades, the strategy used in cancer immunotherapy research was to try to stimulate a stronger immune response against the cancer, but this generally did not work very well. However, the immune checkpoint inhibitors suddenly began to demonstrate marked benefit in many patients.

I will use the analogy of a house where all the lights are off. You keep loading stronger and stronger electrical currents into the power lines, and nothing happens. Then a friend comes along and just flips a switch and all the lights come on. That is what happens with checkpoint

---

[xc]   A monoclonal antibody is an immune protein (i.e., an antibody) that is an exact replica of another antibody. Ordinarily, if the immune system is exposed to an antigen, several different cells will produce antibodies against the antigen. While all the antibodies attack the same antigen, they have minor differences from each other since they came from different cells. For a monoclonal antibody, a company isolates a single antibody-producing immune cell and makes many copies of this cell through "cloning". Since all cloned copies of the cell came from the same parent cell, they will produce identical antibodies. These antibodies can then be collected and used for various functions, such as treating cancers. If another company clones a different cell that produces an antibody against the same antigen, the antibodies from this company's clone will be similar to the original monoclonal antibody but will have some properties that are different. The antibody from the second company will be called a "biosimilar" of the antibody from the first company.

[xci]   Not all patients avoid immunotherapy toxicity, as discussed in Chapter 8.

inhibitors. Rather than trying to "load more power into the power lines" by stimulating a stronger immune system response, the immune checkpoint inhibitors just flip the switch, and this turns on the immune system.

Clinical uses of immune checkpoint inhibitors: In just a few years, immune checkpoint inhibitors have become a major weapon in our war on cancer. Hardly a week goes by without another clinical trial showing benefit. These drugs generally do not work in all patients with a given cancer. However, they are useful in 10–20% or more of treated patients in a wide range of cancers. In some patients with advanced cancers, tumor shrinkage has lasted years. Some researchers have even suggested that at least some patients might be cured by this immunotherapy. I personally do not feel that there is enough evidence to suggest that anyone has been cured, but some patients definitely derive very lengthy benefit.

I have many patients who have experienced this prolonged benefit. One example is a gentleman with advanced, incurable lung cancer who was 80 years old. He appeared to be quite frail and had early Parkinson's disease when I first met him. I told him that I did not think he would tolerate chemotherapy and that the fatigue caused by any chemotherapy might make his Parkinson's symptoms worse. I told him that if he talked to any of my colleagues, they would probably recommend that he not receive any therapy for his cancer. I offered him "best supportive care"[xcii] as an alternative. However, he told me that he very much wanted to at least try chemotherapy, with the option of stopping if side effects were excessive.

With some trepidation, I treated him with a combination of the chemotherapy agents carboplatin plus pemetrexed. Despite his age and frailty, he tolerated this extremely well, and his tumor shrank substantially. After six cycles of the chemotherapy combination, I switched him to "maintenance" treatment with pemetrexed alone.

After about 1.5 years, his cancer began to grow again despite the ongoing pemetrexed. By this time, his Parkinson's was somewhat worse, but he stated that he would still prefer to continue with treatment. We started him on the immunotherapy agent nivolumab, which he again tolerated extremely well. He had an excellent response, with tumor shrinkage.

[xcii] "Best supportive care" means treating patients with incurable cancers with pain medications, low dose radiotherapy, etc., but without any systemic therapy for their cancer.

Over a twenty-seven-month period, he received 47 cycles of nivolumab with ongoing control of his cancer and with excellent tolerance despite his gradually worsening Parkinson's.

We then discussed the pros and cons of carrying on with treatment vs. giving him a break (since some patients can experience ongoing tumor control after stopping nivolumab). We agreed that we would give him a break. At his most recent visit twenty-three months after his last nivolumab, 5.5 years after starting systemic therapy for incurable lung cancer and now 86 years old, he continues to do well, with no evidence of worsening of his cancer.

When he started his cancer journey, his wife of many decades was his highly dedicated primary caregiver. She has recently been admitted to a retirement home with deteriorating health. His main purpose in life is to now help with her daily care. I asked him if he would want to restart the nivolumab if his cancer began to grow again. His response was that he very well might want to do so if his wife was still alive and continued to need his help.

Immune checkpoint inhibitors combined with chemotherapy: For some cancers like lung cancer, adding immune checkpoint inhibitors to chemotherapy appears to work better than chemotherapy alone. Combining immunotherapy with chemotherapy is now a standard treatment for many patients with metastatic disease.

For some malignancies, combining two immunotherapy drugs like nivolumab and ipilimumab which inhibit different immune checkpoints may also be better than using one drug alone. Giving the drug durvalumab after completion of combined chemotherapy and radiotherapy improves the probability of long-term control in patients with locally advanced non-small cell lung cancer. Some of these agents also improve long-term disease control when administered after surgical removal of malignant melanoma and some other malignancies.

Factors associated with increased or decreased benefit from immune checkpoint inhibitors: The higher the proportion of cells in a lung cancer that "express"[xciii] PD-L1, the higher the probability of benefit from PD-1 or PD-L1 inhibiting monoclonal antibodies. However, even some patients with no PD-L1 expression will benefit, and some patients with high expression do not benefit. In addition, the higher the number of mutations in the tumor cells (referred to as having a "high tumor mutation burden"), the higher the probability of benefit. Tumors with a high tumor mutation burden are likely to have more "neoantigens"[xciv] that can be targeted by T-cells.

With most types of therapy, smokers are less likely to benefit than non-smokers. However, the opposite may be true with immunotherapy. Patients who have smoked a lot may have a higher tumor mutation burden in their cancers.

In Chapter 1, I discussed oncogene alterations that can drive tumor cell growth. Tumors with mutated *EGFR* or with *ALK* fusion genes are usually resistant to immunotherapy for reasons that we do not understand. Alterations of some other oncogenes (like *KRAS*) do not appear to influence the efficacy of immunotherapy, and for some oncogene alterations, we do not yet have much information on their impact on immunotherapy efficacy.

Patients who are very weak or debilitated by their cancer are referred to as having "poor performance status." Poor performance status patients are less likely to benefit from immunotherapy (or any other type of therapy) than are patients who are good performance status.

Patients receiving high doses of corticosteroid drugs like dexamethasone or prednisone are also less likely to benefit from immunotherapy, since corticosteroids suppress a patient's immune response. However, if a patient develops severe toxicity from immunotherapy, corticosteroids may be used to treat this toxicity, and this does not appear to have a major negative

---

[xciii] A cell "expressing" PD-L1 means that when a pathologist applies a stain that binds to PD-L1, the stain sticks to that cell (but not to other cells which do not express PD-L1). The pathologist can see this under a microscope. The higher the proportion of tumor cells that stain for PD-L1, the higher the expression. Applying stains to cells to detect presence of a substance of interest is called "immunohistochemistry."

[xciv] An antigen is a molecular substance that can be identified by the immune system. A neoantigen is a new antigen present on cancer cells but not on normal cells.

effect on immunotherapy benefits. Many patients who develop severe toxicity with immunotherapy also have excellent control of their cancer by the immunotherapy.

Patients who receive antibiotics shortly before immunotherapy is started are less likely to benefit from the immunotherapy. We are not certain why this happens, but we all have numerous bacteria in our intestines,[xcv] and these bacteria interact with immune cells lining the intestine. One theory is that antibiotics may interfere with a positive interaction between the bacteria and our immune system. Of interest, very early research suggests that taking stool from a patient responding to immunotherapy and inserting it into the intestine of a patient who is not responding may convert a non-responder into a responder by giving them a more favorable microbiome.[xcvi]

New immunotherapy approaches: In addition to the PD-1/PD-L1 system and CTLA4, there are eleven other immune checkpoints. Work is currently underway to assess the potential impact of inhibiting some of them. There are also several factors that may stimulate the immune system and adding one or more of these to immune checkpoint inhibitors might potentially improve efficacy. Work is also underway on new approaches to activate the immune system in other ways, and some of these are showing early promise. For example, monoclonal antibodies called "bispecific T-cell engagers" have one arm that attaches to a tumor cell and another arm that attaches to a T-cell. They thereby bring the T-cell in close to the tumor cell, potentially allowing the T-cell to destroy the tumor cell.

In another approach called "adoptive cell therapy," T-cells are taken from the body, changed in the laboratory to make them potentially more effective against a cancer, and then infused back into the blood stream. "CAR T-cells" are one example of adoptive cell therapy. CAR T-cells have proven to be highly effective in some patients with leukemias and lymphomas, but this therapy is also quite toxic. We do not yet know if

---

[xcv] It is estimated that we on average have between 200 g and 3 kg of bacteria living in our body. The bacteria living in our body are called the "microbiome."

[xcvi] My wife, Lesley, says "This sounds extremely gross. There is no way I would consider having this done if I were the patient." However, perspectives can change if one has advanced cancer.

CAR T-cells will also prove effective against more common malignancies like cancers of the breast, lung, and colon. Several factors make it more difficult for CAR T-cells to work in these common cancers compared to leukemias and lymphomas.

Conclusions: After a relatively slow start, there has been explosive progress in the immunotherapy field. Over the course of just a few years, the immune checkpoint inhibitors have demonstrated a broader impact across the field of oncology than has any class of drugs since the introduction of the chemotherapy agent cisplatin in the 1970s. While we still have a long way to go, we are making progress.

Further Details and References

Additional details follow.

Immunosurveillance: The immune system may detect and destroy developing tumors in a process termed "immunosurveillance."[496] A type of T lymphocyte called "CD8+ T-cells,"[497] "Natural Killer (NK) cells,"[498] and the STING[xcvii] pathway[499] are among the immune system components that help protect from cancer.

Immunosurveillance may fail for several reasons. These include the protection of tumor cells by some components of the immune system such as "regulatory T-cells" and macrophages.[xcviii] [500] Immunosurveillance also can fail if the immune system does not recognize the "neoantigens"[xcix] that distinguish tumor cells from normal cells, or if immune cells are disabled by "immune checkpoints."

Immunosurveillance and the antigen presenting machinery: There may be numerous changes in a tumor cell that the immune system

---

[xcvii] The term "STING pathway" is short for the "cyclic GMP-AMP synthase/Stimulator of interferon genes pathway."

[xcviii] Regulatory T cells ("Tregs") help reduce the probability that the immune system will damage normal cells. Macrophages secrete substances that help stimulate cell growth in wound healing, etc., but these substances can also stimulate tumor cell growth.

[xcix] An antigen is a distinct molecule that can be recognized by the immune system. A neoantigen is a new antigen (not present on normal cells) that appears due to changes such as mutations in the cell's DNA or due to the cell being infected by a virus, etc. The immune system may identify cells with neoantigens and destroy them.

might potentially recognize, but only a few of these many changes will in fact become distinct neoantigens. Neoantigens are produced when aging proteins are broken down in a cell. In all our cells, old protein is destroyed by machinery called proteasomes as fresh, new protein molecules are produced.[501] The old protein is broken down into short fragments called peptides. Some of the shorter peptides (typically 8 to 10 amino acids in length) will attach to other cellular proteins called human leucocyte antigen (HLA) proteins in the HLA protein's peptide-binding groove.[501] Whether a given peptide will attach to an HLA protein depends on the shape and characteristics of the peptide and on the shape and characteristics of the HLA protein. It's like a key fitting into a lock.

The characteristics of HLA proteins vary widely from one person to another.[c][501] Consequently, the peptides that may be bound by HLA proteins in one person may differ from those in another person.

The HLA proteins are attached to another molecule called beta-2-microglobulin (B2M) to form a protein complex called the MHC-1, a.k.a. the "major histocompatibility complex class I" or antigen processing machinery.[501] The MHC-1 carries any peptides attached to HLA to the cell surface and "presents" them to nearby CD8+ T-cells.[501]

Several other proteins are also important in this process.[501] The CD8+ T-cell does nothing if it recognizes the peptides presented to it as being "normal,". If it recognizes them as being different from what the cell should possess (i.e., as being a neoantigen), this triggers an "adaptive immune response." This response destroys the cells presenting these abnormal

---

[c] The DNA coding for a protein consists of a large number of nucleotides (also called DNA bases). In a given location on the gene, if a nucleotide in one person with normally functioning protein is different from the nucleotide in another individual, the two different "normal" nucleotide options are referred to as single nucleotide polymorphisms, or SNPs. (See chapter 1 for a discussion of SNPs). For the genes coding for HLA proteins, there are an estimated 14,000 SNPs. Hence, there is wide variability between individuals in the characteristics of their HLA proteins. Different gene variants (and hence different SNPs) will be more common in some populations than in others. Hence members of one group of people (e.g., Asians) may be more or less likely to have particular SNPs than are members of another groups of people (e.g., Caucasians).

peptides.[501] In this way, the immune system can recognize and eradicate cells carrying the mutations that may make them cancer cells.

An aberrant cell can go unrecognized if it doesn't present aberrant peptides at its surface. This can happen if a cell has a mutation in its *HLA* or *B2M* genes, or in genes for other molecules required for neoantigen presentation.[501] There also can be "epigenetic" alterations that result in decreased expression of the antigen processing complex. Mutations or other abnormalities of components of the antigen processing complex are extremely common in cancer cells,[501] suggesting that these aberrations may be very important in enabling a cancer cell to escape immunosurveillance.

We each inherit one copy of a given gene from our mother and one copy from our father. The two different versions of a gene in a cell are called "alleles." The two alleles will have differing SNPs. Having two alleles that differ from one another is called "heterozygosity." Consequently, each cell may produce two different versions of a given protein, one encoded by the allele from your mother and one encoded by the allele from your father. For the *HLA* genes, inheriting different versions from your mother vs. your father gives you HLA proteins that can bind a broader range of peptides than if you only had one allele producing one HLA protein type.

However, if the DNA in a cell is damaged, the cell may lose one of its two alleles for a given gene. This is referred to as "loss of heterozygosity" (LOH). LOH for HLA genes reduces the range of peptides that can be bound by HLA and presented to the cell surface where they can be recognized by immune cells. *HLA* LOH is also very common in cancer cells, suggesting that it increases the probability that the cancer cell will escape detection by the immune system.[501]

If a cell is unable to present neoantigens at its surface due to factors such as MHC-I mutations, these neoantigens may in some cases be taken up by other cells that can then present these neoantigens to T-cells.[501] This alternate mechanism consists of the MHC class II system (MHC-II).

MHC-II is also coded by HLA genes and is present on specific cells that are responsible for presenting antigens to the immune system. These antigen presenting cells include dendritic cells, mononuclear phagocytes, and a few others. They take up proteins from their environment and break them down into peptides. The cell's MHC-II system then presents the peptides to immune cells as potential neoantigens.

Lung and gut epithelial cells may also have high expression of MHC-II.[502] Tumors in which the tumor cells have high expression of MHC-II have an increased infiltration of lymphocytes into the tumor. Patient survival is longer than with low tumor cell MHC-II expression.[503]

The bottom line is that abnormalities in the antigen presenting system help cancer cells survive by evading immunosurveillance, and these abnormalities are very commonly found in cancer cells.

Immunosurveillance and immune checkpoints: Immune checkpoints prevent immune cells from attacking and destroying normal cells. In an immune checkpoint, a protein called a "ligand" on normal cells interacts with a "receptor" on the surface of approaching T-cells. This turns the T-cells off. CTLA4 was one of the first receptors recognized as being involved in an immune checkpoint function.[504] A team led by Dr. Jim Allison found that the binding of ligands on tumor cells to CTLA4 on T-cells was also important in inhibiting the immune response to tumors in animals.[505] After the discovery of CTLA4, Dr. Tasuku Honjo and colleagues discovered a second immune checkpoint based on a T-cell receptor they called "programmed death-1" (or PD-1).[506,507] They demonstrated that, like CTLA4, it could mask mutated cancer cells from T-cells.[508] Inhibition of PD-1 in animals improved the immune response to various cancers.[ci][509] It is now known that there are more than a dozen different immune checkpoint systems.[510]

Early immunotherapies: Early clinical trials suggested that stimulating the immune system with factors such as the bacterial product BCG could have an impact on cancers.[511] But these approaches have limited value. My early impression of immunotherapy was that there was substantially more hype than substance. Having said that, some of the early immunotherapies were of unequivocal value. With respect to monoclonal antibodies, some examples where benefit has been demonstrated include rituximab in some lymphomas and leukemias,[492] trastuzumab in some breast cancers[462] and stomach cancers,[479] cetuximab in some colorectal[490] and head and neck[491] cancers, and the antiangiogenic monoclonal antibody bevacizumab in some

---

[ci]   In 2018, Drs. Allison and Honjo shared the Nobel Prize in Physiology or Medicine for their pioneering work on inhibition of the immune checkpoints CTLA4 and PD-1/PD-L1.

lung[493] and colorectal[512] cancers. Interferon has modest activity in malignant melanoma,[513] kidney cancers,[513] and chronic myelogenous leukemia.[514]

Some patients with malignant melanomas[515,516] and kidney cancers[516] achieved impressive prolonged remissions with high dose interleukin-2 combined with tumor infiltrating lymphocytes. While marked activity of this therapy was seen in some patients, the production of the tumor infiltrating lymphocytes in therapeutic quantities remains too cumbersome and expensive to permit this approach to be applied widely. This therapy also can be quite toxic.[517]

Numerous cancer vaccines have also been tested. Most have not proven to be very effective, although sipuleucel-T does prolong life expectancy in patients with metastatic prostate cancer.[518]

Early issues that slowed immunotherapy progress: The probability of major success is low in any attempt to develop a new therapy. From my perspective, there were three main issues that particularly plagued research in immunotherapy. The first was that the animal models that were generally used involved transplantation of a tumor from one rodent to another, and such models probably do not accurately reflect what is happening in humans.[519,520] For example, if you transplant a normal organ from one animal to another, the immune system of the recipient will reject it because the animals are not genetically identical. In the case of the transplanted tumor studies, the donor rodents and recipient rodents were genetically very similar, but they were not necessarily identical.[520] Consequently, if an immunotherapy worked in a rodent, it might be because it was working in a manner analogous to stimulating the immune system to reject a transplanted organ. This is much different from the situation with a cancer patient, where the cancer develops from the patient's own tissues.

Because of this and other reasons, any immunotherapy tested in an animal model might appear to be more effective than it would be in a patient with cancer.[519,520] As a result, these animal models might have a low probability of identifying immunotherapies that would work.

The second problem was that many of the "promising" early clinical trials did not assess the therapy's ability to shrink tumors and prolong life. Instead, they examined the ability of a therapy to create changes in the patients' immune systems. It is much easier to change some aspects of a patient's immune system than to kill cancer cells.

The third problem was that it was widely believed that the immune system would only be effective against very small tumors, since the presence of a large cancer burden can directly suppress immunity.[521] Hence, the vision of many in the field was that surgery, radiotherapy, or chemotherapy would be used to get rid of most of the tumor and that an immunotherapy approach would then be used to mop up the small number of remaining tumor cells that were preventing cure.[522,523] This meant that researchers aimed low when evaluating their new therapies, and aiming low is problematic for many reasons.[524]

Early in the days of Bristol-Myers Squibb's development of the new monoclonal antibody ipilimumab, researchers from the company asked me what patient groups I thought they should target in initial clinical trials. They asked specifically about their wish to target patients with a "low tumor burden." I pointed out to them that if their drug could not shrink a large tumor that chemotherapy could shrink, then their drug probably would also be less effective than chemotherapy in patients with a low tumor burden. If this were the case, then ipilimumab would fail in clinical trials since it would not be as good as currently available chemotherapy. Why use immunotherapy if chemotherapy would probably work better?

Immune checkpoint inhibitors: Ipilimumab targets the CTLA4 immune checkpoint receptor and was the first of the immune checkpoint inhibitors to be tested clinically. Despite my initial skepticism and despite Bristol-Myers Squibb's early temptation to test it primarily in patients with a minimal tumor burden, ipilimumab did in fact prove capable of shrinking advanced malignancies and prolonging life expectancy in at least some patients with metastatic malignant melanomas[525-527] and kidney cancers.[528]

Further clinical trials of immune checkpoint inhibitors demonstrated that monoclonal antibodies targeting PD-1 or its ligand PD-L1 could shrink metastatic disease in 10–30% or more of patients with a wide range of malignancies, including malignant melanoma[529-533] and cancers of the lung,[534-540] kidney,[541-543] bladder,[544,545] head and neck,[140,546] and several others.[cii] [547-564] Only a relatively small number of tumor types show almost no sensitivity to these agents.

Not only may PD-1/PD-L1 inhibitors lead to tumor shrinkage, but they may also significantly prolong life expectancy.[330,532,536,537,539,543,546,565-567] They are superior to our best standard chemotherapy in some patient groups with metastatic lung cancer.[ciii] [330,536,537,539,565,566,568] They also prolong survival when added to chemotherapy in patients with metastatic lung cancer[569-572] or when given following completion of combined chemotherapy plus radiotherapy in patients with locally advanced lung cancer.[573] Very active clinical research is underway to test these agents in a wider range of cancers while also investigating ways to make them more effective.

PD-1/PD-L1 monoclonal antibodies appear to target a wider range of cancers and to be more effective than CTLA4 monoclonal antibodies.[532] They are also less toxic than ipilimumab, with a lower probability of damage to normal tissues.[532] In fact, overall, across a wide range of cancers, the PD-1/PD-L1 inhibitors appear to be markedly more effective than any immunotherapy previously tested clinically, and they are among the most broadly effective systemic therapies of any kind used clinically to date. The introduction of these agents to clinical oncology will probably surpass the impact of the introduction of the chemotherapy agent cisplatin in the 1970s.

Some of the immune checkpoint systems may be more important in protecting normal cells from T-cells very early in life, but their protection of cancer cells does not depend significantly on a patient's age. Because of

---

[cii] Other malignancies that may respond to PD-1/PD-L1 inhibitors in at least some patients include Merkel cell tumor, Hodgkin's Disease, primary mediastinal large B-cell lymphoma, mesothelioma, some types of soft tissue sarcomas, hepatocellular carcinoma, thymic carcinoma and cancers of the anus, colon ("MSI-high" subtype), stomach, esophagus, nasopharynx, ovary, endometrium, cervix, and breast ("triple-negative" subtype).

[ciii] In patients with metastatic lung cancer who have not received previous treatment the monoclonal antibody pembrolizumab is superior to chemotherapy in patients whose tumor cells express PD-L1PD-L1. In patients with metastatic lung cancer who have previously been treated with chemotherapy, the monoclonal antibody nivolumab is superior to second-line chemotherapy even if only a small proportion of the tumor cells express PD-L1PD-L1.

this, immune checkpoint inhibitors may be effective against some cancers without causing unmanageable toxicity.[civ]

Predictors of benefit from immune checkpoint inhibitors: Across a range of malignancies, PD-1/PD-L1 inhibitors typically work to cause tumor shrinkage in about 10–30% of patients. Of interest, chemotherapy can cause a lot of tumor shrinkage in some patients and a bit in some others while having no impact at all in others. Most chemotherapy agents generally appear to work in a "continuously variable" fashion with a range of outcomes across patients. For a given chemotherapy drug in a given tumor type, there generally do not appear to be two distinct groups of patients, with one group sensitive to the chemotherapy and the other group resistant to the chemotherapy. By analogy, if you looked at the height of all people in the population, you would see a range of heights, instead of all patients either being 6 feet tall or else 5 feet tall. Similarly, there is a range in degree of benefit from chemotherapy.[425]

The PD1/PD-L1 inhibitors, on the other hand, appear to work in a "dichotomous" present—vs.—absent way. Analyses that we have conducted suggest that there may in fact be two distinct patient groups, with one group deriving substantial benefit from the agents and the other group deriving little benefit.[425] In this way, they behave more like targeted therapies (see Chapter 6) that are effective in patients having a specific mutation but are largely ineffective in those without the mutation.

Tumors in which a high proportion of tumor cells show expression of PD-L1 and tumors with high "tumor mutation burden"[cv] are more likely to respond to PD1/PD-L1 inhibitors.[574] However, both of these are "continuously variable" factors, with expression varying across a broad range, from being at very low to very high levels, rather than being

---

[civ]   Although not all patients avoid toxicity, as discussed in Chapter 8.

[cv]   "Tumor mutation burden" is the total number of mutations that can be detected in the tumor divided by the number of DNA "bases" assessed. (See chapter 1 for a discussion on DNA bases.) Some malignancies like lung cancer and malignant melanoma on average have a very high tumor mutation burden, while others have a lower burden. It has been proposed that tumors with a very high tumor mutation burden will have more cancer-specific neoantigens that might be recognized by the immune system as being abnormal or foreign.

"present—vs.—absent." Consequently, it is difficult to understand how either of these two factors would explain the "present—vs.—absent" sensitivity to the therapy. Also, some tumors may respond even if they have only very low or absent PD-L1 expression.

We do not yet know why this dichotomous present—vs.—absent pattern happens with immune checkpoint inhibitors. While it might be due to presence of a mutation that makes tumors sensitive—vs.—resistant, no one has yet found such a mutation. Another possibility is that there needs to be a T-cell population already present that targets specific tumor antigens but that is incapable of killing the tumor cells since the immune checkpoint is blocking these T-cells. However, if one inactivates the checkpoint, these preexisting immune T-cells are unleashed. Patients with preexisting T-cells targeting a tumor neoantigen would experience benefit if the T-cells are unleashed by a PD-1/PD-L1 inhibitor while patients without a preexisting population of tumor-immune T-cells would not benefit from checkpoint inhibition.

If a tumor mutation was specifically required for T-cells to work, it might be feasible to discover that tumor mutation and to use it as a test that would enable us to tell who should and who should not receive the immunotherapy. On the other hand, it could prove much more difficult to develop a test to accurately predict who will benefit if the important factor is whether there is a preexisting population of activated immune T-cells just waiting to be unleashed. There are numerous populations of T-cells potentially immune against many different tumor neoantigens, and it will probably be tough to sort out what is truly important.

Resistance to immune checkpoint inhibitors: In Chapter 6, I discussed how some tumors develop resistance to chemotherapy or a targeted therapy agent. In a similar manner, most cancers that initially respond to immunotherapy eventually stop responding due to the development of "acquired resistance." There are several different factors that may underly this acquired resistance.[575] As with targeted therapies, I suspect that the mechanisms underlying acquired resistance will be much more diverse than the factors underlying intrinsic resistance.

Are some cancer patients permanently cured by immune checkpoint inhibitors? Some wonder if at least some patients with advanced metastatic cancers are permanently cured by immunotherapy. Personally, I think this is unlikely. The issue will remain the same as we have seen with chemotherapy and targeted therapies: there are likely to be a small number of resistant tumor cells present at the time of diagnosis that will evade the immune system and will eventually grow out as the dominant tumor cell population. I already know that I have been wrong about immunotherapy before, so with luck, I may also be wrong about this.

Hyperprogression: I mentioned above that some people appear to benefit from immunotherapy while others do not. Some investigators have questioned if immunotherapy might actually stimulate tumor cell growth in some patients, leading to what has been called "hyperprogression."[576] Just as I have doubts that immunotherapy is currently curing any patients, I also doubt that hyperprogression is real. When we apply special analytical techniques called "exponential decay nonlinear regression analysis"[577] to data from patient groups on immunotherapy, the tumors of patients who do not benefit appear to behave the same as if the patient received no therapy at all rather than behaving as if they received something that stimulated tumor growth.

Corticosteroids: Patients receiving corticosteroids such as prednisone and dexamethasone for supportive care have a reduced probability of benefiting from immune checkpoint inhibitors since the corticosteroids antagonize the immune system.[578] However, if corticosteroids are taken as treatment of a side effect of the immune checkpoint inhibitor, they do not appear to adversely impact therapy efficacy.[578]

Antibiotics: Patients who have received antibiotics shortly before or shortly after initiating immunotherapy have a reduced probability of benefiting from the immunotherapy.[579] We do not fully understand why antibiotics reduce immunotherapy efficacy, but we think it might be because bacteria[cvi] in the intestine alter immune function by interacting with immune cells lining the intestine. Antibiotics could change this interaction by changing the bacteria in the intestine.[579]

Smoking: Smokers are more likely to benefit from immune checkpoint inhibitors than are patients who have never smoked.[580] Patients who use

---

[cvi] Microorganisms in the body are called "the microbiome." They are present mainly on the skin and in the intestines, mouth, lungs, and vagina. By some estimates, they outnumber the human cells in the body.

tobacco generally have reduced benefit and increased toxicity from other cancer therapies. However, cancer patients who have smoked heavily tend to have a higher tumor mutation burden than people who have smoked less.[581] This is believed to underlie the increased benefit that smokers derive from immune checkpoint inhibitors.

Oncogene mutations: Patients whose tumors have an *EGFR* mutation[582] or *ALK* fusion gene[583] (see Chapter 6) are usually resistant to immunotherapy. The type of *EGFR* mutation appears to impact immune checkpoint inhibitor efficacy. Patients with *EGFR* exon 19 deletions and *T790M* mutations derive less benefit than those with *EGFR* exon 21 mutations and other alterations.[584,585] It is not clear why these mutations are associated with immunotherapy resistance, but it may in part be because these patients on average have a low tumor mutation burden.[584,586] Mutations of the tumor suppressor gene *STK11* are also associated with reduced immunotherapy efficacy.[587] Most other oncogene mutations appear to have little impact on sensitivity to immunotherapy.

Immunotherapy in patients with "autoimmune diseases": In autoimmune diseases, the immune system attacks normal cells in a patient's body. Examples of autoimmune diseases include systemic lupus, scleroderma, rheumatoid arthritis, Crohn's disease, ulcerative colitis, some types of pulmonary fibrosis (scarring of the lung), and multiple sclerosis. About 20–40% of patients have experienced flares of their autoimmune disease when started on an immune checkpoint inhibitor.[588]

In addition, drugs being used to treat the autoimmune disease might limit the effectiveness of the immunotherapy by counteracting it. However, non-corticosteroid immunosuppressive agents may have less of a negative impact than corticosteroids. It also may be feasible to adjust the immunosuppressive regimen in a way that maintains the potential for benefit from an immune checkpoint inhibitor.[588]

Overall, the presence of a preexisting autoimmune disease is a "relative contraindication" to treatment with an immune checkpoint inhibitor, meaning that I might not offer it to one of my cancer patients as their first treatment. However, after careful discussion of the pros and cons with a patient, I might offer it as a later treatment option if nothing else is working and there is "little to lose."

Immunotherapy in patients with an organ transplant: Patients with an organ transplant are generally given immunosuppressive drugs for the rest of their lives. This is required to prevent rejection of the organ by the immune system. Immunotherapy for a cancer could increase the risk of rejection of a transplanted organ, and the drugs to prevent organ rejection could also reduce any potential anticancer benefits from the immunotherapy. In a report on twelve patients with organ transplants who were treated with immune checkpoint inhibitors, four experienced organ rejection.[589] Consequently, I do not offer immunotherapy to my patients if they have had a heart or liver transplant, since rejection of the transplanted organ could prove fatal. However, I would consider offering immunotherapy as an option to a patient with a kidney transplant if I had no other useful options to offer. If the patient rejected their transplanted kidney, we would at least have dialysis as a potential backup.

How long should immunotherapy continue: In most studies of immune checkpoint inhibitors, the treatment has been continued for two years or less.[590] Some responding patients have worsening of their cancer after the treatment is stopped. Some others have ongoing control of their cancer long after the immunotherapy is discontinued. Some patients who experience tumor worsening after the immunotherapy is stopped will again respond if the immunotherapy is restarted. But others do not.[590] It may take several more years of research for us to better understand the pros and cons of stopping vs. continuing immunotherapy in responding patients.

Conclusions: I have gone from being strongly negative about immunotherapy to recognizing that the immune checkpoint inhibitors are now among the best anticancer agents that we have. There are also other new immunotherapy approaches such as CAR T-cells that have already proven beneficial (although highly toxic) in the treatment of some types of acute leukemia[591] and lymphoma.[592] Several other new immunotherapy strategies are being explored. I am cautiously optimistic that we will continue to see major advances. However, the seventh reason that cancer still sucks is that cancer cells have been able to figure out how to evade the immune system during the initial development of the cancer. Immunotherapy currently works in only a minority of patients, and patients who initially respond to immunotherapy may ultimately become resistant.

— 8 —

For Better or Worse: How Anticancer Therapies Cause Side Effects

The eighth reason that cancer still sucks is that cancer treatments can cause unpleasant (or even dangerous) side effects.

Short Primer

Not only has the effectiveness of cancer therapies improved substantially over the past few decades, but we have also made progress in reducing toxicities. However, we still have a long way to go. Here I will discuss just a few of the more common and distressing side effects of cancer therapies.

Chemotherapy toxicity: Overall, we know relatively little about why drugs are more toxic to some tissues than to others. Different types of toxicity occur with different chemotherapy drugs. It is often stated that rapidly dividing cells (like those in the bone marrow, skin, and the lining of the mouth and intestine) are the most sensitive to chemotherapy, but it is not that simple. For example, the drug cisplatin is regarded as being a particularly toxic drug, but at usual doses it is in fact toxic to only a very narrow spectrum of normal tissues. It causes nausea and vomiting, damage to very specific tissues within the kidney, high frequency hearing loss, and "sensory neuropathy" with numbness in the feet. It rarely causes skin toxicity, mouth sores, diarrhea, or hair loss. It is also less likely than many other chemotherapy drugs to affect the rapidly dividing cells of the bone marrow.

Toxicity of targeted agents and monoclonal antibodies: Like chemotherapy, "small molecule targeted agents" (that are taken by mouth) and monoclonal antibodies (that are administered intravenously) are only effective against a limited range of cancers. They also are generally only toxic to a limited range of normal tissues. For example, inhibitors of the epidermal growth factor receptor (EGFR) such as gefitinib and erlotinib only work well in lung cancers with an *EGFR* mutation. From a normal tissue perspective, they most frequently cause rash and diarrhea, with an occasional patient experiencing nausea, mouth sores, hair thinning, or mild liver toxicity. A small proportion of patients may develop serious lung or liver toxicity. Similarly, most other targeted therapies are better tolerated than chemotherapy, with only mild to modest side effects that vary between agents.

Monoclonal antibodies typically target cell surface proteins. For example, cetuximab targets cells with high EGFR expression and trastuzumab targets cells that have high HER2 expression. Unlike some of the small molecule targeted therapies, the effectiveness of cetuximab does not necessarily depend on a gene mutation.[cvii] By targeting the EGFR expressed on skin and conjunctiva of the eyes, cetuximab frequently causes skin rash, infections around fingernails and toenails ("paronychia"), and dry eyes.

Immunotherapy toxicity: As discussed in Chapter 7, various "immune checkpoints" turn off immune cells that could damage normal cells. New immunotherapy drugs called immune checkpoint inhibitors turn off this defense system, permitting the immune system to attack cancers protected by immune checkpoints. But they also potentially open normal cells to attack by the immune system. Unlike the limited range of cancers potentially impacted by a specific chemotherapy drug or targeted agent, inhibitors of the PD-1/PD-L1 immune checkpoint are active against a very wide range of cancer types.

However, they only work for a subgroup of patients with those cancers. We do not yet understand what makes some cancers sensitive and others resistant. Similarly, they can permit the immune system to attack a wide

---

[cvii]   Although with HER2, the high protein expression is due to the cell having extra copies of the gene, called gene amplification.

range of normal tissues, causing inflammation in those tissues. Despite this, for reasons that are not well understood, most patients have very little toxicity. If they do develop toxicity, it most frequently involves only a single normal tissue. Fatigue, rash, diarrhea, and thyroid dysfunction are the most common. We do not know why most patients with toxicity to immunotherapy only develop it in one or two organs. If we knew the reason, it might be possible to target the toxicity mechanism and permit safer ongoing treatment with the immunotherapy.

Radiotherapy toxicity: Radiotherapy can be toxic to most types of cells that are dividing. It is effective against a much broader range of cancer types than are most individual systemic therapy agents,[cviii] but it also causes toxicity to a broader range of normal tissues. This toxicity to a broad range of normal tissues means that care must be taken to hit as small a volume of normal tissue as possible.

When radiotherapy is given, it goes in one side of the body and out the other side and can damage normal tissues on its way in and out. The bigger the area of cancer to be treated, the larger the volume of normal tissue that is hit by the radiotherapy.

Modern radiotherapy techniques concentrate a greater proportion of the total radiotherapy effect in the tumor itself, but nearby normal tissues still receive a substantial radiation dose. The physicists working in the radiotherapy department calculate the doses of radiation that would hit different normal tissues. The radiation oncologist then uses this information to assess whether the patient would likely tolerate the radiation. If too much normal tissue would be damaged, then the radiation oncologist may decide that it is too risky to give radiotherapy. Alternatively, they might decide to give only a low "palliative" dose of radiotherapy to temporarily improve symptoms, rather than the higher, more toxic dose of radiotherapy that would be needed to potentially cure the patient.

If a tumor is relatively small (typically 3 cm or less in diameter), then "stereotactic radiotherapy" or "CyberKnife" radiotherapy can be used to deliver very high radiotherapy doses. The small tumor size permits

---

[cviii] The range of potentially sensitive tumor types is almost as broad with immune checkpoint inhibitors.

minimization of damage to surrounding normal tissues.[cix] As discussed in Chapter 5, if high dose radiotherapy is given to a larger tumor, then the dose may be divided up, with a small part of the total dose given each day for several weeks. This reduces toxicity to normal tissues.

Radiotherapy can have both short-term and long-term toxicities. The short-term complications mainly involve cells that are rapidly dividing. Short-term complications involving the skin can include redness, pain, and sometimes even blistering or ulceration. Radiation can also cause inflammation of the mouth or esophagus, hair loss, a drop in blood counts, nausea, or diarrhea. While this short-term toxicity can be unpleasant, it is largely reversible at standard radiotherapy doses.

Longer term, chronic toxicity may occur because of damage to blood vessels and to tissues with slowly dividing cells. Radiotherapy can scar these damaged areas, and these scarred areas may have impaired function.[cx] Furthermore, these scarred areas may be easily damaged by trauma, and any damage may not heal well. Long-term radiation-induced damage means that the same area cannot be treated again with high dose radiotherapy. Giving further radiotherapy may cause ulcerated, broken-down areas of tissue with pain, and life-threatening bleeding or infection.

Individual types of toxicity: Almost any tissue in the body may be damaged by at least some types of cancer therapies. I will just discuss a few of the most bothersome examples.

Nausea and vomiting: Nausea and vomiting are among the most feared side effects of chemotherapy. While some chemotherapy drugs cause very little nausea, others can be "highly emetogenic."[cxi] The 1991 movie *Dying Young* had a very graphic and all-too-realistic depiction of how bad chemotherapy-induced nausea and vomiting could be. As in the movie, it

---

[cix] However, even some small tumors cannot be treated by stereotactic radiotherapy or CyberKnife since treatment would involve delivery of too high a dose of radiotherapy to a blood vessel or airway immediately adjacent to the tumor. This could cause a potentially fatal hole in the blood vessel or airway.

[cx] For example, if the esophagus is scarred by radiation, this may interfere with swallowing; scarring of the lung may lead to chronic shortness of breath, etc.

[cxi] A "highly emetogenic" drug is an agent that can cause severe nausea and vomiting.

was frequent for patients to give up on chemotherapy, even if it might cure them, since they could not bear the thought of more nausea and vomiting.

The good news is that 1991 was also the year that the antinausea drug ondansetron was approved. Ondansetron and related drugs revolutionized chemotherapy. It counteracts the nausea and vomiting that can occur in the first 24 hours after chemotherapy. Aprepitant and related drugs help control nausea and vomiting that occur after the first 24 hours. Glucocorticoids like dexamethasone and prednisone, sedatives like olanzapine and other drugs like prochlorperazine can also help. A high proportion of patients treated with chemotherapy now have minimal or no nausea or vomiting. For most who still do get sick, it is not nearly as bad as it used to be.

Before ondansetron became available, a patient who experienced severe nausea and vomiting with their first chemotherapy treatment could become ill at even the thought of chemotherapy. This is referred to as "anticipatory nausea." It is like becoming sick at the thought of a food that had given you food poisoning, or my wife Lesley's strong aversion to scotch since her teen years when she surreptitiously overindulged on her father's favorite.

As a medical student in 1974, I had to mix chemotherapy for my patients on an open table in the kitchen of the oncology ward at the Kingston General Hospital. Patients would swear that they could smell their chemotherapy being mixed from 100 feet down the hallway, as I accidentally sprayed it over the walls and ceiling.[cxii]

Since severe nausea and vomiting are now much less frequent with modern antinausea drugs, anticipatory nausea is also much less of a problem. Ondansetron has been called "the drug that made nurses smile most" since far fewer patients with anticipatory nausea are likely to vomit on the nurse who is trying to administer their chemotherapy.

In the late 1970s, when I was doing my oncology fellowship at MD Anderson Hospital in Houston, prochlorperazine was the only antinausea drug we had, and it was not very effective. If I passed a patient's room and the closed door had a sign saying "Californium," it meant the patient was privately (and illegally) smoking up and was not to be disturbed.

[cxii] We were very naïve back then. Chemotherapy is now mixed by gowned and gloved pharmacy staff under a vented fume hood. It is surprising that any of us survive our misspent youth.

When I returned to Ottawa in 1980,[cxiii] I was also told of compassionate Ontario Provincial Police officers playing Robin Hood by quietly giving confiscated marijuana to chemotherapy patients to try to help their nausea and vomiting. It was a tough time then and is so much better now.

Marijuana and its derivatives unequivocally help some patients with nausea, although most patients do not need it. My older patients who now use these products seem to tolerate them much better than my older patients of years ago. This is in part because CBD oil and other derivatives do not contain some of the toxicity-associated chemicals found in marijuana. It also may in part be due to some of today's older patients having had a long history of using marijuana since youth. They have grown old with it.

Fatigue: Fatigue was unmasked as a major complication of chemotherapy after nausea and vomiting were conquered by ondansetron and related drugs. Before ondansetron, I had assumed that fatigue was a consequence of the nausea and vomiting. I did not appreciate that it would be there even if nausea was controlled. Any anticancer therapy can cause fatigue. It typically begins two to three days after a chemotherapy treatment and often only lasts for a few days. However, in some patients, fatigue can last for weeks or longer. Fatigue can also occur with radiotherapy to even relatively small areas of the body. It is somewhat less common with targeted therapies and immunotherapies but can still be an issue in some patients.

In Chapter 3, I discussed that being as active as possible is the most effective treatment for fatigue caused by a malignancy. The same is true for the fatigue caused by therapy. I often tell patients that they should try to push themselves rather than spending more time in bed after therapy. It may be counterintuitive and easier said than done, but it can be the most effective antidote to therapy-induced fatigue. The muscles may soften up and become weaker if they are not used.

"Chemo brain" or "brain fog": Many patients also experience chemo brain or brain fog (impairment of memory, concentration, or thinking) with chemotherapy. While it is called chemo brain it can also occur with hormonal therapies and radiotherapy. "Cognitive training" exercises can help chemo brain, as can physical exercise and Tai Chi.

[cxiii] I have twice returned from MD Anderson to Ottawa ("the coldest capital city in the world"): the first time in 1980 and the second time in 2011.

Even before any therapy is started, many patients have mild cognitive impairment from the effects of the cancer. For patients with preexisting dementia, I usually recommend against chemotherapy since I am concerned that it will worsen their function. However, the negative impact may not be as great for patients with some other neurological conditions. Specifically, I have been impressed that patients with even relatively severe multiple sclerosis often tolerate treatment well.

Hair loss (alopecia): Chemotherapy is notorious for alopecia, but in fact only a few classes of drugs cause it. Agents causing hair loss include anthracyclines like doxorubicin (Adriamycin), taxanes like paclitaxel (Taxol) and docetaxel (Taxotere), vinca alkaloids like vinorelbine, alkylating agents like cyclophosphamide, and epipodophyllotoxins like etoposide. Cisplatin and carboplatin, most antimetabolites (pemetrexed, gemcitabine, 5-fluorouracil, methotrexate, etc.), and targeted agents do not usually cause hair loss, although some patients receiving them will experience at least mild hair thinning.

If it is going to occur, hair loss typically starts two to four weeks after the first chemotherapy treatment. A patient can go to bed in the evening with a full head of hair but wake up in the morning with hair falling out by the handful. Sometimes this is associated with scalp discomfort. Many patients who know that this is coming elect to have all their hair cut off before it starts falling out.

While some patients lose predominantly their scalp hair, some lose all body hair. One of my patients, the late Canadian columnist Marjorie Nichols, pointed out in her autobiography[cxiv] that the worst sensation was the sides of her nostrils sticking together since they were no longer effectively separated by the usual small, innocuous hairs inside her nose. I think of Ms. Nichols every time my wife implores me to trim once again my own exuberantly growing nasal hairs.

Cooling the scalp while receiving chemotherapy decreases scalp blood flow and can reduce hair loss. However, it may work better in some areas than others, resulting in patchy hair loss that many patients still find distressing. Wearing a wig can help a patient's self-image, but wigs can be

---

[cxiv] Nichols, Marjorie. *Mark My Words: The Memoirs of a Very Political Reporter*; Douglas and McIntyre; 1992.

uncomfortable, particularly on a hot summer day. Furthermore, it has been observed that most women look pretty good in wigs, but most men do not.

Reduced white blood cell count and the risk of infection: Chemotherapy can affect the bone marrow to cause a decrease in white blood cells, red blood cells, and platelets.[cxv] The impact of chemotherapy on red blood cells is usually relatively easy to manage since transfusions can be administered to bring the red blood cell numbers up. The most important bone marrow impact of chemotherapy is on a type of white blood cells called neutrophils. Many types of chemotherapy cause a reduction in neutrophil counts (although not all do). Recovery from a low neutrophil count generally occurs about three weeks after a treatment, and this is the reason that many chemotherapy drugs are typically administered every three weeks.

Many bacteria live in our intestines, on our skin, and in other sites in our bodies. The neutrophils kill any bacteria that get into the blood stream or deep tissues. If the neutrophil count is too low, some of these bacteria may escape and cause a serious infection. If antibiotics are not started within a few hours, these infections could prove fatal. I tell patients on chemotherapy to go to the hospital emergency room immediately if they develop fever or severe chills. If it is in the middle of the night, they should not wait until morning. If it is a weekend, they should not wait until Monday.

I tell patients that if they decide to take chemotherapy, there is about a 10% probability that they will at some point require emergency admission to hospital to treat a serious infection related to low blood counts. I also explain that we can usually bring the infection under control if we treat it rapidly. However, they must understand that there is about a 1% possibility that they could die from such an infection. I often explain that it is like undergoing major surgery: most people come through it fine, but unexpected things can happen. A small proportion of patients will die from complications. The same is true for patients receiving chemotherapy.

The risks are higher for patients who are already very weak due to their cancer or who have other major illnesses. These will reduce their chances

---

[cxv] Platelets promote blood clotting to reduce the risk of bleeding.

of surviving an episode of chemotherapy-related infection. Hence, the decision to initiate chemotherapy is never taken lightly.

It is also important for patients with low blood counts to avoid procedures that may increase the number of bacteria entering the blood stream. This includes any dental work, surgery, a rectal exam, or colonoscopy. Even minor damage to the skin or to the lining of the mouth, or intestines during these procedures increases the risk.

Chemotherapy patients often ask me if they must stay home in isolation to protect themselves from infections. I point out to them that isolation does not accomplish much. The bacteria putting them at risk are generally already in their system. It is important to avoid people with an obvious infection. For example, if they catch an upper respiratory tract virus, the virus itself will not be very dangerous. But we may need to admit the patient and start antibiotics in case there might be a serious bacterial infection. Frequent hand washing does reduce the risk of infections, but isolation is unnecessary. With the use of just a bit of common sense, chemotherapy patients can continue to enjoy their grandchildren.[cxvi]

If a patient develops a serious infection due to low blood counts, administering Neupogen or related drugs with subsequent chemotherapy cycles can help stimulate recovery of neutrophils and reduce the risk of infections. Alternatively, the risk of another infection also can be reduced by just lowering drug doses for subsequent cycles.

There are pros and cons with both approaches, with no clear "right" answer. Different oncologists have different approaches. I prefer to avoid bone marrow stimulants and instead usually use lower chemotherapy doses for subsequent cycles. As discussed in Chapter 6, the dose-response curve for chemotherapy typically begins to flatten at higher doses. Hence, lowering chemotherapy doses may reduce toxicity without substantially reducing efficacy.

During the recent pandemic, patients with some types of cancers and patients on chemotherapy have been at increased risk of dying of COVID-19. These patients may also not get adequate protection from a

---

[cxvi] Things of course changed during the recent COVID-19 pandemic, where far greater caution is needed, with a high importance of social distancing, masks, etc. for everyone, including for people receiving chemotherapy.

vaccine and may be offered an extra "booster" vaccine injection in the hope that it will improve their protection.

Low platelet count and risk of bleeding: Many chemotherapy drugs can also reduce the production of platelets by the bone marrow. This can increase the risk of serious bleeding, although the risk remains low unless platelet numbers are markedly reduced. In patients with a low platelet count, risk of bleeding is increased by platelet antagonists like aspirin, by major infection, and by any procedures (like surgery) that might cause bleeding in any normal person. If needed, platelet transfusions can be given to bring up the platelet count

Damage to cells lining the mouth and gastrointestinal tract: The cells lining the mouth and rest of the gastrointestinal tract divide frequently and can be damaged by some chemotherapy agents. Damage to cells lining the mouth can produce mouth sores, referred to as "stomatitis" or "mucositis." Damage to cells lining the intestines can cause diarrhea. Damage to the lining of the mouth or intestines can increase the risk of serious infections in patients with a low white blood cell count. Mucositis and diarrhea can also interfere with eating and drinking and increase the risk of dehydration.

Numbness, tingling: Platinum agents (such as cisplatin, oxaliplatin and carboplatin), taxanes (such as paclitaxel and docetaxel), vinca alkaloids (such as vinorelbine and vincristine) and a few other agents can cause chronic numbness and tingling in the hands and feet, referred to as a "sensory neuropathy." In my experience, chemotherapy drugs that generally do not cause a neuropathy may nevertheless worsen a preexisting neuropathy in some patients.

At its worst, neuropathy can be painful ("painful dysesthesia"). This pain can have a negative impact on quality of life and may be difficult to control. Numbness of the feet can also impair mobility, particularly at night. When you cannot feel your feet because of a sensory neuropathy, it can be difficult to walk safely if it is too dark to see.

Premature aging: Chemotherapy and radiotherapy can save lives, but they can also cause premature aging of normal tissues that they damage. For cancers that are likely to be fatal if not treated, this premature aging may be a comparatively small price to pay if the therapy cures the cancer and enables survival. On the other hand, we need to think carefully about the impact of "adjuvant" therapies in patients who may already have been cured by surgery. Adjuvant therapies are generally only given if they increase the probability of cure, but due to their potential to cause premature aging, it may be important to only offer these therapies if they substantially increase cure rates.

Impact of strength/fitness ("performance status") and other illnesses ("comorbidities"): In Chapter 3, I discussed how cancer causes weight loss, weakness, and fatigue by wasting energy, impairing appetite, and breaking down muscle mass. The overall fitness of cancer patients may be classified by scales such as the "ECOG[cxvii] performance status." ECOG performance status (PS) 0 patients are asymptomatic. PS 1 patients have symptoms but are fully functional. PS 2 patients have major limitations on normal function due to symptoms. PS 3 patients spend more than 50% of their day in bed. PS 4 patients are completely bedridden. The worse the PS, the lower the probability of benefit from therapy (although even some poor PS patients may benefit), and the higher the probability of major complications.

Advanced age by itself is not a contraindication to therapy. I have had many patients in their 80s who have done extremely well.[cxviii] Having said that, patients who are very frail due to other major illnesses such as congestive heart failure or dementia have an increased risk of experiencing more harm than benefit from chemotherapy and high dose radiotherapy. But some very frail patients will benefit from more easily tolerated targeted therapies and immunotherapy.

Impact of genetic makeup ("pharmacogenetics"): Different individuals inherit different versions of genes that affect a therapy's effectiveness and toxicity. Pharmacogenetics deals with such genetic differences. However, pharmacogenetics information is not usually used routinely to guide therapy. For some drugs (and radiotherapy), everyone is

---

[cxvii] ECOG stands for Eastern Co-operative Oncology Group. Institutions and oncologists have banded together to form several cooperative groups to permit them to collaborate in conducting clinical research. ECOG is one of the oldest and largest cooperative groups. The ECOG PS scale was devised to classify individual patients with respect to their degree of robustness or frailty.

[cxviii] Although I generally reduce chemotherapy doses by about 20% in the very elderly.

given the same dose. For other drugs, dose is based on body size, kidney function, etc., but not on the patient's genetic makeup.

Routinely using pharmacogenetic approaches to individualize therapies might pay dividends. For example, the dose of radiotherapy typically used is the dose that "most" people will tolerate. However, pharmacogenetic approaches would predict that some patients would tolerate much higher doses. Others might be predicted to develop severe complications even with standard doses.

Cancer might be resistant to a standard therapy dose if a patient's genetic makeup results in few side effects with the standard dose of a therapy. While cancers can have gene mutations, many of the cancer's genes will be the same as the genes of the patient in whom the cancer developed. "Resistant" patients with few side effects might also be more likely to have resistant cancers, and *vice versa* for "sensitive" patients. Knowing how much of a therapy a patient would tolerate based on their genetics could help optimize their therapy dose. This, in turn, might maximize the probability that the therapy would be both tolerable and effective.

A much greater exploration of pharmacogenetics is needed to optimize therapy for individual patients.

Further Details and References

I will now go into further detail on therapy toxicity.

Chemotherapy toxicity: While chemotherapy is often regarded as being quite toxic, most drugs are only toxic to very specific normal tissues, in the same way that they are only effective against very specific cancers.[429] Furthermore, a cancer drug will be useful only if the average cancer cell is more sensitive to the drug than are normal tissues. However, in most cases, we have limited understanding of the biological basis of this limited toxicity range across cancer and normal tissue types- and of this relative specificity for cancer tissues over normal tissues.[429] If we knew more, it might help us make the drugs both more effective and less toxic.

Drugs that are retained in human tissues for several months[406,593-595] may be more likely to damage more slowly growing cells like those in the liver, kidney, heart, and sensory nerves than drugs that are retained in human tissues for only a short period of time.[596] However, it will generally only be a very limited range of organs in which the drug causes toxicity, despite being present at prolonged high concentrations in a wide range of tissues. For example, high concentrations of the drug cisplatin are retained in the liver for many months,[406] but it rarely causes liver toxicity.

On the other hand, within a sensitive tissue, drug concentration matters. For example, cisplatin can cause numbness and tingling of the toes and fingers due to sensory nerve damage, but rarely causes damage to other parts of the nervous system. Cell bodies for the sensory nerves reside in structures called dorsal root ganglia. Dorsal root ganglia are less protected from toxins in the blood stream than are other parts of the nervous system. Cisplatin is retained in much higher concentrations in dorsal root ganglia than in other parts of the nervous system. The degree of nerve damage correlates with these cisplatin concentrations.[597]

Toxicity of targeted agents and monoclonal antibodies: As noted previously, targeted therapies like EGFR inhibitors generally only cause mild or modest side effects. About 1–2% of patients may develop potentially life-threatening or fatal interstitial lung disease, or liver toxicity.[598] The greater toxicity of these agents to mutated cancers than to normal tissues is due in part to these agents binding much more tightly to the abnormal EGFR proteins produced by a mutant *EGFR* gene than to the normal EGFR protein produced by a normal *EGFR* gene.[599] In addition, cancer cells with an *EGFR* mutation are "addicted" to its continued activity.[459] If EGFR is blocked, the cancer cell may die. The consequences are generally less severe in a nonaddicted normal cell.

Unlike small molecule targeted therapies that have a special affinity for mutated target protein, monoclonal antibodies target proteins that are expressed at high concentrations on the cell surface. This impacts the pattern of toxicity. For example, the major toxicity of the monoclonal antibody cetuximab is in organs with relatively high levels of its target protein, EGFR. Hence, cetuximab is most likely to cause skin rash, although it can also cause infection around the fingernails and toenails, eye irritation, and diarrhea.[600] Other monoclonal antibodies have differing patterns of toxicity, depending on which organs have high concentrations of their target protein.

Immunotherapy toxicity: In immune checkpoints, a "ligand" on a normal cell or on a cancer cell attaches to a receptor on an approaching immune cell. It thereby turns off the immune cell. Two of the first immune checkpoints to be discovered were the CTLA4 system[504,505] and the PD-1/PD-L1 system.[506-508] Blocking an immune checkpoint using a monoclonal antibody directed against either a ligand or a receptor can unleash the immune system against a tumor. It can also result in the immune system attacking normal cells. However, for reasons that remain unclear, the majority of patients have relatively little toxicity. The most common toxicities seen are rash and diarrhea, but any other organ system also can be affected.[601,602] CTLA4 inhibitors have a different spectrum of organs commonly affected and are more likely to cause serious toxicity than are PD-1/PD-L1 inhibitors.[602] Corticosteroids and occasionally more potent immunosuppressive agents can be used if toxicity is severe.[602]

Immune checkpoint inhibitors have been associated with flares in some patients' preexisting autoimmune diseases, such as rheumatoid arthritis. Many others have not suffered a flare.[603] Consequently, patients with a non-life-threatening autoimmune disease may be considered for immune checkpoint inhibitor therapy. Conversely, caution should be exercised in treating a patient with a preexisting autoimmune disease that is potentially life-threatening, like myasthenia gravis.[603] Risk of flare may possibly be higher with a CTLA4 inhibitor than with a PD-1/PD-L1 inhibitor.[603]

Similarly, some patients with an organ transplant who receive immune checkpoint inhibitors may reject the transplanted organ while others do not.[589] Hence, it might be reasonable to offer immunotherapy to a patient with a kidney transplant (since dialysis can be used if the transplanted kidney is rejected), but much greater caution would be needed in offering it to a patient with a heart or liver transplant.

Radiotherapy toxicity: Radiotherapy may cause either short-term toxicity involving more rapidly dividing cells or late toxicity involving blood vessels and slowly dividing tissues.[604] Toxicity from radiotherapy is increased if a larger volume of normal tissue is hit, if a larger dose is given with each daily treatment, or if a higher total dose of radiotherapy is given over the entire course of treatment.[604] The dose of radiation given with each treatment is particularly important in causing late toxicity. A given total dose of radiation will cause more long-term damage if it is given over just a few days than if it is spread out over several weeks.[604]

Some slowly growing tissues are relatively resistant to radiotherapy. For example, neurons in most parts of the brain do not generally divide much after early childhood. Hence, at least in the short-term, the brain will tolerate a dose of radiotherapy that might not be tolerated by many other tissues. If there are multiple brain metastases, radiotherapy may be given to the entire brain.[605] However, in the longer term, radiation damage to small blood vessels may cause brain shrinkage ("cerebral atrophy") and cognitive decline. Some patients also may develop "radionecrosis" (stroke-like destruction of some areas of the brain due to radiation-induced blood vessel damage).[605] Areas of radionecrosis cause problems not just because functioning brain tissue is lost, but also because they can cause seizures and can be very leaky. This permits fluid from the blood stream to permeate out into surrounding areas of normal brain and impede its function.

As noted in Chapter 5, patients who have some types of autoimmune diseases[cxix] such as scleroderma, related "collagen vascular diseases" and "pulmonary fibrosis" are at substantially increased risk of serious, potentially life-threatening or fatal lung inflammation if radiotherapy is delivered to the lung.[419-421] Consequently, radiotherapy is usually not given to these patients, although occasional patients may be offered it if only a very small volume of normal lung will be hit by the radiation.

Research studies tell us that some patients also may be at increased risk of toxicity from radiotherapy since they have inherited a decreased ability to tolerate this toxicity. For example, variants of some genes related to inflammation susceptibility[422] or related to ability of cells to repair DNA damage[423] are associated with increased radiation toxicity. It is not yet standard practice to test patients for these inherited risk factors, although this could change over the next several years.

Individual types of toxicity: Cancer therapies can damage many different normal tissues.[606] Below is a discussion of a few of the toxicities that can occur.

Nausea and vomiting: As noted earlier, nausea and vomiting are among the most feared side effects of chemotherapy. In the past, patients commonly stopped potentially effective therapy because of

---

[cxix] In autoimmune diseases, the immune system damages normal body tissues.

it.[607] Uncontrolled nausea and vomiting was also a common reason for admission to an oncology inpatient unit.[608] However, the introduction of ondansetron in 1991 substantially improved control of chemotherapy nausea and vomiting[609] and markedly reduced vomiting-induced hospital admissions.[608] Anticipatory nausea induced by even the thought of further chemotherapy[610] is now also very uncommon.

Anticancer drugs vary widely in their propensity to cause nausea and vomiting.[606,607] Cisplatin and dacarbazine are particularly bad, with carboplatin, doxorubicin and cyclophosphamide not far behind. On the other extreme, most "antimetabolites" (such as pemetrexed), targeted agents, and immunotherapy agents cause minimal or no nausea. Since they markedly reduce hospital admissions, modern antinausea drugs are highly cost-effective.[608]

Chemotherapy (or radiation to the abdomen) causes nausea and vomiting by damaging cells in the intestine called enterochromaffin cells.[611] These cells are a warning system that you have ingested a toxin that you should avoid in the future. The highly unpleasant sensation of nausea warns you not to eat the same food again. The vomiting helps rid your stomach of any remaining poison. However, with chemotherapy, the toxin usually enters the system through the blood stream (by intravenous administration) instead of through the stomach, so the vomiting does nothing to solve the problem.

Damage to the enterochromaffin cells leads to the release of chemicals called neurotransmitters, including serotonin and substance P.[612] These neurotransmitters can activate local nerve complexes that directly impact the gut. Even more important, they also interact with specific receptors on the vagus nerve, which transmits signals to areas in the brain stem that are responsible for perceiving nausea and for inducing vomiting.[612] They also travel through the blood stream from the enterochromaffin cells to the brain where they directly interact with receptors in the nausea/vomiting-associated brain stem areas.[612] There is also some local release of neurotransmitters in these brain stem areas by a direct effect of the chemotherapy.[612]

Nausea and vomiting may occur once the neurotransmitters reach a high enough concentration to trigger their receptors. This takes time. Nausea and vomiting usually occur a few hours after chemotherapy administration. Serotonin levels rise over five to six hours. Serotonin is the neurotransmitter that is most responsible for nausea and vomiting within the first 24 hours after chemotherapy administration. However, other neurotransmitters also play a role.[612] Ondansetron blocks the interaction of serotonin with its receptor and substantially reduces the nausea and vomiting that occurs within the first 24 hours.

Delayed nausea and vomiting may occur up to three to five days after chemotherapy administration. Ondansetron is of little benefit against this delayed toxicity. The neurotransmitter "substance P" is the major driver of this delayed toxicity. It increases for a prolonged period of time after chemotherapy, but it also plays a role in the toxicity occurring in the first 24 hours.[612] Aprepitant and related drugs block substance P's "neurokinin-1" receptor and substantially reduce both early and late gastrointestinal toxicity.[609,612]

Corticosteroids like dexamethasone can also reduce both acute and delayed nausea and vomiting through mechanisms that remain uncertain.[613] Prochlorperazine and metoclopramide also help somewhat by blocking dopamine receptors.

Marijuana and cannabinoids can also be effective through their impact on the CB1 receptor,[612] but as discussed previously, they are often not tolerated by older patients. Cannabinoids generally are not recommended for most patients, but they may be of value in some.[48]

Fatigue: As discussed earlier, any type of anticancer therapy can cause fatigue. With chemotherapy, the fatigue typically begins two to three days after a treatment and lasts for a few days. However, in some patients, it can persist for a prolonged time.[614] The fatigue is thought to be caused by the drugs resulting in the production of "reactive oxygen species" or by decreasing antioxidant levels (thereby reducing the body's defenses against high oxidant levels). High oxidant levels impede muscle function.[614] Induction by therapy of pro-inflammatory cytokines[cxx] may also play a role.[615]

[cxx] As discussed in chapter 3, cytokines are proteins released by immune cells in response to the cancer or its treatment. They are an important component of inflammation and play a major role in cancer-related weight loss and fatigue.

Despite radiotherapy only hitting a small part of the body, it frequently causes fatigue that may persist for many months or longer.[616] As with chemotherapy, radiotherapy is thought to cause fatigue by producing pro-inflammatory cytokines and reactive oxygen species that interfere with muscle cellular function.[617]

Fatigue occurs in at least some patients treated with targeted therapies,[618] but is generally less of an issue than with chemotherapy.

Immunotherapy agents that target CTLA4 often cause fatigue.[619] While immunotherapy drugs that target PD-1/PD-L1 appear to be associated with less fatigue than other anticancer therapies, they are nevertheless associated with fatigue in up to 20% of patients.[620]

Of all approaches that have been tested to alleviate therapy-induced fatigue, aerobic exercise has produced the most consistent positive results.[617]

"Chemo brain" or "brain fog": Many patients experience chemo brain or brain fog (impairment of memory, concentration or thinking) with chemotherapy.[621] While it is called chemo brain, it also occurs commonly with hormonal therapies and radiotherapy.[301] Furthermore, some impairment of brain function can be detected in up to one-third of cancer patients even before any treatment is given. This is attributed to the cancer itself. However, the proportion of patients experiencing cognitive dysfunction may increase to as high as 75% during or after treatment.[622] As with fatigue, the symptoms may last for only a short period of time in some patients but may last for months or years after treatment in up to 35% of patients.[301,621]

Chemotherapy may be directly toxic to some types of cells in the brain. It may alter gene expression in brain cells and accelerate brain aging.[621] Reduction in both brain size and function have been seen on MRI scans of patients who have undergone cancer treatments.[301] Just as pro-inflammatory cytokines may play a role in cancer cachexia and treatment-related fatigue, they may also play a role in chemo brain. Worsening and improvement in cognitive function vary with blood levels of inflammation-related cytokines such as tumor necrosis factor-alpha (TNF-α).[301] Genetic factors also may impact the risk of developing chemo brain.[301]

Just as aerobic exercise is a useful therapy for treatment-induced fatigue, exercising the brain with various cognitive training approaches may improve chemo brain.[301] Tai Chi and physical exercise also help, as does the anti-narcolepsy drug modafinil.[301]

Hair loss (alopecia): Some types of chemotherapy drugs (but not all) may also cause hair loss or alopecia. Some patients (particularly older ones) are not bothered a lot by the prospect of losing their hair, but to others it can be psychologically difficult.[623]

Cells in the hair follicle called keratinocytes are responsible for hair formation. If the keratinocytes are damaged by chemotherapy, the hair shaft produced will be thinner than normal. If it is very thin, it may break off at the scalp.[623] Cells called melanocytes add color to the hair, and if they are damaged, the hair will become white. As the keratinocytes and melanocytes recover from the effects of the chemotherapy, growing hair shafts gradually recover their normal caliber, strength, and color.[623] However, some patients who originally had straight hair will have it initially grow back curly (and *vice versa*).[624]

Radiotherapy to the scalp also causes hair loss. With chemotherapy, the hair usually (but not always) eventually regrows. Hair loss with scalp radiation can be permanent in some patients. The risk of permanent alopecia after radiotherapy to the scalp increases with the dose of radiation.[625]

Regrowth of hair with chemotherapy-induced alopecia typically starts about three months after completion of the chemotherapy. However, some patients will note regrowth while they are still receiving treatment.[624] For patients who used a wig after chemotherapy, hair regrowth is sufficient that about half no longer need a wig after one year, and only 15% are still wearing a wig by two years.[624]

Cooling the scalp during chemotherapy administration may partially protect from hair loss by decreasing scalp blood flow (and hence drug delivery).[623] However, scalp cooling substantially reduces hair loss in only 50–70% of patients.[623] Some patients will experience a patchy effect from scalp cooling, with preservation of hair in some areas and loss in others. Many patients do not find this patchy protection to be any less distressing than complete hair loss. Furthermore, some reports have suggested that a tumor may recur in the scalp if cooling reduces scalp chemotherapy concentrations, but the risk of this happening appears to be very low.[626]

Reduced white blood cell count and the risk of infection: As noted earlier, chemotherapy can affect the bone marrow to cause a decrease in white blood cells, red blood cells, and platelets.[cxxi] [606] The most important bone marrow impact of chemotherapy is on a type of white blood cells called neutrophils.[287] About 100 billion neutrophils are produced by the bone marrow every day, but they typically live an average of only 6.5 hours.[cxxii] [627] Neutrophils are renewed and replaced by constantly dividing "stem cells" in the bone marrow. Once a stem cell divides to form new "daughter" cells, these daughter cells go through several additional divisions over about twelve days before eventually producing a mature neutrophil.[628]

Many (but not all) chemotherapy drugs will briefly stop bone marrow stem cells from dividing. During this period when stem cells are no longer dividing, the daughter cells that have already been produced keep on the maturation path toward production of mature neutrophils. But once this supply of daughter cells and their progeny are used up, the neutrophil count drops. When the stem cells start dividing again, they will produce a new crop of daughter cells that will mature. However, this temporary interruption means that there will be a period of a few days to a week or two when the neutrophil count is very low. In most patients, it takes about three weeks from a chemotherapy treatment for the neutrophil count to drop and then come back up again. This is the reason that many chemotherapy drugs are only given once every three weeks. We must wait for recovery of the bone marrow and neutrophils before it is safe to treat again.[606]

Neutrophils are particularly important in fighting bacterial infections. We are each "colonized" by a wide variety of bacteria.[629] Bacteria are a vital part of our everyday life, and we could not survive without them. They play an essential role in food digestion, production of vitamin K, and regulating our immune system. It has been estimated that we each have on average between 200 grams and 3 kg of bacteria in our bodies, primarily in the intestine and on the skin.[630] "Friendly bacteria" and barriers like the skin and lining of our intestines help reduce the probability that potentially

---

[cxxi]   Platelets promote blood clotting to reduce the risk of bleeding.

[cxxii]   This very short survival is the reason that it is not feasible to transfuse neutrophils when they are deficient.

dangerous bacteria will cause serious infections by gaining entry into blood and "deep tissues" in our body.[629]

It is the job of the neutrophils to eradicate any bacteria that do invade. When there are not enough neutrophils, these invading bacteria can get loose and cause very serious, potentially fatal infections.[629,631] "Gram negative bacilli," like *E. coli*, klebsiella and pseudomonas, as well as "gram positive" enterococci are particularly likely to enter the blood and deep tissues from the intestines.

Patients with both a low neutrophil count and diarrhea, mouth sores, or other manifestations of therapy-induced injury to the lining of the digestive tract are at increased risk of serious infection.[632,633] Other types of bacteria called "gram positive cocci" like staphylococci and streptococci are particularly likely to enter blood or deep tissues through breaks in the skin.[cxxiii] [629]

Normally the blood neutrophil count is between 2.0 and 7.5 x 10^9/ Liter (i.e., between 2 billion and 7.5 billion neutrophils per liter of blood). A low neutrophil count ("neutropenia") combined with fever is called "febrile neutropenia." Most patients have no increased risk of infection with a modest reduction in neutrophil counts, but if the level goes down below 1.0 x 10^9/L, the risk begins to rise. The lower the level below this threshold and the longer the duration of neutropenia, the higher the risk of major infection.[634]

The probability of febrile neutropenia occurring varies with chemotherapy type and dose.[631] On average, it occurs in roughly 10% of patients receiving bone-marrow-toxic chemotherapy at some point in their treatment. Older patients are at higher risk than younger patients. Those with "hematologic malignancies" (leukemias and lymphomas) are more likely to develop infections than are patients with "solid tumors"(i.e., the common malignancies of adults, such as breast cancer and lung cancer).[631] At least in part, this is because bone marrow typically takes longer to recover from the very high chemotherapy doses used in hematologic malignancies.

[cxxiii] Breaks in the skin that may predispose to gram positive infections include ulcerating tumors, surgical incisions, intravenous catheters, simple scratches, etc.

Febrile neutropenia is most likely to develop with the first cycle of chemotherapy.[631] Patients will vary in their sensitivity to chemotherapy. Those who develop febrile neutropenia will subsequently have a reduction in their chemotherapy dose or various other preventative strategies. Those who do not develop it with the first cycle have a relatively low risk of developing it with subsequent cycles.[631]

Many patients feel fatigued when they are neutropenic,[631] but the first indication of a serious infection is generally fever and severe chills.[629,631] Fever in a patient with very low blood counts is regarded as an oncological emergency because death can occur within a few hours after fever onset. It is one situation where the physician does not wait for cultures to come back, but instead starts antibiotics immediately.[629,631] With prompt initiation of antibiotics, most patients will recover from febrile neutropenia, but about 8% will die despite treatment.[631]

If about 10% of chemotherapy patients develop febrile neutropenia and if about 8% of those with febrile neutropenia die from it, it means that there is approximately a 0.8% risk that a patient who starts chemotherapy will die from it.

Drugs like G-CSF (including Neupogen, Neulasta and Grastofil) can speed bone marrow recovery.[631] Decreasing the duration and severity of neutropenia reduces the risk of developing serious infections. If febrile neutropenia develops, using G-CSF with later cycles reduces the risk of the febrile neutropenia recurring. However, simply reducing the chemotherapy dose by 20% or so also effectively reduces risk. There is no broad agreement on whether it is better to use G-CSF or simply to reduce chemotherapy doses.

Many of us thought that the approval of G-CSF in 2002 was a breakthrough and that it would make the chemotherapy more effective by permitting higher doses at shorter time intervals. In the test tube and laboratory animals, there can be a steep "dose-response curve"—i.e., the higher the doses of chemotherapy, the better it works.[635]

Some clinical studies suggest that reducing toxic chemotherapy doses may be associated with decreased chemotherapy efficacy,[636] and G-CSF can permit maintenance of higher chemotherapy doses. Despite this, G-CSF use is not associated with a statistically significant improvement in survival for chemotherapy patients.[637] One explanation might be that in common cancers like breast and lung cancer, chemotherapy efficacy increases as the dose increases up to a point. However, as doses get higher, there is little further increase in efficacy despite an increase in toxicity.[436,437,638,639] These cancers behave as if they are running out of factors required for the chemotherapy to kill cancer cells, so further dose escalation has minimal impact on tumor tissues. Despite this, the chemotherapy is increasingly toxic to normal tissues as doses increase.[429,436,437] High dose chemotherapy can be beneficial in "hematologic malignancies" like leukemias, lymphomas, and multiple myeloma and in "germ cell tumors" like cancer of the testicle, but it adds little benefit in the more common "solid tumors" like breast and lung cancer.

Disadvantages of G-CSF include its cost[640] (although it can be cost-effective by reducing hospital admissions, etc.[641]) and potential side effects such as bone pain, which can be severe in up to 5% of patients.[642] In addition, tumor cells may have G-CSF receptors, and G-CSF may thereby stimulate tumor cell growth.[643] This could potentially reduce chemotherapy effectiveness. Furthermore, while the G-CSF partially protects from low white blood cells, it does not protect from other toxicities such as low platelet counts, mouth sores, etc. These other toxicities may limit the ability to give chemotherapy at higher doses or more frequently despite the ability to maintain high neutrophil counts.[640]

Giving "dose-dense" adjuvant chemotherapy may increase the probability of long-term survival in patients with some types of breast cancer who are at high risk of recurrence after surgery. With dose-dense chemotherapy, cycles of full dose chemotherapy are given at 2-week intervals instead of at 3-week intervals, and G-CSF is used to make the blood counts recover fast enough to permit this. This may be an example of G-CSF improving a survival outcome, but the apparent benefit of the dose-dense approach may not be real. In some studies, the dose-dense approach was compared to the agent paclitaxel administered every three weeks, while in other trials, other comparators were used. Benefit of dose-dense treatment was noted primarily in the studies using paclitaxel as a comparator. Dose-dense chemotherapy did not show much advantage when compared to other adjuvant regimens.[644] In other words, the bar may have been set too low by using a paclitaxel regimen as the comparator in some trials.

The bottom line is that some oncologists feel that G-CSF is important and should be used routinely, while others (like me) feel that it is not needed for most patients.

The risk of developing and dying from COVID-19 infection is increased in cancer patients.[645] The risk of death is highest in patients with hematologic malignancies and lung cancer[646] and in patients who have recently received chemotherapy.[647] The impact of chemotherapy on risk is probably more associated with its effect on lymphocytes than on neutrophils. Compared to others, cancer patients mount lower antibody responses to COVID-19 vaccines. Cancer patients may be given additional "booster" vaccine doses to help stimulate greater immunity.[648]

Low platelet count ("thrombocytopenia") and risk of bleeding: Many of the chemotherapy drugs that cause neutropenia may also cause thrombocytopenia. Some are more likely to do it than others.[606] Normally, there are more than 140×10^9 platelets per Liter of blood. Risk of spontaneous bleeding generally does not increase substantially until the platelet count drops to less than $20 \times 10^9/L$,[606] although other factors can increase the bleeding risk. These other risk factors include sepsis[606] or fever,[649] use of aspirin or similar drugs that impair platelet function,[606] use of anticoagulants,[606] and any invasive procedure that involve sticking a large needle or scalpel into deep tissues.

Like the neutrophil count, the platelet count takes about three weeks to recover after most chemotherapy agents.[cxxiv] It is important that the platelet count be recovered sufficiently before proceeding with the next treatment.[606] A persistently low platelet count may be an indication that the bone marrow stem cells have been seriously injured by treatment.

If needed, platelet transfusions may be given for dangerously low platelet levels.[606] Platelets last an average of three to four days,[650] so fairly frequent transfusion may be required for patients with impaired platelet production.

New drugs like romiplostim can speed recovery from chemotherapy-induced thrombocytopenia. It remains unknown if they can improve patient survival,[651] and they are not widely used in chemotherapy patients.

[cxxiv] Although for some drugs like mitomycin-C and nitrosoureas it can take six weeks or more for recovery.

Damage to cells lining the mouth and gastrointestinal tract: Mucositis (i.e., mouth sores) and diarrhea can occur due to chemotherapy or radiotherapy toxicity to the rapidly dividing epithelial cells lining the mouth and intestine. The problem is worsened by treatment-induced inflammation and by the killing of "friendly" gut bacteria that usually help protect against infection by "unfriendly" bacteria.[652]

Mucositis and diarrhea are associated with increased risk of serious infections. Their presence indicates that the barrier of the gastrointestinal tract has been disrupted, thereby facilitating the penetration of bacteria into deeper tissues.[652] When they are severe, mucositis and diarrhea are also associated with an increased risk of dehydration since the patient may be unable to drink enough to keep up with fluid losses.[653] Intake of oral fluids is encouraged, and patients may also be given intravenous fluids at home.

Therapies that <u>interfere with DNA production</u> are more likely to cause mucositis than are therapies that <u>damage</u> DNA.[654] Irinotecan and fluoropyrimidines like 5-fluorouracil often cause diarrhea.[653]

A few therapies that work through mechanisms other than interfering with DNA production may also cause diarrhea or mucositis, including radiotherapy, some targeted agents,[653] and immunotherapy approaches like ipilimumab and the PD-1/PD-L1 inhibitors.[655]

Agents such as loperamide may help control diarrhea,[653] as may reducing the therapy dose. High dose corticosteroids or potent immunosuppressive drugs may be needed to treat immunotherapy-induced diarrhea.[656] Sucralfate[654] and mouthwashes such as salt and soda, chlorhexidine or a combination of lidocaine, Benadryl and Maalox[657] may help with healing of mucositis.

Numbness, tingling: The platinums, the taxanes, the vinca alkaloids and occasionally other anticancer agents can cause a sensory neuropathy, with numbness and tingling of the hands and feet. The brain is protected from toxins by a lining between blood vessels and brain tissues called the "blood-brain barrier."[597] Most of the rest of the nervous system is also protected from toxins by a similar barrier, apart from the nerves responsible for sensation ("sensory neurons"). A single sensory nerve cell has very long projections called axons, with a short axon going from the cell body to the spinal cord and a much longer axon going to a distant part of the body such as the ends of the toes or fingers.

The cell body of a sensory neuron is in a structure close to the spinal cord called the "dorsal root ganglion." Unlike most of the rest of the nervous system, sensory neuron cell bodies in the dorsal root ganglia have little barrier to protect them from toxins in the blood.[658] Neurotoxic drugs like cisplatin reach higher concentrations in the dorsal root ganglion than in any other part of the nervous system.[597] That is probably the main reason that toxin-induced sensory neuropathies are much more common than toxin-induced damage to the brain or to motor neurons (nerves that make muscles move).

The longest nerves (those that supply sensation to the toes) are usually the ones that demonstrate symptoms of damage first, so that numbness in the toes often begins earlier than numbness in the fingers. The numbness is initially in a "stocking and glove" distribution, rather than first developing higher up on the legs, on the trunk, etc.

Different drugs have different time patterns for developing neuropathy. With taxanes and vinca alkaloids, numbness with or without pain will often develop within a few days of drug treatment. The numbness and pain may then improve prior to the next treatment three weeks later. However, the severity of residual symptoms may gradually increase over the course of several treatments.

On the other hand, cisplatin may not cause any numbness until several weeks after the last treatment. The numbness may then gradually increase over a few months, then stabilize, and then gradually decrease. However, in some patients, it may be persistently severe. With most drugs causing a neuropathy, the oncologist may determine whether it is safe to give more by the severity of numbness currently present. But this is not an option with cisplatin since the numbness may only develop long after the last dose of cisplatin has been administered.

There are no effective agents to prevent or reverse this nerve damage and reducing or limiting drug doses is the major tool used to minimize problems. While no drug will reverse the damage, analgesics like pregabalin may help reduce neuropathic pain.

Premature aging: Chemotherapy, radiotherapy, and other cancer therapies can save lives and improve life expectancy and quality of life, but they may also be associated with premature aging. Among other things, they can cause accelerated brain aging[621] and altered function of the heart, lungs, thyroid gland, muscles, skin, eyes, and other organs.[659] They can also cause infertility, sexual dysfunction, osteoporosis, impaired tissue regeneration, chronic inflammation, frailty, and an increased risk of developing other cancers.[659] Survivors of childhood cancers have a shorter life expectancy than their peers and an increased risk of chronic illnesses.[659]

Various factors contribute to premature aging. For example, a cell's chromosomes have a bit of extra DNA at the end called a "telomere." Each time a cell divides, a bit of the telomere is chopped off and not replaced. After several cell division, when the telomere reaches a critically short length, the cell will stop dividing or will die. The increased cell divisions required to repair therapy-induced damage means that available telomere length is used up more rapidly.[659]

In addition, when DNA is damaged by therapy, a gene called *TP53* may initiate changes that force the cell to kill itself through a process called "apoptosis." Alternatively, *TP53* may force a damaged cell to undergo a process called "cellular senescence" in which the cell is still alive but no longer able to divide. Loss of stem cells through apoptosis, senescence, or exhaustion of telomere length will reduce the number that are available to renew aging or damaged tissues.[659]

Not only are senescent cells no longer available to divide, but they can also cause inflammation and other adverse consequences.[659] Therapy generates "free radicals" (highly charged molecular particles that can damage tissues) and induces "epigenetic" changes (gene alterations that permanently inactivate the gene). These can increase tissue damage or decrease ability to repair damage.[659]

Various strategies have been tested in animal models to reduce or reverse this premature aging. They include therapies that might restore telomere length or eradicate senescent cells.[659]

Impact of strength/fitness ("performance status") and other illnesses ("comorbidities"): ECOG performance status (PS) 0 patients have no symptoms, while at the opposite end of the spectrum, PS 4 patients are bedridden. The worse the PS, the less likely the patient will derive major benefit from any type of treatment for their cancer. Having said that, even poor PS patients may have at least some improvement in cancer symptoms[660,661] and life expectancy.[441]

Poor PS patients also have a higher risk of major complications from treatment, including death from chemotherapy-induced febrile neutropenia.[660] Consequently, the usual recommendation for poor PS patients is that they not take chemotherapy or high dose radiotherapy since treatment is more likely to cause major harm than to produce meaningful benefit.

Poor PS patients with the right target may be offered targeted therapy and poor PS patients with high PD-L1 expression may be offered immunotherapy. These therapies tend to have fewer side effects than chemotherapy. However, the probability of benefit will be lower than in good PS patients.[662,663]

Impact of genetic makeup ("pharmacogenetics"): We have more than 20,000 protein-coding genes in each of our cells.[664] In addition, different people may have different versions of a given gene. (See Chapters 1 and 15 for more detail.) These different versions of a gene are referred to as "single nucleotide polymorphisms," or SNPs. These SNPs contribute to people having different eye color, hair texture, etc. But there also are several different SNPs for genes that make proteins responsible for drug activation, for drug breakdown, for repair of DNA damage, etc.[665] Consequently, a particular dose of a drug will have different levels of toxicity and efficacy in different patients. For example, specific SNPs are known to be associated with the risk of developing severe diarrhea with the chemotherapy agents irinotecan[666] and 5-fluorouracil.[667] SNPs for different genes related to inflammation are associated with the degree of lung toxicity with radiotherapy for lung cancer.[422] SNPs for a variety of different genes are associated with efficacy and toxicity of cisplatin in the treatment of lung cancer.[668-670] We are in the very early days of using this information to help personalize therapy choices. I expect it to be much more broadly useful within the next decade or two.

Conclusions: In addition to the side effects I discuss above, cancer therapies also can cause issues with the heart, lungs, kidneys, liver, and just about any other tissue in the body. The eighth reason that cancer still sucks is that cancer therapies can still cause substantial toxicity. We have made a lot of progress on this front, and these therapies are generally much less toxic than they were just a few years ago. However, we still have a long way to go.

— 9 —

Do I Have a Cure for You!
Complementary and
Alternative Therapies

The ninth reason that cancer still sucks is that unsubstantiated claims are made about potential benefits of complementary and alternative therapies.[cxxv] My stepdaughter Jenika Alvarez played a very important role in making this book happen. She tells me that at least some people are going to be very unhappy with me for what I say in this chapter. But here I go!

Short Primer

It's devastating for a patient to be informed that they have incurable, metastatic cancer. For many, this news is followed by advice from family and friends regarding alternative therapies. However, for most alternative therapies, there is no reliable indication that they work. Many claims of efficacy are unsupported by solid evidence, and many are based on very shaky logic.

[cxxv] For the rest of this chapter, I will use the term "alternative therapies" to refer to both complementary and alternative therapies. I define complementary therapies as unproven drugs or practices used in addition to standard therapies while alternative therapies are used instead of standard therapies.

Alternative therapies are alternative therapies rather than standard therapies since they have not been reliably demonstrated to shrink cancers, or to prolong life expectancy or to make patients feel better. Either there is no evidence to support their benefit, or the evidence is inconclusive or unreliable.

Alternative therapies can give false hope. Desperate patients may spend large amounts of money on something that has no value, or they may give up standard therapies that could help them. I have seen numerous examples.

Many physicians would gladly prescribe and promote an alternative therapy that was demonstrated to be beneficial. Therapy does not have to come at a high price from a major pharmaceutical company. For example, we promote eating a healthy diet, exercise, a full night's sleep, avoiding smoking, moderation in alcohol consumption, etc. Cheap, low dose aspirin is routinely used to help reduce the risk of heart attacks and strokes. We will use anything that helps improve outcomes for our patients. The important thing is not how much something costs or who profits from it. Instead, there are two important questions: how effective is it, and how reliable is the evidence supporting its use?

When a patients asks me if I am OK with them using an alternative therapy, I tell them that I generally have no problem with it. I advise against using high dose antioxidant vitamins since they might antagonize a therapy's effects, and alternative therapies can occasionally be toxic.[671] But most are not. I have no objection to people using them, provided they understand that I do not advise patients to use them. Some may help with some symptoms, but there is little credible evidence that any of them help control cancers.

Further Details and References

I outline further details below.

Approximately 40% of patients with advanced malignancies try alternative therapies.[672] When patients ask me about them, I explain that alternative therapies are alternative (rather than standard) because there is little or no credible evidence supporting their use. I point out that most of the claims supporting them are unsubstantiated. Furthermore, in most cases, people who claim to have benefited from them either also received effective standard treatment at the same time (and it was the standard therapy that actually helped them) or else had not been reliably diagnosed as having an advanced cancer.

As one example, one woman told me that she had cured herself of cancer using only an alternative therapy. When asked for details about the type of cancer, she stated that she felt sick, so she was "certain" she had cancer. She never had a scan or biopsy to confirm it, but instead started the alternative therapy, began to feel better, and concluded that she had cured herself of the cancer that she was convinced she had. Many people feel sick for many reasons and begin to feel better as the illness abates. One cannot claim that they have had cancer (and have cured themselves of it) unless they have had scans and a biopsy to prove it. But she convinced many of her friends that she had cured herself of cancer. Her friends did not understand that they were being misled.

As another example, an acquaintance convincingly told my wife, Lesley, that he had cured himself of a major cancer using an alternative treatment. Lesley suggested I talk to him to broaden my horizons. When I did talk to him, I learned that he had been treated with standard radiotherapy for a type of localized cancer that is highly curable by radiotherapy. He agreed with me that the radiotherapy was actually what cured him, but he continued to tell anyone else who would listen that his cure was instead due to the alternative therapy that he took. He "forgot" to mention that he had also received high dose radiotherapy.

The above are examples of great stories that make the teller the center of attention. Harmless, right? No, not harmless at all! Desperate patients with incurable cancers hear such stories and may pay a fortune for an alternative therapy that has virtually no chance of helping them. It is fantastic when one can give true hope. It is a cruel, self-centered, selfish hoax to give false hope by misleading others with unfounded stories of "miracles."

Many alternative therapies come from plants. For thousands of years, humans have been deriving medical benefit from plants in their environment. A recent "systematic scoping review" summarized outcomes with naturopathic therapies for a range of illnesses.[673] Most of the studies assessed used "retrospective" analyses or "prospective cohort" approaches.

Either of these methods might suggest benefit from a therapy that could then be tested properly in a randomized controlled trial (RCT), but the initial analyses themselves would not be considered reliable enough to draw any conclusions. There would be far too many ways in which other factors could have impacted the outcome.

A small number of RCTs were included in the review. Some of these suggested possible benefit in a few areas like some types of musculoskeletal pain, cardiovascular risk factors and anxiety. The only cancer study in this review was a small retrospective analysis and such an analysis would not permit any meaningful conclusions.[673]

Several of the chemotherapy drugs that we currently use (e.g., paclitaxel, docetaxel, irinotecan, vinorelbine and etoposide) are derived from plant sources.[674] For example, etoposide was synthesized from a chemical found in a folk remedy for cancer, the mandrake root or May apple.[675] Hence, if someone is using a plant product as an alternative therapy, I cannot say there is no possibility that it will have meaningful anticancer effects, but I can tell them that any benefit is likely to be very limited.

Through evolution, plants have developed many toxic compounds to protect themselves from insects and animals. If the insect or animal eats the plant and the plant has a toxin, then the insect or animal will not return. If you saw the 1992 movie *Medicine Man* starring Sean Connery, you will have seen a scientist working in the jungle trying to extract effective new drugs from local plants.

The story in *Medicine Man* has a basis. For example, I have been told that a jungle plant with relatively few insect bites in the leaves will have an increased probability of expressing toxic chemicals. There is a possibility that these toxins may prove beneficial as an insecticide or as a chemotherapy agent. Animals may also discover plants that help improve their health. Scientists found potentially effective new antimalaria compounds by observing what plants were eaten by chimpanzees in Uganda.[674]

Even if a plant has a useful substance in it, there may be several problems with just using the plant "as is." The first is that in addition to the useful chemical, there may be several other chemicals that are toxic. The second is that the concentration of different chemicals in the plant may vary depending on weather, soil conditions, etc. Hence, some plants may only have very low concentrations of the useful chemical while others may have very high, toxic concentrations. One may not know the right dose to maintain both efficacy and safety. As a result, using a highly purified, quantifiable chemical extracted from the plant is likely to be more reliable than using the plant itself.

Currently, while there are several effective chemotherapy agents that have been developed from plants, there are no direct plant products that have been demonstrated to be effective in metastatic cancers. A good naturopath can give reliable guidance to clients on toxic plants or combinations to be avoided and benefits that might possibly be experienced with others.

However, the advice in some cases may not be fully reliable. For example, a few years ago, I asked one naturopath if he told patients that he might cure their cancers with his alternative therapies. He replied that he told them "I will not *necessarily* cure you." If I were a patient, I would interpret this statement as implying that a cure was possible, but not guaranteed, when in fact a cure would be very highly unlikely.[cxxvi]

There are some plant products that may be helpful in other ways. In Chapter 1, I discussed the importance of fruits, vegetables, nuts, and whole grains in possibly preventing cancer, but there is no evidence that they can cure a patient or slow down cancer growth once a cancer is established. As I discuss in Chapters 3 and 8, marijuana derivatives also may have at least modest benefit in cancer-related pain and appetite loss and in chemotherapy-induced nausea and vomiting.

Some patients who had tried both marijuana smoking and synthetic marijuana derivatives have told me that the smoking worked better, although I am unaware of any formal, properly conducted clinical trials to directly compare them. However, I anticipate a lot of additional useful research, now that medical marijuana is legal and widely available in many jurisdictions. What is learned from clinical research on marijuana and its derivatives may also provide insights on how we should approach research on other plants with potential medicinal value.

While marijuana does help some patients with nausea, vomiting, or pain, there is no evidence supporting claims that it has any clinically

---

[cxxvi] Of course, there are also some physicians or other health care providers who may overstate a therapy's potential benefit. This is not a problem restricted to the occasional naturopath. Many naturopaths give reliable advice.

useful anticancer activity.[48] Many people use marijuana, and if it had any meaningful anticancer activity, we would be seeing patients whose cancers suddenly shrank when they used it. I have never seen this happen. On the other hand, I had one recent patient whose lung cancer symptoms were rapidly worsening despite treatment. When I told her that the scans confirmed that her cancer was growing, she exclaimed, "How can that be possible? I am taking marijuana!" Her local cannabis merchant had lied to her, telling her that marijuana was known to be effective against cancer. He was charging her $850 per week for false hope based on a product with no proven anticancer benefit.

Similarly, various other approaches such as acupuncture, "mind-body interventions," etc., may help a number of symptoms related to cancer or to therapy toxicity,[676] but there is no indication that they do anything that directly helps control the cancer.

There are many alternative therapies for which there have been unsubstantiated claims of benefit. Examples include laetrile, Essiac, antineoplastons, high dose vitamin C, and many others.[676] As discussed in Chapters 4 and 5, antioxidants like vitamins A, C, and E might actually antagonize the beneficial effects of radiotherapy and some chemotherapy agents. Furthermore, while fresh fruits and vegetables may help reduce the risk of developing cancer, antioxidant vitamins as medications are either of no value or else increase the risk of developing cancer, instead of reducing it.[cxxvii] These vitamins may promote the growth of established cancers rather than inhibiting them.

Various types of logic are used to back up claims for alternative therapy approaches. For example, diabetics are at increased risk of cancer.[677,678] A cell growth factor related to insulin may drive growth of some types of tumor cells,[678] and diabetics with cancer do better overall if their diabetes is being treated with the drug metformin than if it is being treated with insulin.[677,678] Furthermore, cancer cells generally use glucose much less efficiently to produce energy than do normal cells, so they need more glucose to survive and grow than do normal cells.[295] Consequently, it has been claimed that markedly reducing carbohydrates in the diet might

---

[cxxvii] See Chapter 1.

reduce cancer growth.[679] However, there is very limited research or clinical experience that supports this claim. The notion that you can deprive cancer cells of glucose by markedly restricting carbohydrate intake also ignores the fact that cancer cells enhance breakdown of normal tissues like muscle as a source of glucose for the cancer,[680] as discussed in Chapter 3. If you eat less sugar, the cancer can still get a ready supply of it by inducing breakdown of your normal tissues.

In assessing whether a therapy is beneficial, clinical researchers look at different types of evidence. The gold standard is evidence that a therapy can prolong life expectancy, although there are also some other "endpoints" that are accepted as evidence.[430] However, life expectancy is highly variable across patients and across trials, so unless something is highly effective, the only way to reliably test if something prolongs life expectancy is to perform a randomized clinical trial, where by random selection (equivalent to the flip of a coin) some individuals receive the new therapy and others do not. This helps balance for unknown factors and for the usual variability in life expectancy across patients.[cxxviii]

To prove that low sugar diets are useful in cancer patients, one would need to do a randomized clinical trial where half the participants are assigned a low sugar diet and the other half are assigned a regular diet. Furthermore, to give reliable results, such a trial would need to be very large. Linus Pauling (who won both the Nobel Prize in Chemistry and the Nobel Peace Prize) felt strongly that high dose vitamin C was effective in cancer and several other illnesses. He worked with physician Dr. Ewen Cameron to demonstrate this. In their assessment of high dose vitamin C in patients with advanced cancer, Cameron and Pauling compared 100 patients receiving vitamin C

[cxxviii] In Chapter 10, I discuss situations where randomized clinical trials are not needed, and some examples of how they might mislead us. However, there are no examples of alternative therapies that are so effective that a randomized controlled trial would not be needed, and no examples where I think a randomized clinical trial led us to mistakenly discard an alternative therapy that was actually beneficial.

to 1,000 "historical controls."[cxxix][681] They concluded that high dose vitamin C substantially prolonged life expectancy.

However, studies using historical controls do not do a good job of predicting what will happen in a randomized trial. There may be several unrecognized important factors that differ between the treatment group and the historical controls.[682] Subsequent randomized clinical trials of vitamin C in cancer have all been negative,[401] suggesting that Cameron and Pauling were misled by flawed historical controls.

Most alternative therapies are not even backed by formal trials using historical controls. Instead, they are based on observations that may draw erroneous conclusions due to underappreciated factors that one may not notice. As an analogy, if you go to a car race and the fastest cars there are all green cars, you might conclude that painting a car green will make it faster. However, before you can claim this to be true, you need to formally test this in a randomized trial where it is decided by random selection if a given car is painted green vs. (for example) black. If you don't know a lot about cars, you may not have noticed that all the green cars were Ferraris, and all the other cars were Volkswagens. Your perspective also may be influenced if you have just written what you hope will be a million-dollar best seller on the "Miraculous Benefits of Painting Cars Green."

It also might be claimed that a new therapy improves patients' quality of life by making them feel better. This would be a reasonable objective for a new therapy, but again, it is important to prove this in a randomized trial. If your conclusions are to be based on how patients feel, then not only must the trial be randomized. It must also be "double-blind," where neither the participants nor the researchers analyzing the data know whether an individual patient is receiving the new therapy or a control therapy.

After such trials are completed, the code is broken to reveal which participants received the new therapy and the data are analyzed to see

---

[cxxix] As noted above, in randomized trials, one takes a group of patients and the therapy a given patient receives is decided by the equivalent of the flip of a coin. In studies using "historical controls," one looks for patients seen previously that have similar characteristics to the patients you are treating with the new therapy. The outcome (such as life expectancy) seen in this earlier group of patients is then compared to the outcome of the patients who are now receiving the new therapy.

if the new therapy made a difference. It is essential to use this approach in assessing if a new therapy makes people feel better because of the very powerful "placebo effect." If you believe that a new therapy will make you feel better and if you know you are receiving the new therapy then there will be an increased probability that you <u>will</u> feel better, not because the new therapy is actually doing anything, but instead based on the placebo effect offered by hope.[683]

Pauling claimed that high dose vitamin C made cancer patients feel better,[684] but one cannot accurately draw this conclusion without a formal double-blind randomized comparison of high dose vitamin C to other approaches. A few alternative therapies such as acupuncture have been convincingly demonstrated to help some symptoms such as nausea,[676] but benefit has not been demonstrated for most other alternative therapies.

As I discuss further in Chapters 10 and 12, one does not <u>always</u> need a randomized trial to conclude that a therapy is beneficial.[430] One can conclude that a new treatment is effective if it causes tumor shrinkage in a high proportion of patients. With only a very few exceptions, cancer therapies that prolong life expectancy and improve quality of life also can directly shrink cancers. To shrink the cancer, you need to kill cancer cells. Generally, this killing of cancer cells is required if you are going to prolong life expectancy. If the cancer shrinks, the patient will often also feel better due to the associated reduction of cancer-related symptoms.

In the early days of immunotherapy (see Chapter 7), it was proposed that the immunotherapy would not need to shrink cancers for it to be effective,[523] but most immunotherapies that could not shrink cancers proved to be of little benefit. While it is not widely appreciated, even effective "antiangiogenic" agents that are designed to slow cancer growth by decreasing new blood vessel formation can cause tumor shrinkage. For example, among fifty lung cancer patients given the antiangiogenic agent bevacizumab prior to surgery, all but three showed at least some tumor regression.[685] As one poetic oncologist[cxxx] quipped, "If tumors don't shrink, then the treatments stink!"

[cxxx] My apologies, but I do not know which oncologist first said this.

If an alternative therapy (given on its own) is demonstrated convincingly to cause tumor shrinkage in patients with a defined type of cancer, this would be of substantial interest. It would be sufficient to stimulate further formal trials.

On the other hand, an apparent increase in the proportion of patients with tumor shrinkage when the new therapy was added to an established therapy would <u>not</u> necessarily be sufficient to stimulate much interest. The increase in response rate might all be due to the combination of the known benefits of the established therapy and the variability in outcome one sees based on patient selection.[430] Adding the new therapy to an established therapy would require a formal randomized comparison (established therapy alone vs. new therapy combined with established therapy) to conclude that the new therapy is sufficiently beneficial to become part of standard treatment.

The FDAs "breakthrough drug" approaches permit approval of some new therapies based solely on their ability as single agents to cause tumor shrinkage in a high proportion of patients. The logic is that this demonstrates that the drug can be effective against the targeted cancer.[686] If another new therapy also could cause tumor shrinkage in a high proportion of patients in this situation, you might still need a randomized trial to determine whether one was more effective than the other. But you could at least be reasonably confident that both new agents could provide benefit.[cxxxi] This generally is not the case with alternative therapies.

Conclusions: In summary, the ninth reason that cancer still sucks is that many proponents of alternative therapies claim benefits that have not been substantiated by reliable evidence. While some of these alternative therapies are cheap and harmless (and there is even a small possibility that an occasional one will be beneficial), others can be expensive or toxic. They also can divert patients away from or antagonize the effects of effective standard therapies.

Hope can be a very positive force for patients in difficult circumstances, but false hope and misinformation are destructive forces that help no one other than those selling them.

[cxxxi] I have heard Dr. Richard Pazdur (Director of the FDA Oncology Center of Excellence) state that he very much wants to approve effective new agents, but he has no interest in approving an expensive placebo.

— 10 —

Oncology Myths and Legends: Unfounded "Facts" That Impede Us

Our cultural view of cancer is rife with mythology. The negative impact of this mythology is the tenth reason that cancer still sucks.

Short Primer

Life is chaotic, yet we all function best if there is at least <u>some</u> order. Too much order is stifling and counterproductive, but the "right amount of order" helps promote efficiency, safety, and happiness. In virtually all areas of our lives, this order is promoted by societal rules, regulations, guidelines, laws, conventions, common understanding, or data. These are supplemented by the beliefs and conclusions arising from our own personal observations and from the mythologies entrenched in our culture. These deep-rooted cultural mythologies and personal beliefs and observations are a vital source of wisdom that can be essential to progress and stability. But they can also mislead. I will discuss a few illustrative examples of cancer myths that can have negative impacts.

I have divided the examples into myths common among the general public, those common among non-oncology healthcare professionals and administrators, and those common among oncologists and cancer researchers.

Myths common in the general public: In the previous chapter, I discussed complementary and alternative therapies. "Myth" is often the major factor driving their use. Other examples include:

Conspiracy myths: Some claim that the cure for cancer already exists, but Big Pharma and others are suppressing it.[687] This whole book is about the major challenges that we face in making cancer advances. With these major challenges, we do not need conspiracies to explain why progress is so painfully slow.

The myth of the major breakthrough: Then there is the myth of the "major breakthrough." The announcement in the press of a major breakthrough typically means that some scientist has developed a new approach that appears to be effective in the laboratory. However, very few of these major breakthroughs trumpeted by the press are very important on their own. Huge numbers of scientific papers are published every year, each presenting a bit of new knowledge. Progress is typically made when, over time, we slowly put together several of these individual bits of information to guide us in setting up the experiments that lead to the next small step in a very long process.

For example, in Chapter 7, I discussed the discovery of the CTLA4[505] and PD1/PD-L1[506] pathways. In retrospect, these were incredibly important in the development of effective immunotherapy. But neither of these was recognized as a "major breakthrough" when first announced. If they had been, I would not have believed it, considering all the earlier major breakthroughs in immunotherapy that had gone nowhere. When you see a major breakthrough in the press, it often means that a scientist's institution has decided to put out a press release labelling the publication as such. The public would not be interested unless it was a major breakthrough, and positive publicity raises the institution's profile.

It is relatively easy to kill cancer cells in test tubes and mice, yet such discoveries may be called a breakthrough. While such statements may help raise the institution's profile, they also may distract from other important projects that are more promising. In this way, the frequent news stories about "major breakthroughs" can indirectly impede progress while giving patients false hope.

The myth that cancers grow more rapidly "after being exposed to air": The two myths that I hear most commonly from cancer patients are that eating sugar will speed cancer growth (discussed in Chapter 9 on alternative therapies) and that undergoing surgery will "make the cancer grow faster by exposing it to air." This latter myth is based on the observation that some patients deteriorate rapidly and die shortly after cancer surgery. This happened frequently prior to the modern age of CT, MRI, and PET scans. Because they did not have scans able to determine the extent of the cancer, a surgeon would perform "exploratory surgery" where they would find that the cancer was much more extensive than initially appreciated. Being unable to remove the cancer, the surgeon would then close the incision and the patient would continue to have rapid tumor progression. Life expectancy would be short since that was what was destined to happen anyway. It was not shortened by "exposing the cancer to air."

Myths among non-oncology healthcare professionals:

The belief that cancer systemic therapies are of little benefit: Some senior healthcare professionals and administrators have limited understanding of the potential benefits of anticancer therapies. One former senior administrator (who played a role in the hospital's policy and resource allocation decisions) recommended to me that I have the medical oncologists just tell cancer patients that treatment will not help them, and that we strongly recommend that they not take treatment. The patient's care would then be transferred to a separate palliative care program and the hospital would avoid the costs of delivering care to the patient.

This administrator was quite surprised when I told him that systemic therapies more than triple the average life expectancy for even relatively resistant malignancies like metastatic lung cancer and pancreatic cancer, and that these systemic therapies have unequivocally been shown to improve quality of life in the average patient (despite their potential toxicity). Making a cancer shrink even slightly can be a much more potent pain killer than morphine and the government's decision to fund these treatments is based on very solid evidence.

While a government agency funds the anticancer drugs we give our patients, other parts of the patient's care (such as essential scans) are funded through the hospital's global budget. There are many strains on this hospital budget. It is essential that all administrators overseeing these resources understand the importance of patient access to them.

The myth that wait times are not important: While many people in the healthcare system understand the importance of dealing with cancer rapidly, a few do not. I have had former senior people push for oncologists to spend less time expediting cancer patient care and more time doing non-oncology undergraduate teaching as part of their contribution to the medical school. Some proposed that oncologists should spend more time caring for inpatients who were too debilitated to be candidates for any treatment rather than trying to offer faster care to outpatients who could benefit from therapy. Two former senior people specifically questioned whether shortening cancer wait times was of any real value apart from earning a bonus for the hospital's Vice President overseeing cancer services. If it could take several months to get a rheumatology or neurology consult, why should cancer be any different? Just a few influential people with misguided opinions like this can impede timely access to care. Patients with small, slow-growing cancers may not benefit a lot from speed, but speed is essential for patients with more rapidly growing cancers.

Myths among oncologists: There are also various myths among oncologists. Some are gradually abating but others have been remarkably durable. Some members of the public will be unhappy with what I said in Chapter 9. Some oncologists will be unhappy with some of the things I say in this chapter. *C'est la vie!*

Myths around therapy resistance: When I started my oncology training in 1976, we were taught that if you have treated a patient with a given regimen and then stop the treatment, you cannot go back to the same treatment. It was believed that the patient's tumor would now have mutations that made it permanently resistant to the initial treatment.

However, we now know that this is not necessarily the case, particularly if the patient was responding to the treatment at the time it was stopped. While restarting previous therapies became common practice for some malignancies, the situation was different for others. For example, in 2017, I had a patient with metastatic non-small cell lung cancer who had responded to the drug pemetrexed. She decided to take a break from the therapy. When I tried to restart her on the pemetrexed when her cancer began to grow again several months later, the Ontario government cancer agency told me that this was not permitted, since she would be resistant. I appealed this decision. They let me restart her therapy, and she responded again. She is currently continuing to respond to pemetrexed five years after restarting it.

The myth that one needs a randomized controlled clinical trial (RCT) to prove that a new therapy is beneficial: Some oncologists and payors believe that to determine whether a new therapy is beneficial, you must compare a new therapy to a "control"[cxxxii] in an RCT. In RCTs, the equivalent of a flip of a coin is used to select whether an individual patient will receive the new treatment or not. However, contrary to the opinion of some "believers," an RCT is not necessary if a new drug results in tumor shrinkage in a high proportion of patients with resistant tumors that would usually not respond to standard approaches.[cxxxiii] The fact that the therapy can cause tumor shrinkage in a high proportion of these patients is all the proof of benefit that is needed.

RCTs can take years to complete, they require large numbers of patients, and they generally are very expensive. For uncommon malignancies, it can be very difficult to find the required patients to conduct an RCT. Hence, requiring such trials greatly delays access to effective new drugs. It also greatly increases the cost of development of the drug.

RCTs <u>are</u> generally needed to answer some specific questions. For example, is it better to give two drugs in combination with each other, rather than giving one for as long as it works and then switching to the other? Does a drug that does not cause tumor shrinkage on its own nevertheless improve outcome if added to a standard therapy? Are there advantages in giving a new therapy as first-line therapy as opposed to only giving it after another therapy? Are there important differences in efficacy of two drugs that work in the same way? Is the probability of cure improved by adding a new therapy to a potentially curative approach like surgery or radiotherapy? Does a therapy prolong survival or improve quality of life?

---

cxxxii A "control" consists of an older established treatment, a placebo, or "best supportive care." "Best supportive care" consists of treatment of the patient's symptoms using pain medications, low dose radiotherapy and similar approaches, with no systemic therapy that might shrink the cancer.

cxxxiii As discussed in Chapter 12, drugs yielding high response rates in tumor types that would be expected to be resistant to standard therapies may be approved based on the FDA's "breakthrough therapy designation."

One major problem in Canada is that some payors may say that they will only pay for the new drug if it costs less than $50,000 per "quality-adjusted life-year" (QALY) gained."[cxxxiv] One can only determine how many QALYs a therapy is adding if one has done an RCT. As I discuss further in Chapter 12, there are problems with using QALYs gained as the basis of funding.

Additional statistical myths: In RCTs, statistical calculations assess the probability that any observed differences between a new therapy and an older therapy[cxxxv] are real, as opposed to being due to chance alone. These statistical calculations generate a "p" value. The results of the study are designated as being "statistically significant" if p is less than 0.05. P less than 0.05 means that there is less than a 5% probability[cxxxvi] that any observed difference between the therapies is due to chance alone. Conversely, it means that there is more than a 95% probability[cxxxvii] that the difference is real.

If a new therapy is demonstrated to be better than an older standard therapy, with a p value less than 0.05, it usually will be approved for marketing as an "effective" therapy. On the other hand, if the p value is greater than 0.05, the new therapy may be declared to be ineffective, and it may be discarded.

This excessive reliance on the p value can create problems. The p value does not give an indication of the size of the benefit derived from a treatment. It instead just tells us how certain we can be that there is a real difference in benefit between a new and an old treatment.

The more patients included in the trial, the higher the degree of certainty. Another way of saying this is that the more patients included, the greater the "statistical power" of the study. The greater the statistical power, the greater the ability to detect a difference between the two therapies being compared in the study.

---

cxxxiv This means, $50,000 for every year of life added by the new therapy compared to older therapies, with an adjustment such that life-years added don't count if the patient is suffering from a lot of therapy side effects or cancer symptoms.

cxxxv Or compared to placebo or best supportive care.

cxxxvi 0.05 is the same as 5%.

cxxxvii 100% minus 5%.

To illustrate this, let's say that we have two therapies. We will call one Old and we will call the other therapy New "1". For this example, let us assume that we treat 1,000 patients on each arm of the trial, and the average survival with New "1" is 3 months longer than with Old, with a p value of 0.0001. This would mean that there was only one chance in 10,000 that the observed difference between New "1" and Old was due to chance alone. There would be a very high degree of certainty that New "1" is better than Old. Based on this, New "1" probably would be approved for marketing

On the other hand, if there were only 100 patients on each arm of the trial and the difference in average survival between New "1" and Old was again 3 months, the p value might only be 0.10. The much smaller number of patients (which would give much lower statistical power) would mean less certainty in the outcome. Hence, the weak p value. New "1" might now be discarded based on the weak p value, despite the size of the gain being the same as in the large study.

Let's take it one step further. If another drug, New "2" is compared to Old and there are 5,000 patients on each arm of the trial and the difference in average survival is only 1 week, the p value might nevertheless be 0.0001. The high patient numbers give a high degree of certainty in the benefit of New "2" despite the size of the benefit being very small. Based on this p value, New "2" could be approved for marketing.

On the other hand, in a comparison of New "3" to Old with only 100 patients per arm, let's say that the difference in average survival is 6 months. The low patient numbers might mean that the p value is only 0.06 despite the large size of the difference in survival.

The p value of 0.06 would mean that there was a 94% probability that New "3" was better than Old. In fact, there would be a high likelihood that it is <u>substantially</u> better than Old. It also would probably be better than New "2". Despite this, there would be a high risk that New "3" would be discarded as being ineffective since the p value was not less than 0.05

Historically, the reason for using a p value less than 0.05 as being the cutoff for deciding a therapy was beneficial was completely arbitrary. Culturally, it became accepted as the way we make decisions, but there is no scientific basis to conclude that we should have a cut point at all, or that p less than 0.05 is the "right" cut point. Many statisticians now question our over reliance on p values for decision making. So far, not enough oncologists and payors share the statisticians' concerns. It makes our lives easier if we use a simple p value cut point to guide our decisions, but "simple" is not necessarily the same thing as "right".

The myth of the impact of the blood-brain barrier: The blood-brain barrier is a membrane-like system that limits the ability of toxins to cross from the blood stream into the brain. The blood-brain barrier limits penetration of systemic therapies like chemotherapy into the normal central nervous system. There is a widespread belief that the blood-brain barrier also limits penetration of systemic therapies into tumors in the brain and that most systemic therapies will be of only limited benefit in brain metastases. However, the blood-brain barrier is largely disrupted in brain metastases. This allows chemotherapy drugs to reach high concentrations in brain metastases (despite only low concentrations in the normal central nervous system), and systemic therapy agents that do not cross the intact blood-brain barrier nevertheless may be effective against brain metastases.

Further Details and References

Further details and documentation are outlined below.

The myth that cancers grow more rapidly "after being exposed to air": As discussed earlier, many people have told me that they have heard that exposing a cancer to air during surgery will result in an explosion in tumor growth. This is not the case. This perception is based on the deterioration of patients who would have progressed rapidly whether or not surgery had been done.

Having said that, surgery can somewhat accelerate growth of a cancer, but it has nothing to do with exposing the cancer to the air. There are several ways in which surgery can stimulate tumor growth or increase the risk of metastases.[688] For example, I discussed the concept of Gompertzian tumor growth in Chapter 5: as a cancer becomes larger, the growth rate slows due to a shortage of nutrients and other factors.[387] If one removes a large primary tumor then metastases that are initially too small to be seen may grow faster.[689] That is why in patients at high risk of recurrence it is important not to wait several months after surgery before starting "adjuvant" chemotherapy.[cxxxviii] The now rapidly growing micrometastases may be more sensitive to chemotherapy than they were prior to the removal of the primary, but if one waits too long to start the chemotherapy, the micrometastases become larger and, therefore, less likely to be eradicated by the chemotherapy.[cxxxix] The probability of long-term survival may decrease with longer delays in starting adjuvant therapies.[690,691]

Acceleration of growth of micrometastases also may happen if radiotherapy is used to treat the primary tumor instead of surgery.[692] Hence, there is no indication that how one tackles the primary tumor is determinative: it is simply the biological impact on micrometastases of removing the primary tumor. The important thing, one way or another, is to eradicate the primary tumor if possible. The patient may then be cured if there are no micrometastases or if the micrometastases can be eliminated by adjuvant systemic therapies. Overall, the potential benefits of surgery greatly outweigh the potential risks for the average patient who has a safely resectable cancer, no definite distant metastases, and no major health issues that would make surgery particularly dangerous.

The myth that wait times are not important: As noted above, some important decision-makers in the healthcare system do not understand the importance of speed in the diagnosis and treatment of cancers. In fact, speed is very important. Across Canada, oncologists on average have shorter wait times than most other specialties.[693] Short wait times are extremely important for patients with the more rapidly growing malignancies like lung and pancreatic cancer. These patients can deteriorate and die if therapy is not started promptly.

For example, using a methodology called population kinetics[577] to extrapolate from clinical trials, we calculated that among patients with metastatic non-small cell lung cancer, about 4% of the patients will die in each additional week that therapy is delayed.[424] Therefore, if it takes two weeks from the time a family doctor puts in a consult until a lung cancer patient is seen by a medical oncologist, there would be about an 8% probability that the patient would die while waiting. Many of the patients

---

[cxxxviii]   See chapters 4, 5, and 6 for a discussion of adjuvant chemotherapy.
[cxxxix] See Chapter 6 for a discussion on why rapidly growing and small tumors are more sensitive to chemotherapy than are slowly growing and large tumors.

who survive the wait time deteriorate so rapidly that even if they are still alive, they are no longer strong enough to consider therapy. This is one reason that fewer than 25% of Ontario patients with metastatic non-small lung cancer receive systemic therapy[427] despite this therapy being funded by the government healthcare system. At our institution, we undertook a successful "transformation" initiative to identify and eliminate bottlenecks that slow down diagnosis and access to treatment for patients suspected of having lung cancer.[694]

One reason that some administrators question the importance of speed in dealing with cancer is because even some oncologists feel that wait times are not very important. Hospital administrators who must balance budgets might prefer to listen to oncologists who say that wait times are unimportant.

Some oncologists may feel that wait times are unimportant because the impact of wait times varies substantially across tumor types. Oncologists who treat primarily lung or pancreatic cancer may have a much different view than do oncologists who treat primarily breast cancer. Hormone-sensitive breast cancers in elderly patients can be quite indolent, and this is the reason why fewer than 20% of women who develop breast cancer will ever die from it.[cxl] This is much lower than the 87% of pancreatic cancer patients and 74% of lung cancer patients who will die from their malignancies.

In Chapter 2, I discussed the impact of delays in diagnosis and pointed out that screening is likely to have a much greater impact in high-risk individuals than in low-risk individuals. The same holds true for wait times. The higher the risk of rapid cancer growth, the greater the importance of

---

[cxl] In Canada in 2017, there were approximately 26,500 new cases of breast cancer and about 5,000 deaths. Since the incidence of breast cancer is not changing rapidly, dividing the number of deaths by the number of new cases gives a reasonable rough estimate of the proportion of all patients who develop breast cancer that will eventually die from it (19%). This does somewhat underestimate the true proportion who will die from it since the denominator (number of new cases) is increasing due to growth and aging of the population, and the patients dying now would in many cases have been diagnosed several years earlier, when the denominator was smaller. Nevertheless, it does give a reasonable estimate of the probability that a breast cancer will eventually prove fatal.

short wait times, and the greater the potential problem if anyone in the hospital's leadership team is not strongly behind rapid access to care.

Myths around therapy resistance: It was once believed that if a cancer had been exposed to a therapy it would be resistant to that therapy if it was reintroduced in the future. However, over time it was noted that some patients with colon,[695] ovarian,[696] breast,[697] non-small cell lung,[698] small cell lung,[699] and other cancers who had been given a therapy break responded positively to the original treatment if it was restarted when the cancer began to grow again. Patients are particularly likely to respond to restarting a therapy if they were still responding at the time the therapy is interrupted.[700]

The myth that one needs a randomized clinical trial (RCT) to prove that a new therapy is beneficial: RCTs are regarded as the "gold standard" for proving that a new drug is effective.[430] I have heard one senior oncologist state, "There is nothing smarter than a randomized clinical trial." But I disagree with his view. While there are some situations where RCTs are helpful and important, as I discussed earlier, there are other situations where they are unnecessary or misleading.[430]

RCTs are often unnecessary if a drug results in major tumor shrinkage in a high proportion of patients. This is particularly likely to be the case if the patients would be expected to be resistant to other therapies. This is the basis of the FDA's "breakthrough therapy" designation.[701]

RCTs may not be needed if a high proportion of patients respond to the treatment because tumors rarely shrink spontaneously. Tumor measurements can be imprecise, and some tumors may appear to shrink by 10% or more simply because of measurement errors.[702] However, "response" to an anticancer therapy is defined as a decrease in the tumor diameter by 30% or more.[703] It is very uncommon for measurement error to give this large an apparent change. So a 30% reduction in measured tumor size is a reliable indicator that the tumor probably really did shrink with the therapy.[430]

A high response rate may be due to chance alone if only a few patients are treated. Hence, the more patients treated and the higher the observed response rate, the greater the degree of certainty that the drug is active.

But a breakthrough therapy designation may be granted based on relatively small trials. For example, crizotinib was granted breakthrough status for lung cancers with *MET exon 14* alterations based on results from just 18 patients. Of the treated patients, 8 (44%) responded and another 9 (50%) had stable disease.[704] When the study was expanded to include 65 patients, the response rate was still 32% (with 95% confidence intervals of 21% to 45%). Another 45% of the patients had stable disease and only 6% had tumor progression.[705] Hence, the initial small study gave a reasonable estimate of the level of activity seen in the later, larger study.

While there can be greater certainty in the efficacy of a drug with larger patient numbers, it takes longer to find the patients and conduct the study. By using the results from a smaller number of patients, we accept a decreased level of certainty in exchange for faster drug access. Delays in access to an effective drug can translate into suffering that could have been avoided and into many years of life lost.[706,707]

In some cases, responses may last only a few weeks. Very brief responses are not of much value. Consequently, duration of response is also taken into consideration. In the crizotinib example above, the median duration of response was 9.1 months, with some responses lasting more than two years.[705]

While non-randomized trials can be a reliable indicator that a single drug is effective, they are less reliable as an indicator that a combination is more effective than a single agent. This is because the standard drug in the combination generally has some degree of efficacy, but the efficacy level could be quite variable due to patient selection or chance. Consequently, efficacy would have to be markedly higher with the addition of a new agent for one to be confident that the apparent increase in benefit is not due solely to patient selection or chance. RCTs (old drug alone vs. old drug combined with new drug) can help to determine a genuine increase in efficacy with the addition of a new drug to an old one.

From my perspective, many RCTs risk misleading us since the question being asked in the study doesn't make much sense.[430] For example, two drugs, A and B, with different mechanisms of action may be compared to one another. Since the two drugs work in different ways, they would probably be effective in different patient different subpopulations. However, we are often unable to identify these subpopulations reliably. If drug A is effective in a subpopulation that constitutes (for example) 40% of the entire population and drug B is effective in a different subpopulation that constitutes 20% of the entire population, then drug A will be identified by the randomized trial as being the better drug.

Drug B may well be discarded. Drug A will become the new standard of care for the entire population, despite being ineffective in 60% of the population, including the 20% that would have responded to B, had B not been discarded.[430] The more rational approach if both drugs may work, but in different patients, would be start with the more effective drug A first, but then to offer drug B if drug A doesn't work. This is a much better option than just throwing drug B out since it works in fewer patients.

There are numerous instances of two drugs with different mechanisms of action being compared to each other. Examples include the PD-1 inhibitor nivolumab being compared to the chemotherapy drug docetaxel in lung cancer.[536,537] Such comparisons are biologically irrational and can lead to effective drugs being discarded.[430] In practice, what now happens in advanced lung cancer is what should happen: the immunotherapy drug may be tried first, with a change to a chemotherapy drug if the immunotherapy drug is not effective, or *vice versa*. This is the biologically rational way to use two drugs with differing mechanisms of action, but this approach is not assessed in most RCTs.

In another example, let's assume that a new drug, B, is being added to a standard drug, A. In RCTs, the combination of drugs A+B will be compared to drug A alone to see if the combination is more effective. However, it would be more rational to instead test whether the combination of drugs A+B is more effective than giving the drugs sequentially, with one followed by the other.[430] The usual study design (drugs A+B vs. drug A alone) assumes that drug B has no value unless it adds to drug A. This will not be correct if it has already been demonstrated to cause tumor shrinkage in some patients when given alone.

In many RCTs, when patients on the control arm experience tumor progression, they may cross over to the experimental therapy being assessed. This can make the control arm therapy look better than it really is. This in turn can lead to the incorrect conclusion that the new therapy being tested is no better than the control arm therapy, and the new therapy may be discarded.[708] As I discuss further in Chapters 12 and 13, some studies try to get around this by forbidding crossover, but many of us[709] feel that forbidding crossover is unethical if we already know that the new drug can help patients.

There are also several other ways in which RCTs can mislead.[430] RCTs are used extensively in the assessment of new cancer therapies. The information they add is often very helpful. But some are misleading or irrational or unnecessary.[430]

The myth of the supremacy of "statistical significance" as a benchmark for research outcomes: The "p value" for a study is an indicator of the probability that an observed difference or association is real, rather than being due to chance alone. Many years ago, statistician Ronald Fisher managed to convince much of the world that the p value calculated using his preferred statistical methods should be the deciding factor in whether we accepted an outcome of a study as being important. However, some authors have questioned this reliance on the p value.[710]

The problem with the p value is that it is a measure of precision or certainty rather than being a measure of size of an effect. As Ziliak and McCloskey put it, the p value fails to take into consideration the "oomph" of a study result, and this causes problems.[710] For example, if RCTs of a new drug "magic potion 1" compared to standard therapy included several thousand patients and showed a one month improvement in average life expectancy, the 95% confidence intervals for the difference might be 0.9 months to 1.1 months.[cxli] The very high patient numbers mean that the difference between the two therapies can be measured with a high degree of precision.

Because of this high degree of precision, the p value might be a very impressive 0.0001. This p value would mean that there is only one chance in 10,000 that the observed difference is due to chance. Because of this highly significant p value, magic potion 1 probably would be approved for marketing.

Conversely, consider "magic potion 2" which is designed to hit an uncommon target that is only present in a small proportion of all patients. In initial studies, it causes tumor regression with symptomatic improvement

---

[cxli] Meaning that there would be a 95% probability that the "true" difference in average life expectancy with the two therapies was between 0.9 months and 1.1 months.

in 80% of treated patients, and average survival of treated patients is two years longer than would be expected with standard approaches.

However, because it only works in patients with an uncommon target, it is difficult to accrue patients to trials. The RCT comparing magic potion 2 to standard therapy can only accrue 100 patients. In the trial, the average life expectancy with magic potion 2 is two years longer than with standard therapy, but because of the small number of patients accrued, the 95% confidence intervals for the difference between the two therapies indicate that, on the one hand, survival with magic potion 2 could possibly be 1 day shorter than with standard therapy, but on the other hand, it could be up to four years longer.

In this case, because of the very wide 95% confidence intervals that come with low patient numbers, the p value might be 0.051. This p value would mean that there is only a 94.9% probability that magic potion 2 is better than standard therapy. Despite this problematic p value, there would be a high likelihood that magic potion 2 really is substantially better than standard therapy. It is probably also better than magic potion 1, which only improved survival by 1 month. Nevertheless, magic potion 2 might be discarded since it did not meet the rigid criterion of giving a p value less than 0.05.

In the example of magic potion 2, there would be a 5.1% probability that it is no better than standard therapy. On the other hand, since the observed average life expectancy was two years longer than with standard therapy, there would be approximately a 50% probability that magic potion 2's true average survival is two years or more greater than standard therapy.

Since the upper limit of the 95% confidence intervals for magic potion 2's average survival is four years longer than standard therapy, there would be about a 5% probability that the true average life expectancy with magic potion 2 is four years or more longer than with standard therapy.

Many clinicians, regulators and payors have historically had a fervent faith in the paramount importance of a trial's p value. But many statisticians now question it.[711-714] The p value is one of several bits of information that can be helpful in assessing a drug's efficacy, but blind faith in the p value as the ultimate proof of an approach's value can cause a lot of harm.[710]

The p value is a continuous variable. As discussed in chapters 1 and 7, dealing with factors that are "continuous variables" can create problems.

Using cut points to assess impact of a continuous variable can be helpful in assessing biological properties: for example, are outcomes in patients with a lot of something different than patients with only a little bit of the same thing. But using cut points to make therapeutic decisions can be problematic. In the use of p values, it is assumed that a clinical trial with a p value of 0.049 is a winner, while a trial with a p value of 0.051 is a loser. This is biologically and clinically irrational.

Life is easy if we have simple rules to follow: we keep a treatment with p less than 0.05 and discard a treatment with p greater than 0.05. But this simplicity can result in some pretty stupid decisions.

The myth of the impact of the blood-brain barrier: Perhaps no other myth is so firmly entrenched in medical oncology lore than the myth of the impact of the blood-brain barrier.

The blood-brain barrier is very real. It protects the brain from many toxins (including chemotherapy) circulating in the blood stream. It prevents the toxin from reaching the brain's neurons. There has been a long-standing belief that the blood-brain barrier also protects tumors in the brain from systemic therapies like chemotherapy and targeted therapies,[715-726] that brain metastases will therefore be resistant to systemic therapies,[718,720,723,725] that treating patients with systemic therapies will achieve little,[720,723,725-727] that life expectancy is particularly bad if there are brain metastases,[717,718,728] and that we either need drugs that cross the blood-brain barrier or new ways of delivering drugs to get more into the brain.[715,717,718,721,724] There are several problems and negative consequences associated with these mistaken beliefs.[729]

Neuro-oncology is the specialty responsible for the treatment of primary brain tumors. But most medical and radiation oncologists (and not just neuro-oncologists) may be involved in the management of patients with metastases that have spread to the brain from other sites. Early in my oncology training, I was interested in neuro-oncology, but when I went to my first neuro-oncology meeting in the late 1970s, I was the only medical oncologist there. All the other participants were neurologists, neurosurgeons, radiation oncologists specializing in brain tumor therapy and basic scientists. A central theme of the early neuro-oncology meetings I attended was that all cancers except for brain tumors are highly curable by chemotherapy. The message of the meetings was that brain tumors were highly resistant to chemotherapy since they were protected by the blood-brain barrier.

The only problem with this is that it was not true. None of the common malignancies could be cured by chemotherapy because of a host of resistance mechanisms in those tumors. Neuro-oncologists looked at what was happening in curable malignancies like childhood acute lymphoblastic leukemia, Hodgkin's disease, and testicular cancer. They incorrectly assumed that it also happened everywhere else except for brain tumors. They brushed off my attempts to correct this misperception.

While only low concentrations of chemotherapy are achieved in normal brain and in the cerebrospinal fluid (CSF) bathing the brain, the blood-brain barrier is largely disrupted in brain tumors.[729] That is the reason that these tumors show up so well on scans. We assessed 15 different chemotherapy agents and found that all 15 achieved high concentrations in human brain tumors, even if they reached only low concentrations in normal brain and CSF.[729]

With respect to the widespread belief that systemic therapies are ineffective against brain metastases,[715,719,720,722,723] there is a growing recognition that systemic therapies can in fact be useful.[719] For various types of cancer, the response rates are actually very similar for brain metastases as they are for metastases outside of the brain.[730,731]

When a drug does work well against brain metastases, it is generally claimed that this effectiveness must be due to its ability to cross the blood-brain barrier.[718,719] In fact, it is usually just because it is a good drug. A case in point is the use of alectinib in lung cancer patients with what is called an "*ALK* fusion gene." The drug crizotinib is more effective than chemotherapy in lung cancer patients with the *ALK* mutation,[465] but it has been claimed that it is relatively ineffective against brain metastases.[722] However, it is in fact capable of shrinking brain metastases,[732] and the relative benefit with crizotinib compared to chemotherapy is as great in patients with brain metastases as it is in patients without brain metastases.[733] If it could not get into tumors in the brain, then it should not be better against them than chemotherapy.

The newer agent alectinib is superior to crizotinib against *ALK* mutated lung cancer, including in patients with brain metastases.[732] It has been claimed that this superiority against brain metastases is due to a better ability to penetrate into the brain. However, the more likely explanation is that it is just a better drug than crizotinib and will work better against *ALK* mutated tumors whether they are outside or inside the brain.

Over the past few decades, a lot of effort has gone into developing new drugs specifically for their enhanced ability to enter the brain. There also have been numerous attempts to change the blood-brain barrier so that drugs could enter the brain better. None of these has had much impact. The important factor is the effectiveness of the drug: not whether it enters the normal brain.

With respect to the belief that survival of patients with brain metastases is particularly short,[717,718,728,734] and that this is partly because systemic therapies penetrate the brain poorly,[715,717,718] this is again not entirely correct. The survival with brain metastases from (for example) lung cancer is the same or better than with metastases to the liver, bone, adrenal glands, and subcutaneous tissues.[735-738]

A publication may demonstrate that survival time is shorter with brain metastases than without brain metastases and imply that the blood-brain barrier is responsible for the difference. But there is a trick here: the lung cancer patients who do substantially better than others are those who only have metastases involving the lung or the outside lining of the lung or heart. In the current lung cancer staging system this latter group of patients is classified as being the "M1a" group, and their life expectancy is substantially better than for patients with a single metastasis outside of the chest ("M1b") or with multiple areas of spread outside of the chest ("M1c").[739]

When life expectancy in patients with brain metastases is compared to those without brain metastases, most patients with brain metastases will have M1c disease (multiple brain metastases or metastases that also involve other organs). They will be compared to a group that includes a lot of patients with better prognosis M1a disease. The most important factor with the brain metastasis group is not that the brain metastases *per se* give a poor prognosis but that they are being compared to a group of patients enriched with patients with better prognosis M1a disease. As noted above, on average, when outcomes for patients with brain metastases is compared to other patients with M1c disease, patients with brain metastases have similar or better survival.[735-738]

Some authors who comment on the poor survival of patients with brain metastases also include in their assessment patients who only develop the brain metastases late in the course of their disease.[740] There is increasing recognition that the limited apparent benefit of systemic therapy in patients with brain metastases may in some cases be because these patients may already have developed resistance to systemic therapies before brain metastases developed.[741,742] In this situation, the poor outcome is not due to protection of the brain metastases by the blood-brain barrier. It occurs instead because the tumor cells had become resistant to systemic therapies before they had ever spread to the brain.

If one did a similar comparison of patients with liver, bone, adrenal, or subcutaneous metastases to those without them, one would get roughly the same result as when one compares those with brain metastases to those without them. However, no one usually does that. The Holy Book of oncology says that the poor outcome with brain metastases is due primarily to the blood-brain barrier, but there is no Holy Writ to explain why this also happens with other tumor sites, so no one pays much attention.

There are a variety of biological reasons why brain metastases might differ from other tumor sites in their response to systemic therapies.[743] The role of the blood-brain barrier has probably at best been overstated, if it has any impact at all.[729]

There are at least four major negative consequences of this blind blood-brain barrier worship. The first is that a lot of resources have been expended on developing therapies that get more drug into the brain. We might have been better served if these resources had instead been used to develop more effective drugs. When effective new agents are developed, they work against brain metastases despite not penetrating the normal central nervous system.

The second negative outcome is that patients with brain metastases may be excluded from clinical trials of potentially effective new agents based on the belief that the new therapy would be ineffective against brain metastases.[718,719,744] Approximately 10–40% of patients with cancer will develop brain metastases at some point.[717,719,722,724,725,740] Exclusion of patients with brain metastases means that accrual to clinical trials is slowed since there are fewer available patients.[744] Slow accrual to clinical trials is a major problem. Large numbers of years of life are lost since it takes too long to complete the clinical trials that are needed for approval of effective new drugs.[706,745]

The third major problem is that patients with brain metastases who are excluded from clinical trials of potentially effective new agents are denied access to therapies that might help them.[744]

The fourth major problem is that for some trials, patients with brain metastases are eligible if they first have radiotherapy to their brain and are demonstrated to then have stability of the brain metastases for four weeks or more. The net consequence of this is that patients with small, asymptomatic brain metastases and (for example) large liver metastases are sent first for brain radiation that they do not need. Systemic therapy that could control their liver (and brain) metastases is delayed. The very fastest that one could then get these patients on systemic therapy on the trial (having given one week of radiation then having observed the patient for four weeks to make sure the brain metastases were stable) would be five weeks.

My population kinetics calculations indicate that 18% of the patients would have died from their liver metastases within this period, without receiving the systemic therapy that they need.[424] Many others would have deteriorated to the point that they were no longer eligible for either the clinical trial or for any other standard systemic therapy that might have helped them.[424]

Even outside the clinical trials realm, I have seen patients have their essential systemic therapy delayed while they first receive brain radiation for small asymptomatic brain metastases. Brain radiation is a very important part of the treatment of patients with large, symptomatic brain metastases. But many patients with small, asymptomatic brain metastases often do not need it until late in the course of their illness.

Overall, the myth of the impact of the blood-brain barrier involves a lot of smoke and mirrors, and it costs lives.

Conclusions: In summary, there are several oncology myths and legends, and this is the tenth reason that cancer still sucks.

— 11 —

Mired in the Mud: The High Cost of Caution

Cancer research has made tremendous progress. However, the eleventh reason that cancer still sucks is a procedural and regulatory swamp that slows progress, shortens the lives of cancer patients, and drives up the cost of cancer therapies. This chapter will discuss why this has happened and the toll it takes. In the following chapter I will cover some of the particular contributing factors and solutions.

Short Primer

I will start here with a brief overview and will then follow in the next section with additional detail (for those interested) and some documentation to support my views.

New cancer treatments must be approved by regulators before they can be marketed. Approval of a therapy requires clinical trials demonstrating its efficacy and safety. Over time, these clinical trials have become more tightly regulated, increasingly complex, and very expensive. Consequently, it takes longer to conduct the trials required to acquire the data needed to approve a new therapy, and the process has become very expensive.

This translates directly into deadly delays. Thousands of years[cxlii] of life are lost that could have been saved had the therapy been available sooner.

[cxlii] One life-year is one person being alive for one year.

In addition, delayed access to effective therapies means that patients whose symptoms could have been improved continue to suffer.

The costs of the clinical trials must eventually be recouped from sales after the drug is marketed. If a company cannot recoup the costs, then the company will go bankrupt and new drug development will stop.

Companies also need to make a profit. If a company has no prospect of making a profit, no one will invest in it and it will have no funding to conduct the research required to develop new therapies. Higher clinical trials costs mean that companies must charge more for the therapy if they are to both cover the costs of developing the drug and also make a profit.

Major reform is urgently needed if we are to speed access to new therapies and bring down high drug costs. This reform must be done in the context of appropriate clinical research regulation.

Further Details and References

I will now go into more detail and provide some references as documentation.

Inevitably, without regulation, some researchers or companies would put their personal interests ahead of the interests of trial participants and of society.[746]

Clinical research is governed by statutory laws, but statutory laws can never foresee and include all possibilities. To deal with this reality, custodians or regulators oversee the fine details through "administrative law" or regulation.[746] In other words, they interpret the statutory law and attempt to apply it to the reality of the clinical research situation. However, this interpretation can shift over time and can lead to unintended consequences. Corrective actions are periodically required to address these.[746]

From my perspective, we need to pay more attention to the dense bureaucratic web in which clinical research is enmeshed. Investigators attempting to carry out clinical trials must traverse a morass that is reminiscent of the World War I Flanders battlefield where soldiers fought their way through sticky, thigh-deep mud that held them back and pulled them down.[747] I will describe several aspects of this more fully in the next chapter.

Tight, but not strangling clinical research regulation is needed. Bad things can happen without it. While the integrity of most researchers is unquestionable, there are, nevertheless, numerous examples of falsification of research findings[748] or alleged misrepresentation of data.[749] Researchers and companies are under intense pressure to produce positive results. In addition, researchers can draw incorrect conclusions from research results. Falsified data and erroneous interpretation can cause harm. Stringent oversight helps reduce the risk.

There are also well-known examples of serious drug toxicity that was not disclosed. For example, the analgesic Vioxx was associated with an increased risk of fatal heart attacks.[750] The responsible pharmaceutical company allegedly knew about this but hid the data.

In addition, while most clinical researchers have the best interest of their patients at heart, there are notorious exceptions. Examples include cruel experiments conducted by Nazi researchers during the German Third Reich.[751] Or the Tuskegee study, in which African American men with syphilis were not told that they were infected and were observed without treatment to assess the natural history of the disease.[752] In a New York nursing home, residents were injected with cancer cells to see whether their bodies would reject them.[751] In another New York study, mentally disabled children were injected with the hepatitis virus in a vaccine development study.[751] Hypervigilant regulation is now in place to help ensure that clinical trial participants give fully informed consent and that their safety and well-being are a top priority.

The regulatory process also plays a paramount role in protecting the public from drugs that are harmful or ineffective.[753]

Although strict regulation is essential, it also slows clinical research while increasing its cost.[745] The net result is that huge numbers of years of life are lost due to long delays in approval of effective new drugs.[706] Furthermore, new drugs must be very expensive if research costs are to be recouped.[745]

By the early days of the twenty-first century, it took on average 12–15 years to take a drug from discovery to approval, compared to just 8 years in the 1960s.[754] Much of this 12–15 year period is the time required to set up and conduct the needed clinical trials.[cxliii] [755,756] Regulatory compliance slows this process.[745,757] Thousands of patients may be denied a drug that prolongs life, reduces cancer symptoms, and improves quality of life.

A person's attitude toward the regulatory process and the safety of a treatment can change radically when they are suddenly facing certain death and suffering from rapidly worsening symptoms that are poorly controlled by even the most potent analgesics. Consider this striking finding. Slevin and colleagues asked healthy people and cancer patients under what conditions they would be willing to try a toxic treatment. The average healthy person wanted a 50% probability of cure, a two to five year increase in life expectancy, or a 75% probability that their cancer symptoms would improve.[428] The average cancer patient would take a highly toxic treatment if it had a 1% probability of cure, a twelve month increase in life expectancy, or a 10% chance that their cancer symptoms would improve.[428] For a less toxic treatment, the average cancer patient required the possibility of just a three month increase in life expectancy or a 1% chance that their symptoms would improve.[428]

So, what is the potential impact of delays in drug approval? To assess this, we examined 27 clinical trials published between 2000 and 2015 in which a new drug improved outcome in patients with advanced cancers to an extent that was statistically significant.[cxliv] [706] The gain in survival with these therapies was generally very modest. The average was a four-month prolongation of life for malignancies that typically were associated with a life expectancy of a few months to a year or two with standard treatments.

A four-month gain in life expectancy may seem minuscule to someone who is healthy. But it may be very important to a patient who has just

---

[cxliii] In a recent analysis, it took on average 31.2 months from the discovery of a new compound until initiation of human testing, then on average 95.2 months to conduct the clinical trials required for approval.

[cxliv] I am indebted to my son Andrew for this study. While he was in medical school, he did the majority of the research underlying this study.

been diagnosed with incurable cancer. Therapies that prolong lives even modestly also often improve cancer symptoms[440] and quality of life.[cxlv] [758]

For each of the 27 trials, we multiplied the gain in survival with the new therapy by the estimated number of patients dying from that cancer worldwide. For each therapy, there was an average of 80,000 years of life lost for every year of delay in drug approval. On average, more than 1 million years of life were lost in the time between drug discovery and eventual marketing.[706] This finding highlights the importance of trying to find ways to safely speed access.[cxlvi]

When we submitted our paper on this for publication, we expected that reviewers would be concerned by the high negative impact of approval delays. However, the reaction we did receive surprised me. We had to submit the paper to nine different scientific journals before eventually finding one that would accept it. For the journals rejecting it, some reviewers stated that the years of life-years lost was so large that it could not possibly be true, even though anyone could use the readily available public data to verify our calculations. Other reviewers stated that it did not help anyone to know these findings since the problems causing the delays were unfixable, despite there being a wide range of potential solutions.[430,524,745,757,759]

In general,—and sadly—I find that older oncologists feel that we have become so tightly entangled in the current regulatory web that escape is all but impossible. Younger oncologists assume that current approaches are based on some immutable "truth," with no other possible options.

[cxlv] There is a common misperception that even if a cancer therapy prolongs average life expectancy, it will generally reduce quality of life because of side effects. However, where this has been assessed, the overwhelming majority of cancer therapies that prolong survival are associated with either improved quality of life or no net change in quality of life, since the improvement in cancer symptoms more than compensated for therapy toxicity. This is a major reason that patients stay on cancer therapies despite the side effects.

[cxlvi] While we used the total number of patients dying worldwide of the cancer in question, we recognized that many patients would not be treated even if a drug were available. They might be too frail to consider the treatment, or their healthcare system might not agree to pay for it. But even if only (for example) 1–10% of patients could be treated with the new therapy, the burden of avoidable suffering and the number of life-years lost due to approval delays would still be very large.

Some reviewers called the paper an unfair criticism of the FDA (US Food and Drug Administration), even though the paper stressed that it was not the FDA's fault, and the FDA alone could not fix it. We are *all* complicit in what has happened.

Still other reviewers stated that it would not make any difference if the drugs were available faster since many patients could not afford the drugs anyway. An op-ed we submitted to the lay press was rejected by several newspapers before one finally agreed to publish it.[760] Again, there was virtually no reaction. Society has become numb.

Stalin observed that "If one man dies it is a tragedy. When thousands die, it's statistics."[761] Not thousands, but millions are dying of cancer worldwide. Louise Binder, a Canadian AIDS advocate has recently become a cancer advocate. She has explained our apathy by contrasting AIDS vs. cancer patients. AIDS patients were mad, but "cancer patients are just too polite."[cxlvii]

Most cancer patients and their families do not understand how bad the situation is until they are standing alone, facing the barriers that prevent them from accessing promising new therapies. Many cancer physicians go to great lengths to try to obtain these new therapies for their patients. At the same time, most physicians are fatalistic and do not feel that it will ever be possible to fix "the system."

However, there have recently been various fruitful initiatives to try to speed patient access to effective new agents.[762,763] These included the passage by the US Congress in 2012 of the Food and Drug Administration Safety and Innovation Act (FDASIA). This act enabled "breakthrough therapy" designation to expedite approval of a promising new drug.[701]

Typically, the "gold standard" required for drug approval is a large, randomized trial that demonstrate a statistically significant improvement in survival time. But these studies take years, and there are several ways in which they can yield misleading answers.[430] However, with the breakthrough drug approach, a therapy could be approved rapidly if it gives a high "response rate," with major tumor shrinkage in a much larger proportion of patients than would be expected with standard therapies.

[cxlvii] Personal communication.

Response rate in early non-randomized clinical trials is a reliable predictor of the drug being confirmed to be effective in later, larger trials,[764] and response rate correlates strongly with survival.[cxlviii][765,766]

Several promising agents have now been approved as breakthrough drugs in far less time than if a randomized clinical trial had been conducted. For example, the lung cancer agents crizotinib and ceritinib were approved in just 6.0 and 6.4 years, respectively, from the date of patent application.[430] This is much better than the usual twelve to fifteen years required for approval. Approval of dabrafenib for malignant melanoma, osimertinib for lung cancer, and ixazomib for myeloma occurred just three to four years from the date they were patented.[767]

Other encouraging US initiatives have included the 2016 21st Century Cures Act that facilitates clinical research.[768,769] However, some groups have opposed this legislation.[770]

The 2015 White House Cancer Moon Shot Task Force was also a positive step. This initiative focused on areas that included improving collaboration and data sharing between different cancer research groups, opportunities for strategic new investments, and improving research access to patient data.[771]

Some groups have advocated "Right to Try" approaches. With Right to Try, patients with incurable cancers who are ineligible for a clinical trial of an experimental drug may, nevertheless, access the drug. Access may be permitted even if there is very little clinical data on toxicity or efficacy.[772]

I very much understand the sentiment behind this approach, but I and others[772] have concerns.

[cxlviii] High response rate in non-randomized trials is a good indicator of a drug's effectiveness if the drug is given alone as a single agent. Tumors do not usually shrink on their own. Hence, if you treat patients with a new agent and 60% of them (for example) have tumor shrinkage, you know that the drug is effective. On the other hand, if a standard therapy on average causes tumor shrinkage in 20% of patients and the rate of tumor shrinkage goes up to (for example) 50% when you add the new agent, you cannot be certain that the apparent improvement is due to the addition of the new agent. It might, on the other hand, simply be due to patient selection or some other unrecognized factor. Hence, a randomized trial is generally needed to determine if adding a new agent to an older therapy truly improves outcome whereas a randomized trial may not necessarily be needed if the agent is effective when given alone.

One concern is that without the tight regulation associated with clinical research, misleading hype may be used to coerce patients into trying a therapy that has very little possibility of being helpful. It is also unclear who will pay for the medication. It is one thing if a company provides a drug free of charge. However, will the company be able to do this, and will they still have enough of the drug and resources available to conduct clinical trials? Some companies might decide to just sell the therapy at a high price to Right to Try recipients without doing the expensive clinical trials required to prove efficacy. This could be too tempting for some companies and may mean that drugs that are really of no benefit can be sold at a large profit with no clinical trial to confirm the drug's effectiveness.

I am also concerned that if patients can access the drug in this manner, it may become impossible to recruit patients to the clinical trials that are necessary to prove drug benefit and safety. Instead of Right to Try approaches, we must streamline the clinical research that is essential for rapid drug assessment and approval.[cxlix]

The Tufts Center for the Study of Drug Development estimated that by 2013 it took an average of $2.9 billion to take a new drug from discovery to approval for marketing.[756] This compares to $4 million in 1962 and $231 million in 1987.[773] The major part of this increase is the cost of doing clinical research. The costs of the early laboratory work to discover the drug and to test it in animals are markedly lower than the clinical trial costs.[756]

The very high cost of clinical research is driven largely by the cost of complying with the regulationsthat are intended to protect the rights and safety of patients participating in the clinical research.[757] This regulation is extremely important, but it needs to become much more cost-effective. To put this into context, a new cancer therapy may be judged to be too expensive if it costs more than $250,000 per life-year saved.[774] Some Canadian agencies are trying to reduce drug prices to the extent that they cost no more than $50,000 per life-year saved. However, complying with

---

[cxlix] There are numerous approaches that can be used to streamline clinical research. I discuss some of these in Chapter 12.

increasingly stringent clinical research regulation is much more expensive. By my calculations, it costs millions of dollars per life-year saved.[757]

To gain approval for one specific drug, it generally costs far less than the $2.9 billion calculated by the Tufts Center. The $2.9 billion figure was calculated by dividing the amount drug companies spend on research and development by the number of drugs approved.[756] However, only about 12% of all drugs that make it into clinical trials are eventually approved.[756] Many others do not make it into clinical trials at all.

Cancer drugs are less likely to be approved for marketing than are other drugs.[775] Only 3–8% of cancer drugs entering clinical trials are approved.[775-777] Yet, it takes longer to do clinical trials in cancer than in other diseases,[775] and the complexity of cancer clinical trials is rising rapidly.[775] Large amounts of money may have already been spent on a failure before it is discarded. When a drug succeeds, the revenue from its sales must pay for its own development in addition to the costs of many discarded failures.

The fact that therapies are becoming more "personalized" adds to the problem. For many drugs, the number of patients that may be treated with them is rapidly shrinking.[759,778] In the "old" days, clinical research might have led to the development of a drug that could be used to treat, for example, most patients with advanced lung cancer. This is a lot of patients, but only a small proportion of the patients treated would benefit from the drug. With personalized therapy, only a small percentage of all patients with lung cancer would have the target required to make them a candidate for the drug, but a high proportion of those treated would benefit.[759]

For example, the ROS1 fusion gene is present in about 1% of patients with lung adenocarcinoma. Only this small number of patients would be candidates to receive an approved ROS1-specific therapy. However, unless a drug achieves the breakthrough drug designation, the cost to gain approval for a drug that can only be used to treat a small number of patients will be the same as for a drug that can be used to treat many patients. Hence, the price of the drug for each treated patient will have to be high if the research investment is to be recouped.

Personalized therapies have a higher probability of being approved than do non-personalized drugs,[775,778] so there is a lower risk of failure. Despite this, the development costs for each drug remain very high.

The overall costs of cancer care are rising rapidly.[767,779,780] The cost of new anticancer medications is rising by more than 10% per year.[775] In 2015, cancer drugs accounted for 10.2% of total annual drug costs in Canada and 11.5% in the US.[767] By 2018, it cost an average of $149,000 to treat a patient for a year with a new anticancer medication.[775] Not only are new anticancer drugs more expensive than older agents, but the total population treated is larger. The number of cancer cases is increasing, and new drugs are becoming available to treat patients who were previously untreatable. In addition, the efficacy of these new agents is increasing, so individual patients are being treated for longer.

High drug development cost is not the only factor driving high drug prices, but it is a very important one. We risk bankrupting healthcare systems if we continue to pay these rapidly rising drug prices.[781] Many healthcare systems already limit access to new drugs.[775] Very high cancer care costs also put individual patients (particularly Americans) at risk of personal bankruptcy.[782]

We have only a few viable options to deal with this. New drug development and progress against cancer will stall if healthcare systems and patients decide not to buy these effective but expensive therapies. The same will happen if drug prices are legislated to a level too low to permit recouping development costs. No investor or drug company could survive investing in new drug development if the development costs—including the cost of the many failures—cannot be recouped.

As I discuss in more detail in Chapter 13, the only rational way that I can see to solve this problem is to make drug development much cheaper.[745,757]

There are many reasons why drug development costs must be brought down.[524,745,757,759] If it costs a large amount to develop a single drug, then these resources are not available to test other new agents. Hence, high costs directly limit the number of effective new agents developed. In addition, if drug development costs are very high, then only large companies have deep enough pockets to succeed. High drug development costs thereby limit the competition that could potentially help bring down drug prices.

Furthermore, if costs of doing clinical research are very high, then clinical research is severely limited unless it is funded by a large drug company. This means that these companies drive the clinical research agenda. Pharmaceutical companies have made an incredibly important contribution to our progress against cancer, but the current funding model for clinical research has disadvantages. Moreover, the high costs of new drug development discourage risk-taking by investors. These high costs raise the temptation to back development of a "me too" drug that works in the same way as other currently available drugs, rather than embarking in bold new directions.

The same things that could make approval of effective new agents much faster could also make it much cheaper. Breakthrough drug designation is one way to do this for drugs giving high response rates, but we need much faster, cheaper access for other agents as well. For example, we cannot reliably predict who will respond to new immunotherapy agents, so they do not meet the criteria for breakthrough drug designation. Nevertheless, they provide a very marked benefit in a subset of treated patients.

Conclusions: In summary, the eleventh reason that cancer still sucks is that it takes too long and costs too much to develop effective new therapies. But there is hope. Tight regulation of clinical research and new drug development is very important. However, we need careful reassessment and corrective action of this essential regulation. Its negative impact on access and costs for effective new therapies takes too high a toll. It is never possible to make regulation perfect,[746] but we can make it better. The next chapter discusses a few possible options.

— 12 —

Speed Bumps on the Autobahn: The Many Barriers to Progress

There are major issues with clinical research speed, efficiency, and costs. The twelfth reason that cancer still sucks is that these issues delay access to drugs that can save lives and alleviate suffering—and they play a major role in driving up drug prices.

This chapter is a call to action. No matter what country we live in, we must all keep up the pressure on our government representatives and the pharmaceutical industry to fix these issues.

Short Primer

We will again start with a brief overview. Human nature and the Darwinian principle of survival of the fittest mean that rational decisions by individuals frequently have adverse consequences for the collective.[783] Countries craft constitutions and enact laws and regulations to control the adverse consequences of this reality.

The same logic applies to clinical research. In Chapter 11, I briefly discussed a few examples of the bad things that have happened historically in clinical research, such as Nazi experiments on prisoners during World War II and the Tuskegee study, in which treatment was withheld from African American males infected with syphilis. These examples and numerous others resulted in the creation of strict rules to govern clinical research.

The history of clinical research is like a highway plagued with numerous accidents. To reduce risks, regulators have added speed bumps every few feet along the clinical research highway. This reduces accidents, but markedly slows research traffic. It takes much longer and costs far more for researchers to get to their destination. It also increases the probability that they will give up in frustration and turn back. For example, high research costs[784] and delays in trial initiation[785] both increase the risk that a trial will be abandoned before it can be completed.

Conditions like acne, obesity, depression, arthritis, bacterial infections, or irritable bowel syndrome are diseases that can be cured or controlled. Vaccines or therapies for nonlethal conditions must be extremely safe and so warrant extensive testing and cautious regulation. In these cases, safety speed bumps are important even if there are too many.

But the risks and rewards are much different for lethal diseases like metastatic cancers. We need a different approach for developing therapies for lethal diseases. It should entail processes and regulations that effectively protect data integrity, patient rights, and safety while simultaneously enabling and encouraging speed.

I suggest a model based on the Autobahn approach, which minimizes speed bumps and provides safety according to road conditions. Germany's Autobahn is famous for having long stretches with no speed limit, but nevertheless having relatively low traffic fatality rates.[786,787] The key is that they have excellent roads, excellent cars, well-trained drivers, and simple, effective regulations.[787] For example, a driver cannot pass on the right and must generally be in the right-hand lane unless passing. It only has speed limits when road conditions dictate them.

My wife Lesley and I rented an Audi to drive across Europe in 2016. We drove the Autobahn at an average of 140–150 km/h (85–90 mph) and would speed up to 160 km/h (100 mph) or more to pass. We would then smoothly merge back into the right lane as a Mercedes or BMW traveling 240 km/h (145 mph) rapidly passed us. This was faster than we would comfortably drive on North American roads, but we felt completely safe doing so. The right conditions had been created to maximize safety. Similarly, we need a regulatory/research Autobahn for development and dissemination of therapies for lethal diseases.

The press is much more likely to criticize regulators for approving a "bad" drug than for being too slow in approving a "good" drug.[788] Therefore, regulators are pressured to proceed slowly and carefully. Regulators can only permit drug development to move faster if society allows them to do so.

Pressure and political action are essential to make regulators move faster.[788] For example, despite breast cancer being on average much less deadly than lung cancer, news coverage of breast cancer is far more extensive. The net result is that the average time taken for the FDA to approve a breast cancer drug is much shorter than for approval of a lung cancer or prostate cancer treatment.[788]

The dramatic impact of political pressure brought by AIDS patients and supporters is well known.[789] AIDS activists were very successful in attracting both wide press coverage and political support. Media coverage and patient group pressure markedly accelerated drug development and approval.[790]

Their success raised concerns that the FDA had been inappropriately transformed from "a consumer protection agency into a medical technology promotion agency."[791] However, this activism had a major impact on the speed of true progress.[789]

As I write this in August 2021, the US and Canada are administering vaccines for COVID-19 that were developed at breakneck speed. COVID-19 kills about 2% of those infected. Conversely, many cancers kill more than 40% of patients and some kill more than 90%.[1,792] Cancer is responsible for about 30% of all deaths in Canada (far more than heart disease or any other factor)[792] and it causes 22% of all deaths in the US.[cl 1] Between the start of the pandemic and the time of writing this, approximately 27,000 Canadians died from COVID-19, while approximately 140,000 died from cancer in that same time frame.

Part of the urgency to fight COVID-19 is its impact on economic productivity and healthcare costs. But cancer also has a huge impact.[793] We[794] and others[795] have argued that we must have the same urgent approach in our fight against cancer as we have with COVID-19.

[cl] In the US, heart disease is slightly ahead of cancer, causing 23% of all deaths.

As outlined in Chapter 11, it takes far too long and costs far too much to make lifesaving therapies available to dying patients. This must stop. We need better solutions. Clearly, the clinical research missteps in the past cannot be condoned. However, the measures enacted to prevent further missteps are overkill. They are extracting much too high a price.

Several people have told me that this problem is unfixable. I am convinced, however, that there are many achievable things we can do to speed clinical research.[430,524,745,757,759,760] The rest of this chapter outlines a few of them.

Solutions: There is no one "big thing" that would result in promising clinical research findings getting to the clinic faster and more efficiently. Removing the biggest speed bump from a highway with a thousand speed bumps would have almost no impact. The remaining 999 smaller speed bumps would still bring things to a near standstill.

We must address <u>all</u> the speed bumps. Taken together, they are deadly, causing the loss of countless lives. We need a consensus that speed is extremely important in the development of new therapies for lethal diseases, and that current approaches unnecessarily slow progress. Currently, there is no such consensus.

Here is a list of some things that need to be addressed quickly:

Clinical research oversight

- We need to separate the oversight for clinical research and drug approval for lethal diseases from the oversight for nonlethal diseases.
- The US must deal with regulatory gridlock. Simply put, US regulatory bodies are tripping each other up. Currently, some regulatory bodies that try to foster reform are slowed by other regulatory bodies that get in their way.
- We need privacy regulations that help promote progress rather than blocking it.

Clinical trial design

- We need small studies aiming for large gains rather than large studies aiming for small gains.
- Irrationally rigid "eligibility criteria" inappropriately restrict patient participation in many clinical trials. If the patient cannot satisfy the eligibility criteria, then they are denied treatment with the new therapy. This slows progress while markedly limiting the number of patients who can access promising treatments.

Clinical trial activation

- It takes too long to assess toxicity and pharmacology of a new drug in animals before human trials begin. Much of this animal data adds little value.

- There are currently several different players that review a trial before it can begin to accrue patients. These include government regulators, institutional ethics review boards, and others. These current extensive review processes take far too long—and often add little value.
- Trials of new "targeted" agents often select patients based on their tumor expressing a specific biomarker. However, the regulations that govern the development and use of these biomarkers slow progress by being inappropriately stringent.

Study conduct

- The "protocol" that forms the blueprint for a clinical trial was once regarded as a guideline for the trial. However, it is now regarded as a strict and very narrow path from which the investigator must not deviate. There are many situations in which this rigidity is irrational and slows progress without adding meaningful levels of safety or value. For example, the eligibility criteria may stipulate that a patient's kidney function must be normal for them to enter the trial. It would generally be perfectly reasonable for a patient with slightly abnormal kidney function to also participate, but very safe, rational deviations from the protocol are not permitted. An investigator may initially try to create a protocol that is flexible, but in most instances, this is forbidden by either the regulators, study sponsors, or Institutional Review Boards.

- Trials now demand a huge amount of very detailed, expensive documentation. Much of this documentation is of no practical use. The documentation requirements are now so extensive and onerous that it is very difficult to run a trial without a very rich source of funding. Pharmaceutical companies are generally the only source available for this funding. Consequently, this necessity for rich funding to collect vast amounts of unimportant data means that pharmaceutical companies inevitably drive the clinical research agenda.

- Like many clinical investigators, I once found it exciting and satisfying to participate in clinical research. However, the thrill is now gone. Clinical research has instead become a masochistic exercise in frustration. Clinical investigators are essential to do the clinical research necessary for progress. With the rise of disenchantment, the number of clinical researchers is rapidly dwindling. This trend must be reversed, or progress will halt.

Drug approval and funding

- Once the trials have been completed and an application for approval of the drug has been submitted, review times for this application need to be shortened.

- Canada needs to align itself with either the US or with the European Medicines Agency so that a drug approval application to them also constitutes a de facto application for approval in Canada. Otherwise, delays in applications to Canada create delays in access to effective new therapies for Canadian patients.

- Drug funding processes in Canada (and many other countries) need to be reformed drastically. The interval from drug approval to funding for a drug must be markedly shortened.

Further Details and References

I will now go into further detail on the factors slowing progress.

Statutory law and regulation (or "administrative law") both play roles in oversight of clinical research and new drug development. As Garrett Hardin explained in "The Tragedy of the Commons,"[746] it is never possible to cover all nuances and possibilities when drafting statutory laws. Consequently, regulators or "custodians" are entrusted with interpreting these statutory laws and adapting them for application to a wide range of circumstances.[746]

Over time, the regulation evolves as it is applied to an ever-expanding series of events. The diverse ways in which the regulations are applied can—and often do—differ from the initial intent of the statutory law and create unintended consequences. Hardin points out that periodic "corrective feedback" is needed.[746]

The regulatory speed bumps I described earlier have been created by evolution of clinical research regulation.[524,745,757,759] There was a valid reason for the creation of each new speed bump, but many of these speed bumps have had a negative impact that extends far beyond fixing the problem for which they were intended. This is the consequence of what the late Emil J Freireich referred to as "applying a general solution to a specific problem."[796]

In oncology, we are taught to adopt approaches that are evidence-based. Regulators carefully monitor and audit compliance with these speed bump regulations,[797] so we have evidence on *compliance*. We know how long it takes to bring drugs to market and can calculate the number of years of lost life during this time.[706] Hence, we can calculate the *harm* that is done by the speed bumps. We can also calculate the rapidly rising expense of bringing a drug to market.[756,773] Therefore, we also have evidence on the *financial costs* of these speed bumps.

What we do not have is evidence of their *benefits*. As clinical research regulation has become more and more stringent and expensive, the proportion of patients dying from toxicity on phase I clinical trials has only dropped from 0.8% to 0.5%.[cli] When we divided the high costs of this augmented regulation by the number of lives saved, we calculated a cost of millions of dollars per year of life saved.[757] In contrast to this, an effective new drug that is used to treat an advanced cancer is deemed to be too expensive if it costs more than $50,000 to $250,000 per year of life saved.[757] It is always good to make things as safe as possible, but this comparison suggests that from the perspective of years of life saved, hypervigilant clinical research regulations are neither cost-effective nor efficient.

While scientific fraud in cancer is relatively uncommon, some of the speed bumps probably do reduce the risk of fraudulent or inaccurate data. This helps ensure that approved therapies are both effective and reasonably safe. The speed bumps likely help to ensure that investigators respect patient privacy and obtain proper informed consent. However, there are no data that show how well the speed bumps do this. More importantly, there has been sparse assessment of other approaches that might work as well (or even better) at a much lower cost, and with far less negative impact on the time required for approval of effective new drugs.

We accept the current speed bumps as a matter of faith. [798]This must stop. The price we are paying for this blind faith is much too high.[524,706,745,756,757,759,773,796] We have let ourselves become locked into what Thomas Kuhn would have called an unworkable paradigm.[798]

These shortcomings are not the fault of regulators. They play the hand they have been dealt.[clii] However, regulation is a very inexact process.[746] It inevitably "drifts" over time as it attempts to deal with an ever-widening realm of real-life issues. This drift eventually begins to cause problems. Hence, periodic "corrective feedback" is essential.[746]

[cli]   Phase I trials are among the riskiest clinical trials since before starting the trial, investigators have little information on what dose is safe and what kinds of toxicity might occur in humans.

[clii]   In fact, without the superb leadership and commitment of people like Dr Richard Pazdur and his colleagues at the FDA and similar regulators at Health Canada, the situation would be much worse. They have done an outstanding job of facilitating access to effective new drugs despite the bureaucratic and political challenges they face.

In recognition of this, corrective feedback has indeed been applied periodically to regulation overseeing research and new drug development. As I discussed in Chapter 11, this corrective action in the US has included passage by Congress in 2012 of the Food and Drug Administration Safety and Innovation Act (FDASIA).[701] The establishment of the White House Cancer Moon Shot Task Force in 2015 was a further corrective measure, as was passage by Congress of the 21st Century Cures Act in 2016.[768,769]

Each of these corrective feedback activities and the response of regulators to them had positive effects. They have been a key enabler of rapid approval of several effective breakthrough cancer drugs. However, they tackle only a few of the numerous speed bumps that impede us.

We need to do much, much more. For lethal diseases, we need "progress-centered regulation" with the paramount objective being the marked acceleration of progress.[745] Yes, safety, data integrity, patient informed consent and privacy are vital. But rapid progress is equally important. We have lost sight of this.

Would speeding drug development increase profits for Big Pharma? Maybe. There are some—like those who opposed the US 21st Century Cures Act[770]—who would argue that making life easier for Big Pharma would be a sell-out. But faster, cheaper progress is essential for all of us. We must not be "crabs in a barrel"[cliii] just out of spite against Big Pharma! This spite can kill our loved ones.

Below I have listed just a few of the speed bump issues that must be addressed. I expect that many of you will be amazed and perhaps even dismayed at the labyrinthine speed bump bureaucracies as you read through this section.

Separate oversight for development of new therapies for lethal diseases: It would help if oversight for the development of therapies for lethal diseases and nonlethal diseases were separate and distinct from each other. The FDA has a highly effective Cancer Center of Excellence to oversee oncology efforts, and much greater levels of risk are accepted

---

[cliii] The term "crabs in a barrel" comes from the legend that if you put one crab in a barrel, it will readily climb out. However, if you put several crabs in the barrel, the crabs below will grab and pull back any ahead of them that try to climb out. Booker T. Washington reportedly used this analogy in his speeches.

for oncology drugs than for the drugs used to treat many other illnesses. However, I suspect that oversight of the development of cancer therapies is still ultimately impacted by many of the same threats[788] and constraints that affect other areas. Safety is a paramount concern when using an anticancer drug as an adjuvant treatment (where the intent is to reduce risk of recurrence in a patient who may already have been cured by surgery), as well as when assessing a therapy for nonlethal diseases like acne and arthritis. But the processes that are so essential for protection of safety of patients with nonlethal diseases can markedly slow the progress of research in lethal diseases where the consequences of delay are orders of magnitude higher.

By far the greatest threat that patients with a lethal disease face is the threat from the disease itself. It is about to kill them. Toxicity is important, but there are ways to monitor this that could have a much smaller negative impact on progress.[759] This is already a key consideration for agents approved as breakthrough drugs.[701]

Regulatory gridlock: In the US, many different agencies and regulations directly or indirectly affect clinical research. These include Institutional Review Boards (IRBs), the FDA, the Office of Human Research Protections (OHRP), the National Cancer Institute (NCI), NCI's Cancer Therapy Evaluation Program (CTEP), the Center for Medicare and Medicaid Services (CMS), CMS's Clinical Laboratory Improvement Amendments (CLIA), the US Department of Health and Human Services, the Health Insurance Portability and Accountability Act (HIPAA), the Department of Veterans Affairs, the Internal Revenue Service (IRS), the Office of the Inspector General (OIG), the US Patent and Trademark Office, and the Joint Commission on Accreditation of Health Care Organizations (JCAHO).[757] Space does not permit a full discussion of the impact of each body, but the essence is this: each may make changes that appear small and simple, but that can have a serious adverse ripple effect on others. Conversely, if one agency makes a change to try to facilitate progress, this can be thwarted by even modest tightening of regulations by a different agency. There are too many hands in the pot.

Communication barriers between these agencies further complicate the process. I asked friends in NCI's CTEP why they could not meet with the FDA and other federal agencies to try to solve some of the barriers to clinical research. I was astounded when they told me that it is illegal (punishable by firing and jail) for employees of one federal agency to "lobby" another federal agency. Any attempted discussions could be interpreted as illegal lobbying. How can progress be made if agencies that unintentionally block each other cannot even discuss it?

Harmonized regulation: I am now back in Canada but am spending a lot of time in this book talking about things that happen in the US. One might expect that one country could make rapid progress even if speed bumps slowed progress in another country. However, this is not the case because research oversight rules have been "harmonized" internationally. Initially, the Helsinki Accord and Good Clinical Practice (GCP) guidelines[799] were widely adopted. These were well-thought-out responses to abuses exemplified by Nazi human experimentation and the Tuskegee syphilis experiment. However, over time, interpretation of these approaches has evolved. This has spawned hundreds of unhelpful rules slowing almost all aspects of clinical trials. They have been adopted internationally, so no country has an easy way to escape them. While there were good intentions, this has spiraled out of control and has become an impediment.

Regulatory fundamentalism: When a problem arises, rational, well-thought-out corrective rules are formulated. With every rule, there is a specific intent, but over time, this intent is often forgotten. There is a gradual emergence of what I call "regulatory fundamentalism," whereby interpretation of the rule evolves in a way that forgets the original intent.[757] The US Constitution's Second Amendment is a good example. It gives all citizens the right to bear arms. The intent was to allow citizens to protect themselves should a heavy-handed government try to deprive them of their freedom. However, it is now interpreted by many as giving everyone the right to own a gun—even a military assault rifle—with which they may potentially terrorize their neighbors.

Regulatory fundamentalism spawns "interpretation creep" in which interpretation of guidelines slowly moves further away from the original intent. In our analogy, many US states now have enshrined the right for people to carry guns openly. In some states, even feeling threatened by a neighbor is legal justification for shooting them.[cliv]

In clinical research, this interpretation creep makes it increasingly difficult to comply with any given regulation. In addition, people start worrying that even though they are complying with today's interpretation, they may run into difficulty if today's actions do not comply with tomorrow's updated interpretation.

Houston's MD Anderson Cancer Center is a phenomenal institution that has been responsible for many advances in cancer treatment. I had many great years working there. But the "Office of Compliance" is a particularly prominent department at MD Anderson, staffed by lawyers and administrators whose job it is to ensure that the institution and its employees are fully complying with all government regulations. When I was there, the medical staff received a steady barrage of threatening emails about the potential for us to be fired or jailed if we went astray of a series of federal or state government mandates.[clv]

When an MD Anderson physician would ask the FDA (for example) why they were now interpreting a regulation in a particular way, the FDA would often protest that this was not, in fact, their interpretation. The Office of Compliance would then point to the need for risk minimization. Their approach was not based on current directives from the FDA and other regulators but on how they predicted the regulators might retroactively interpret the regulations next year.

Such a highly risk-averse approach slows progress. The real barriers created by fundamentalist reinterpretation of regulation wording are problematic enough. This fear of future regulation reinterpretation makes it even worse.

Privacy regulation: Making private healthcare information public has resulted in embarrassment, social isolation, or discrimination for many people. They may be denied insurance or a job. The information can be misused in family disputes. Individuals may be targeted by those preying on their vulnerabilities or by those trying to sell them a new "cure."

[cliv] My apologies to my Houston friends and former neighbors for using this example. My "Canadian" is coming out in me again.

[clv] I discuss this further in Chapter 14.

Respect for privacy is extremely important. But it can be taken too far. For example, privacy regulation such as HIPAA makes it difficult to contact patients who might be candidates for a trial. It also increases clinical research costs, the complexity of documentation, and the administrative burden associated with clinical research.[800] The burden of complying with privacy rules delays many trials and can lead to complete abandonment of others.[801] It also seriously impedes the review of health records to try to learn more about a disease. The rules are well-intentioned, but they directly slow progress against lethal diseases.

The details of how privacy regulations are applied must be modified. They must minimize the negative impact on progress against lethal diseases while continuing to protect individuals from harm. The US Congress has taken an initial step in this direction through the 21st Century Cures Act, but it remains unclear how helpful this will be.[802-804]

Overall, it is ironic that with the flimsiest of consent processes, Google, Facebook, etc., have been free to sell much of our private information to anyone who wants to pay for it,[805] while privacy rules create major barriers to clinical research that could save lives. Patients facing death have repeatedly told me that progress against their disease is much more important than their healthcare privacy.

Study size: The bigger a study, the easier it is to achieve "statistical significance." This is the magic of "statistical power." In a big study, a very small, clinically meaningless gain may be statistically significant. Conversely, a large gain may be observed in a small study, but because of the small study size, this large gain does not achieve statistical significance.[430] Based on this, a new therapy giving a small gain in a large study may be approved while a drug giving a large gain in a small study may be discarded. Hence, it is tempting to conduct very large, expensive studies to maximize the probability of a result that is statistically significant. Consequently, we end up with very expensive drugs (since the research costs associated with the large trial must be recouped) but with only a small medical gain. Instead, we need faster, less expensive, smaller studies aiming for large gains.[524] This can speed progress while bringing down drug development costs and fostering development of highly beneficial drugs. This approach will dismiss therapies that give only small gains in favor of other therapies that yield large gains.

The use of small studies that aim for large gains is precisely the approach taken when agents are assigned breakthrough drug designation,[701] as discussed in Chapter 11.

Preclinical toxicology: It takes months or years to conduct expensive testing to define a drug's toxicity in animals prior to human trials. However, most of this effort is a waste. This testing generally predicts obvious toxicity that we would watch for anyway in patients—or it misses important toxicities—or it predicts toxicity that is not a major problem in humans. There are very few instances where preclinical toxicology predicted something that was both unexpected and important.[524,757]

The only preclinical toxicology needed for anticancer agents is a small, rapid, inexpensive assessment to determine the dose of the drug that kills 10% of treated rodents. Once this study has been completed, it is safe to then start trials in humans using 1/10th of this dose.[806] Clinical trial initiation—and hence progress—could be faster using this approach.

Preclinical pharmacology: Preclinical pharmacology tests how a drug is metabolized and eliminated in animals prior to initiation of human trials.[757] Yet animal pharmacology studies are generally a poor predictor of the drug's pharmacology in humans. In general, animal preclinical pharmacology studies add little useful information. But there are exceptions. Limiting preclinical pharmacology to these exceptions would speed up progress and drive down research costs.

The exceptions include studies to determine how much of the drug is absorbed into the blood stream if given by mouth (since drugs with high absorption may then be administered orally). Other helpful studies determine how the drug interacts with different components of the "cytochrome P450" system (which may break down a drug). Under some circumstances, comparing the pharmacology of several related drugs can help choose one to bring into clinical trials.

If one sees unexpected toxicity in early clinical trials, doing further focused toxicity and pharmacology studies in animals could lead to a better understanding of why the toxicity occurred, but doing these studies before clinical trials generally does not help much.

Study review and activation: Clinical trial design and activation begins after preclinical evaluation. It takes an average of about two years to bring a clinical trial from initial concept to study activation in trials run by cooperative groups.[clvi] For each trial, this entails hundreds of time-consuming, costly steps (many of which add little value).[807] Each of these hundreds of steps is a potential bottleneck— and it only takes one bottleneck to slow everything down.

For businesses, various "business process management" strategies have proven useful in improving processes.[808] Similar strategies to speed clinical trial activation have also been beneficial. For example, there are often delays between several necessary sequential steps before a trial is activated. The time to trial activation can be sped up significantly if several of these steps are undertaken in parallel, rather than requiring one step to be completed before the next is begun.[809] We need much more of this.

Approval by Institutional Review Boards (IRBs): A trial must be approved by a center's IRB before it can be activated. The IRB assesses the ethics of the study and determines whether the informed consent document is accurate and adequate. This review process can be prolonged and arduous. For example, one researcher reported that for a simple "observational" study,[clvii] interactions with the IRB accounted for almost 17% of all study costs. There was an exchange of more than 15,000 pages of materials, but with almost no impact on patient safety or trial procedures.[810] An analysis of IRB reviews of 103 studies at another center reported that it took a median of 131 days for IRB approval, and one study took 631 days[811]—nearly two years before a patient could receive potential benefit!

I served as an "alternate" on one of MD Anderson Cancer Center's three IRBs. If a regular IRB member was away, I would step in to participate. What I found most striking was the very high frequency of trials turned

---

[clvi] A cooperative group is a formal association of several different institutions and investigators to carry out clinical trials together. Highly successful ones that have been conducting important cancer clinical trials for several decades include the Eastern Co-operative Oncology Group (ECOG), the Southwestern Oncology Group (SWOG), and the Canadian Clinical Trials Group (CCTG), but there are also several others.

[clvii] In an observational study, data are gathered on study subjects, but no therapy is given as part of the trial. There is very little possibility of harm coming to a patient from participation in such a study.

down due to very minor comments on wording in a consent form. To comply, the investigator would change the wording and resubmit it for the following month's IRB meeting. Commonly, there would be a new requirement for one or two more words to be changed. As a clinical trials investigator, I had protocols sent back to me up to four times, each time asking for one or two new minor changes.

In their quest for perfect wording on a protocol consent document, IRB members forget that patients desperately need new treatments. Such delays can translate into thousands of years of life lost worldwide because they delay broad, international access to the drug.[706]

IRB members often feel that they have not done their job properly unless they find at least one minor fault with a trial that they are reviewing. We pay a very high price for this narcissism.

In the past, this same prolonged, review process would play out at the IRB of every institution participating in the trial. Frequently, a step mandated by one IRB would be specifically disallowed by another.

Research ethics is not an exact science—in fact, it isn't a science at all. However, we have made some progress over the past few years. For example, many jurisdictions now permit use of "central IRBs," whereby approval by a specified central IRB enables many different institutions to activate a trial. In the province of Ontario, Canada a cancer trial may be activated at participating centers across the province if it is approved by the central Ontario Cancer Research Ethics Board (established in 2004). This has substantially reduced the time required for clinical trial approval.[812]

However, we need to go further: in Canada, for example, it is legal to use a central IRB based in your own province, but you cannot activate a trial approved in a different province or country. This needs to change so that any trial only needs to be approved by a single accredited IRB.

Just-in-time trial activation: We live in an era of personalized medicine. Drugs increasingly target uncommon mutations. Often, a center might put time, effort, and expense into activating a targeted trial and then see no eligible patients with the required mutation.[clviii] On the

---

[clviii] Not only is it expensive to do all the things required to activate the study, but it is also expensive to keep it open due to ongoing regulatory requirements for each study.

other hand, it doesn't make sense to wait for a patient with the mutation before activating a trial. There would be a very high probability that the patient would deteriorate or die in the months that it takes to have the trial approved.

The physician might refer the patient to another center that is already participating in the trial. However, travel to another city to receive the therapy could be a major burden for the patient and their family.[clix] And at the end of the day, there would be no guarantee that the therapy would be effective.

We need "just-in-time" trial activation. If the trial is already approved at another site, an accredited investigator needs to be able to immediately activate the trial at their site based on this prior approval.[745,759,813] Initial experience with "just-in-time" trial activation indicates that this is a feasible strategy.[813]

Exclusion of patients from trials: Fewer than 5% of adult patients with cancer ever make it onto a clinical trial in North America,[814] despite 70% being willing to do so if participation were an option.[815] In some cases, lack of participation is because the patient's local oncologist is not an approved investigator for the trial, and it is not feasible for the patient to travel to a center that offers the study.[clx]

In other cases, patients who need to start the therapy immediately may not be able to participate in a trial because it takes too long to go through the procedures to enter a patient on an open study. For example, with metastatic non-small cell lung cancer, we have calculated that 4% of remaining patients will die every week that therapy initiation is delayed.[424] Many others deteriorate rapidly. Several factors contribute to delays for standard therapies.[816] On average, it takes weeks longer to initiate treatment on a clinical trial than it takes to start a standard therapy.[817] Many patients cannot afford to wait this extra time.

Many other patients are "ineligible" for a trial, based on a long list of exclusion criteria. For example, 13–23% of patients with advanced lung cancer have brain metastases at the time of diagnosis.[818] Many clinical

---

[clix] For some studies, the patient must move to the other city if they are to participate in the trial.

[clx] "Just-in-time" trial activation could fix this issue.

trials exclude patients with brain metastases.[819] This exclusion is based on the deeply entrenched myth that patients with brain metastases do much worse than other patients due to anticancer drugs not crossing the "blood-brain barrier." However, as I discussed in more detail in Chapter 10, survival of advanced lung cancer patients with brain metastases is no worse than those with metastases to other sites such as liver and bone.[735-738] As for the blood-brain barrier, drugs generally reach high concentrations in brain metastases even if they do not reach high concentrations in the normal brain,[729] and drugs that reach only low concentrations in the normal brain may nevertheless provide effective palliation against brain metastases.[730,731]

I have heard concerns that drugs might prove toxic to the brain if a patient has brain metastases, but there is absolutely no evidence to support this concern.

Overall, there is no rational reason to exclude most patients with brain metastases from a clinical trial if they satisfy other eligibility criteria.[744] Nevertheless, large numbers of patients are inappropriately excluded.

Prior malignancies are common in patients who develop other types of cancer. For example, about 15% of patients with advanced lung cancer will have had a different cancer previously.[clxi] [820] Many clinical trials exclude patients who have had a prior malignancy. The exclusion is based on unfounded[820] concerns that such patients will do worse than patients without a prior malignancy. Also, some investigators are concerned that if the cancer worsens on treatment, it might be difficult to differentiate which cancer is progressing—the cancer being treated or the earlier cancer. However, in practice, it is rarely difficult to tell which cancer is worsening. Exclusion from trials of patients with prior malignancies unnecessarily slows progress.[744]

Patients with "poor performance status"[clxii] are also often excluded from trials.[744] With chemotherapy, poor performance status patients do

---

[clxi] For example, smoking can cause cancers of the lung, bladder, head and neck, and several other types, and a patient may develop one smoking-related cancer after having been treated for an earlier one.

[clxii] "Poor performance status" means that the patient is quite sick due to severe cancer symptoms. For example, a "performance status 3" patient spends more than 50% of the day in bed and a "performance status 4" patients is completely bedridden.

have an increased risk of serious treatment toxicity.[660,821] Despite this, they may experience improvement in cancer-related symptoms.[440,661] They may also experience improvement in quality[660,822,823] and length of life.[441,660,824] It may make sense to exclude poor performance status patients from trials of a particularly toxic therapy, but for less toxic therapies there is little reason to exclude them.[744]

Abnormal lab values may also exclude patients from clinical trials. For example, the study protocol may say that the serum creatinine must be 1.2 mg/dL[clxiii] or less. The purpose of this is to prevent increased drug toxicity. A patient with abnormal organ function may be unable to properly clear the drug from their system. Years ago, a patient who did not meet eligibility criteria—but was close (e.g., creatinine 1.3 mg/dL)—would be granted an exception by the "principal investigator".[clxiv] The reason? There is no valid clinical or biological reason to expect that a patient with a creatinine level of 1.3 mg/dL would experience more toxicity than a patient with a serum creatinine of 1.2 mg/dL. Despite this, it is now generally forbidden to permit such a patient to participate.[825] This is irrational and it means that many patients who could safely participate in a trial are excluded.

Some investigators have tried to write clinical trial criteria that permit reasonable discretion in accepting low-risk patients who do not meet strict exclusion criteria. Occasionally they have been successful, but more frequently, they have not. IRB's, drug companies sponsoring the studies, or regulators have insisted that the protocol be written with very strict eligibility boundaries. These boundaries are strict and clear but are also ludicrous pseudoscience.

Some trials exclude patients if their tumors were biopsied by fine needle aspiration (FNA) rather than by a core biopsy. This makes no sense since diagnostic and molecular tests conducted on tissues obtained by the two methods appear to be equally reliable.[826] Requiring a repeat biopsy exposes the patient to risks from the procedure, and also exposes

---

clxiii For many clinical laboratories, this is the upper limit of normal for a creatinine level in the US. Canada and other countries use different "units" and hence have normal values substantially different than this.

clxiv The principal investigator is the person who oversees the study.

them to the risk of rapid deterioration or death as a consequence of therapy initiation being delayed to accommodate the unjustified repeat biopsy.

It is perfectly reasonable to remove excessively stringent exclusion criteria. Studies could accrue patients faster. We could get reliable answers and faster drug approval. In addition, many patients who are currently denied access to promising new experimental therapies would be able to access them.

There have been several calls to action.[744] It is important that we act.

Requirement for "CLIA certification" for biomarker assessments in clinical trials: The Clinical Laboratory Improvement Amendments (CLIA) requires that a US clinical laboratory can only accept human blood, tissues, etc. for testing, if it has first been certified for that test by the Center for Medicare and Medicaid Services (CMS).[827] Similar regulators oversee clinical laboratory operations in other countries. CLIA certification indicates that the lab has undergone extensive quality control assessments to ensure that its results are reliable.

Patients are selected for a "personalized therapy" trial if their cancer has a specific "biomarker" that suggests that the patient would have a high probability of benefit. The required biomarker is generally a genetic mutation or a highly expressed target protein. For example, high expression of the protein Her2/neu was used to select patients for early breast cancer clinical trials of the antibody trastuzumab (Herceptin). These trials demonstrated that trastuzumab could be effective for these patients.[462]

In the early days of "personalized therapy" trials, patients might be selected based on an experimental test. The test would be done in a research laboratory that would not have undergone the careful vetting of a CLIA certified lab. This raised concerns that the results might not be reliable. For example, in the early Herceptin trials, 18% of patients had different HER2/neu testing results from a local community laboratory than confirmatory testing done in an experienced, high volume central laboratory.[828] Some patients called positive by the community laboratory were called negative by the central reference laboratory, and vice versa. In another example, research laboratory gene expression profiling methods used at Duke University to assign lung cancer patients to a specific arm of a clinical trial were found to be seriously flawed.[829]

Because of such concerns, it was determined that biomarker testing used to select patients for a clinical trial had to be done in a laboratory that is CLIA certified for that biomarker.[830] Furthermore, in many cases, the test for the biomarker had to be granted an "Investigational Device Exemption" (IDE) by the FDA before the clinical trial could be started.[830,831]

This may sound like a good idea, but there are two problems.[829] The first is that the initiation of the trial is delayed at least several months until the test has been formally approved.[830-832] In some cases it can take one to two years or longer. Up front delays in trial initiation mean that completion of the trial and approval of the therapy for general use will also be delayed. For effective drugs, such delays in drug availability can translate into large numbers of lives lost.[706]

The second problem is that resources are limited. Fewer tests can be developed and assessed when one adds the high expense of IDE and CLIA certification to other test-related costs. This means that investigators must guess up front which test(s) they should develop.

We are often wrong when we guess which tests will best predict drug benefit. For example, in the early days of the development of EGFR[clxv] inhibitors, investigators thought that drug benefit would be predicted by the expression of the EGFR protein or by an increase in the number of copies of the EGFR gene in tumor cells.[clxvi] Both of these biomarkers are associated with increased benefit from EGFR inhibitors.[833,834] However, drug benefit is more effectively predicted by the presence of an activating mutation in the EGFR gene.[835,836]

The more expensive a test is to launch, the fewer the biomarkers that can be tested. I feel that, instead of this approach, the best way to comprehensively test for drug benefit is to initially run a series of low-cost tests. One can later use CLIA certified tests to confirm efficacy of the marker that performs best.[829]

Protocol deviations: A specific "protocol" governs the conduct of each clinical trial. Several bodies (such as the trial sponsor, relevant

---

[clxv] EGFR: epidermal growth factor receptor

[clxvi] Increase in the number of copies of the gene is referred to a "high gene copy number" or "gene amplification."

regulators, and the local IRB) must formally approve the protocol before the trial can start.

The protocol is a "blueprint" for the trial. It includes lists of inclusion and exclusion criteria that determine which patients will be permitted to participate. It also has criteria for when specific follow-up tests should be done and whether it is OK to proceed with another cycle of the therapy. For example, the protocol may state that patients need to have normal kidney function to be eligible for the trial, that they should have follow-up blood work done on days 7, 14, and 21 after each treatment, and that the next cycle of treatment can only be given if the absolute neutrophil count is more than 1.5×10^9/L.

Each trial also has a designated "Principal Investigator," or PI, who oversees the trial. As discussed in the earlier section entitled "Exclusion of patients from trials," it used to be possible to call the PI for permission to include a patient who did not satisfy all the inclusion criteria. This is no longer permitted.[825]

Some "study deviations"—such as doing the blood work at a time different from the day mandated by the protocol—are permitted if properly documented. Others are not. Entering a patient who does not meet all eligibility criteria would be a "study violation," and this would not be permitted.

The rigid, written protocol trumps the judgment of experienced clinician investigators. Moreover, a study violation would be a black mark against the investigator, who might be forced to discard all the scientific information gained from that patient, however valuable. In many cases, study violations are not material—they do not affect the validity of the data or violate the essential purpose of the study. Why do we value knowledge and experience in most other areas of life but not in clinical research?

It often doesn't make sense to have rigid cutoffs. In the real world, is it safe to drive 60 mph (100 km/h) on the highway but reckless and dangerous to drive 61 mph (101 km/h)?

Protocol rigidity does not improve safety or study reliability. It certainly does not reflect how the new therapy will be administered by practicing oncologists if it is approved. Instead, such rigidity slows trial accrual, reduces the ability of patients to access therapies that might help, and increases the cost and complexity of conducting the trial.

Study rigidity can also make life difficult for participating patients. A colleague shared with me a story of a patient in tears because the trial she was on mandated that she come in on day 7 for blood work. Her son's wedding was on day 7, and rigid study rules threatened her ability to attend her son's wedding. An excessive requirement for exact compliance with a protocol is in fact unjustifiably coercive.[825]

Study amendments: Once a study starts, investigators typically discover some aspects of the protocol that are problematic. Hence, changes or "amendments" are required for a high proportion of protocols. For more than half of all clinical trials, the amendments are major.[837]

Reasons for these amendments vary. They could be due to issues that arise between the time the study is approved and entry on the study of the first patient—or because of new information about the therapy—or because it proves difficult to find patients who exactly match the rigid eligibility criteria—or due to experience with the first several patients.[837]

The important point is that the accrual of patients must be put on hold while the IRB, regulators, study sponsors, and investigators assess these protocol amendments.

On average, this amendment approval process takes more than two months,[837] but I have seen several instances where it has taken a year or more. The time required to complete a clinical trial can be substantially lengthened if an amendment is required,[837] and having the amendment approved can cost hundreds of thousands of dollars.[837] In most cases, the reviewers mandate no major alterations to the proposed protocol amendments. Hence, rather than taking months or years, the amendment approval process needs to be reengineered so that in most cases it takes only days.

Excessive documentation requirements: Clinical research costs are driven up by the requirement for massive amounts of documentation. Data collection and other study-related procedures require research personnel such as research assistants, nurses, monitors, and others.

These requirements have rapidly become very extensive.[757] By 2013, study complexity had increased to the point that it cost on average

$42,000 US per patient entered in a phase III[clxvii] clinical trial, with some studies costing almost $90,000 US per patient.[838] Pharmaceutical company sponsors often hire a Clinical Research Organization[clxviii] (CRO) to oversee collection of the required data. CRO involvement magnifies the problem.[839,840] CRO data requirements are often rigid and excessive. This is partially for liability reasons. But many CRO personnel also do not have the clinical knowledge required to know which pieces of data are important and which are not. So their objective is to collect all of it, whether it is important or not.[839] Researchers spend a lot of time, effort and expense collecting data of minimal value from the perspective of either safety or efficacy.[839-841]

After data are submitted to the CRO, the researcher in many cases will be required to do a time-consuming "fix" of numerous minor details.[839,840] For example, I have received "urgent" requests to clarify whether a patient's mild fatigue actually got better on day 5 or day 6 after treatment. The reason cited was patient safety! Of course, this is unimportant to patients whose major concern is that they will be dead within a few weeks unless the therapy works.

However, any minor reporting errors or delays may be interpreted as indicating that the investigator is an unreliable clinical researcher. The investigator may be publicly shamed by the CRO[839] or in unbalanced, overblown, sensationalized reports by the friendly local newspaper.[842] The extent of data collection is so great that there is little possibility that no deficiencies will be identified if an investigator undergoes a routine audit by the FDA,[797] Health Canada, or regulators in other countries. Every empty data box may be regarded as a deficiency that could threaten patient safety, no matter how trivial the omission.

As Ayn Rand wrote in her novel *Atlas Shrugged*, "One declares so many things to be a crime that it becomes impossible for men to live without breaking laws."[843] Clinical researchers carefully collect a myriad of unimportant details[841] under a coercive, threatening cloud.[839,840]

There are multiple factors driving this massive collection of minimally important data. One is that it means increased CRO profits. "Several

---

clxvii A phase III clinical trial is a large, randomized trial.
clxviii Alternatively called a "Contract Research Organization."

reports have called into question whether ethical and professional standards in research conduct are at times secondary to economic considerations", with the prospect of increased profit at times driving CRO demands for more data.[840] Additionally, this increased detail may look good to the sponsor, and may make the CRO more likely to win a future contract.[839] The FDA may put pressure on the sponsor to collect these data, and the sponsor in turn may put pressure on the CRO.[839] The net result is that huge numbers of data boxes must be filled. Meticulous attention is paid to whether these boxes are filled, but little attention is paid to whether the data in the boxes are of much real importance.

These data, much of which will never be used, must be stored for years.[clxix] This storage is expensive. The investigator is also required to sign and date every lab report on every patient and to comment on whether any abnormality is or is not clinically significant. In routine practice and in the conduct of clinical trials, most lab reports have at least one abnormal value, and most of these abnormal values are unimportant.

The situation is reminiscent of the week I started premedicine at Queen's University in Kingston, Ontario in the fall of 1968. Senior students submitted us to hazing rituals as part of our initiation. One of my classmates ran afoul of the hazers. As his punishment, he had to count all the hundreds of holes in the ceiling tiles in his room in residence. He could have rapidly counted the holes along one side and the end of one tile, multiplied these to estimate the number of holes in that tile, and multiplied this by the number of tiles in the ceiling. After fifteen minutes he could have reported his estimate of the number of holes in the entire ceiling. Alas, he was told that he had to document his work. He spent much of the next week standing on a chair putting a checkmark beside each hole to document that he had counted it.

And that is where we are now in clinical research: spending billions of dollars worldwide each year documenting hole after empty hole.

The problem will not get fixed unless regulators, sponsors, investigators, IRBs, CROs, patient advocates, healthcare payors, and others decide what data are required—and what is not needed and unlikely to be used. It is

---

[clxix] In Canada, huge piles of paper must be stored securely for 25 years after the drug is approved.

tempting to demand as much data as possible, but there is a huge cost to this, and much of the information is never needed or used.[841]

The huge cost to this data collection and storage is a big part of the cost of developing a new therapy. This means that fewer new therapies can be developed with available resources, and the price of marketed drugs must be very high to pay for these excessive clinical research costs.

Reporting of toxicity: A lot of the detailed data collection deals with therapy toxicity. Toxicity information is very important. The problem centers on how this detailed information is disseminated to clinical trial investigators. Because serious safety data have been hidden in the past (e.g., Merck allegedly not disclosing the risk of heart attacks with the analgesic Vioxx[750]), many companies now send regulators and investigators large amounts of "toxicity" data labeled as "Serious or Unanticipated Events"— whether or not the label is appropriate. The problem is that regulators and investigators can be buried with voluminous reports that are not useful.

Any adverse event a patient experiences while on a trial will generate an adverse event report. Often, the event reported is simply tumor progression, is highly unlikely to be related to the drug, or is an already well-known drug side effect.

Typically, an initial form on the event is then followed by a flurry of follow-up forms. Regulators and investigators may then receive several copies of the same form. Regulators hire large numbers of people just to track and assess the forms received and investigators rapidly become overwhelmed and numb.

The company cannot be accused on hiding any data, but the data they send out is not helpful. Any important toxicity information may be easily lost or overlooked in a vast sea of irrelevant reports.

I was listed as a participant on a trial of the immunotherapy drug atezolizumab. I was not the principal investigator for our center, and I did not end up entering any patients on the study. Nevertheless, between August 2018 and October 2020 I received more than 1,000 emails from the sponsoring company detailing reports of "serious events" with the drug. Very few of these reports identified a previously unrecognized atezolizumab side effect. The sponsor judged the overwhelming majority to be neither unexpected nor an issue caused by atezolizumab. A few were thought to have been caused by atezolizumab but were already well-known immunotherapy side effects.

Nevertheless, as an investigator, the sponsor charged me to decide whether I needed to submit each report to my local IRB. I was also required to archive a copy of each submission in my study file. Let me stress that the sponsor of this trial is not necessarily any worse than any other company sponsoring clinical research, and I am only using this as one illustrative example.

There have been exploratory initiatives to try to lessen this problem. Dr. Richard Pazdur recently outlined the FDA's approach to reduce the deluge of unhelpful data sent to them. They requested that companies only send them a summary of adverse event data that they feel are an important toxicity, rather than letting them know every time a patient stubs her toe.

However, uptake on this request has apparently been slow.

Compliance vs. progress: Clinical research projects may be audited by study sponsors, regulators, or IRB's. Deficiencies are usually identified. [797] In many cases, the finding of these "deficiencies" will necessitate the time and expense of corrective action.

The results of the trial may be discarded if the deficiencies are judged to be major. For example, one of my MD Anderson colleagues was given a $50,000 grant to review patient charts for clinical information that was of potential scientific importance. All the patients signed a consent form to have their data reviewed, but the IRB later decided that the wording of the consent form was not specific enough to cover all aspects of the data collected. The IRB told my colleague to discard all the data collected: one year's worth of scientific work that could have contributed to progress.

Fully compliant investigators may—and often do—publish completely negative and unhelpful data. But very important data may have to be discarded over a minor wording issue. Has compliance usurped progress as the primary, most important objective of clinical research?[825]

Unquestionably, noncompliance can cause serious harm, but I am concerned that we have gone too far. We may not be properly weighing the negative impact of regulatory actions.

Funding of clinical research, intellectual property, and contracts: In my early days as an oncologist, I could run a clinical trial without any formal funding. I would personally write the protocol, guide it through the

IRB approval process, collect all needed data from patient medical records, analyze, and publish the data. However, that changed with the widespread adoption and increasingly stringent interpretation of Good Clinical Practice[799] regulations over the past two decades. These regulations require the collection of so much additional detailed data that study personnel have become essential. That in turn means that funding is essential.

A few clinical trials can be funded directly or indirectly by research grants.[clxx] But most require funding by the pharmaceutical company that makes the drug. That creates several problems. The first is that pharmaceutical company funding increases the risk of bias because you are not completely insulated from the interests of the company.

The second issue is more complex. It is often notoriously difficult to resolve patent issues if a trial involves combining agents belonging to different companies. They often have competing or overlapping interests.[844]

The third issue is that the center or institution must sign a contract with the sponsoring company.[845] This usually takes several months, but can literally take years, as company lawyers and institutional lawyers argue over the fine wording in the contract. Furthermore, the amount of time required to negotiate a contract is significantly longer if the trial involves a Clinical Research Organization (CRO).[846]

Some institutions have succeeded in simplifying the process with selected sponsors by agreeing upon a master contract that might apply to multiple different studies that they worked on together. Where these have been implemented, they have markedly shortened the time it takes to complete contract terms.[846]

Just formulating the institution's contract requirements can be daunting. In the early days of clinical research, it was understood that different groups would be playing a role, usually at no added charge. Most people wanted to contribute to research progress, and research participation was expected in academic centers. However, as potential participants became overloaded with ever-increasing demands, they first began to prioritize specific studies.

[clxx] It takes a lot of work to write a grant application to support a clinical trial, and the probability that the application will be successful is usually small. Writing a grant application to fund the study also means that it will usually take 1–3 years or more from the time you initiated work on the idea until you can enter your first patient.

To improve the priority attached to their study, some sponsors began to offer payment for the support. This then became the understood norm for pharmaceutical company studies. Over time, multiple different groups (IRB's, hospital pharmacies, radiologists, and others) began to charge for their contribution. This meant that in each participating center, there were separate negotiations with each participating subgroup in the building of the final contract. This can take a lot of time.

Not only does this take time, but it also increases the cost of clinical trials. Further increasing clinical trials costs is the fact that most institutions also charge "overhead." For every dollar that the sponsor must pay the institution, this overhead is added on. It is typically a 25% surcharge on the overall budget costs for the trial,[847] but some institutions may charge far more.

It also changed motivation and the ability to get others to co-operate. It is one thing to be internally motivated by an interest in research, and another to be motivated by financial compensation. In his book *Drive*, Daniel Pink quotes examples of young children who love art.[848] They are motivated by this love. However, if someone then pays them to do the art, the price required to keep the child interested becomes higher and higher as the promise of payment has killed their internal motivation. They lose their inner drive to create art for the simple love of doing it. Eventually, no payment amount will be high enough to motivate them.

Similarly, it becomes increasingly difficult to keep people interested in doing clinical research when the driving force has changed from internal motivation to external rewards. Whether these external rewards are direct financial compensation or academic promotion, they begin to lose their appeal, and it becomes harder and harder to get the collaboration needed for successful clinical research.

Use of real-world data: While clinical trials have typically been used to demonstrate that a new agent is effective, there has been a push to use "real-world data." This means that the drug's efficacy and toxicity would be determined in standard practice. Such real-world data can certainly be helpful in confirming that a drug is effective, but problems can arise if the real-world data are used as the main "proof" of benefit. There are too many potential biases at play.[849] Apparent benefit of the agent may be due to the selection of patients or could be due to other things one is doing while taking the new drug. For example, in Chapter 1 I discussed real-world data suggesting that various vitamin supplements and aspirin might reduce the risk of developing cancer, but randomized clinical trials have generally failed to confirm this.

I favor formal clinical trials to establish efficacy and safety. Real-world data, especially if gathered in a rigorous way, can be a useful supplement.

The shrinking pool of clinical investigators: Clinical trials require physician clinical investigators. However, the number of clinical investigators is rapidly declining. The drop in <u>experienced</u> investigators is particularly concerning.[850] When I started my oncology training in 1976, participating in clinical research was exciting and intellectually rewarding. However, it has become progressively more challenging.[524,745,757,759,825,829,851] Investigators have described the experience of activating and running a clinical trial as "traumatic" and "nightmarish," and young investigators are increasingly deciding not to put themselves through this.[851]

I have also seen respected colleagues threatened, denigrated, and intimidated by overzealous newspaper reporters,[842] IRBs, and regulators. Minor errors in documentation have spawned accusations of major misdeeds, and deviations that have minimal or no importance have been portrayed as huge threats to patient safety.[842] Researchers have committed suicide when facing overblown accusations,[852] and others have told me they had contemplated it.

If we want to maintain progress against cancer, we must fix this.

Drug approval processes: There are other speed bumps on the road to access to new therapies. Once companies have proven that their drug is safe and effective, they must apply to regulatory agencies for marketing approval.

Average times to complete the reviews vary by jurisdiction. They are all too long. The FDA takes 10.6 months and the European Medicines Agency (EMA)[clxxi] takes 12. For Health Canada, it's 12.9 months.[853] These times are close to the goals set by the different agencies (10 months for a standard review and 6 months for a "priority review" for the FDA[854] compared to

---

[clxxi] The EMA reviews medications for the European Union members, as well as Iceland, Norway, and Liechtenstein.

12 months for a standard review and 7 months for an "accelerated review" for Health Canada[855]).

These review times may not seem long. Nevertheless, they translate into a substantial number of years of life lost.[706,707] Any improvement could pay dividends. The experience with COVID-19 tells us that faster is feasible: the FDA granted Emergency Use Authorization for the Pfizer COVID-19 vaccine just 21 days after submission of the application.[856]

Canada faces an additional problem. Companies typically apply for Canadian approval <u>after</u> applying to the FDA and the EMA. For anticancer drugs approved between 2008 and 2013, Canadian approval was sought an average of more than 10 months after the US and more than 8 months after the EMA.[857] The delay in approval of new medications in Canada compared to the US and Europe may be growing over time.[853,858] The average delay in Canada was 1.3 years for medication approved from 2012 to 2019.[858]

Companies delay applying to Canada since it is a much smaller market than the US or Europe.[clxxii] The approval request process differs for different countries. It is costly and time consuming to put together an application for each country. Countries also charge hefty fees to process the application.[clxxiii] [859-861] Consequently, companies will apply first in jurisdictions where they are likely to earn higher profits.

The EMA has made European countries more attractive. An EMA approval permits marketing to more than 500 million people in twenty-seven different countries. One option for Canada with its relatively small population would be to partner with the US or Europe, and thus receive applications sooner.

There has been a move in that direction for anticancer drugs. "Project Orbis" was launched by the FDA Oncology Center of Excellence in 2019 to provide "a framework for concurrent submission and review of oncology products among international partners," including Canada and Australia.[862] The first year led to thirty-eight approvals of cancer therapies,

---

[clxxii] The population of Canada is about 38 million people, compared to about 333 million in the US and more than 500 million served by the EMA.
[clxxiii] In 2021, the application fee was $2,875,842 USD for the FDA, €296,500 for the European Medicines Agency and $437,009 CAD for Health Canada.

including eight by Health Canada.[863] It is too early to assess the impact of this initiative on drug access in Canada and other countries, but it is an encouraging start.

Drug funding processes: After the drug is approved by regulatory agencies, the next speed bump is the drug funding process. The US sidesteps this issue. By the Medicare Prescription Drug, Improvement and Modernization Act of 2003, US Medicare automatically accepts the drug price set by the manufacturer, without negotiating.[864-866] As I discuss in Chapter 13, this plays a major role in the US having the world's highest drug prices,[867] but it also means that Americans get the fastest access to effective new drugs.

In contrast, after the EMA approves the drug for marketing, each individual country negotiates a price with the drug company. The average time from EMA approval to reimbursement is 16.6 months, ranging from 4.2 months for Germany to 27 months for Poland.[868]

Once a drug is approved by Health Canada, Canadian patients may access it if they have private insurance or if they have the resources to pay for it. However, most patients instead must wait until the drug is funded by their provincial healthcare plan. Some drugs are provided free by the company through a compassionate release program while negotiations with government payors are underway.

For a provincial government healthcare plan to pay for an anticancer drug, the Canadian Agency for Drugs and Technologies in Health (CADTH) must first conduct a Health Technology Assessment (HTA) of the drug.[clxxiv] Once a company has prepared and submitted their CADTH application[clxxv], CADTH takes about seven months to complete their review.[869]

CADTH assesses the drug's efficacy, but also calculates the cost per "quality adjusted life-year" (QALY) gained.[clxxvi] CADTH may approve the drug, but generally makes their approval conditional on the price being

---

[clxxiv] The province of Quebec has its own separate funding processes.

[clxxv] With an application fee of $74,000.

[clxxvi] One life-year is one patient living for one year. A QALY adjusts this for the impact of disease symptoms and therapy side effects.

brought down to less than $50,000 per QALY. To do this, some companies would have to reduce the price of their drug by as much as 80% or more.[870]

This puts a disturbingly low value on a human life. This value of $50,000 per QALY gained is the same value that was advocated in the early 1970s.[871] Because of inflation, a 2021 Canadian dollar was worth less than 20% the amount of a 1975 dollar.[872] By not adjusting the cost per QALY for inflation, CADTH has devalued a Canadian life by 80% compared to 50 years ago. What basis do they have to do this?

Adding to the problem is the fact that calculation of a QALY requires calculation of quality of life (QOL). QOL assessments can be imprecise, flawed and biased.[873] Furthermore, a drug must have undergone a randomized trial comparing it to a standard therapy. Randomized clinical trials can be misleading in several ways,[430] and if a drug is so good that it was granted expedited approval without a randomized trial, there will not be any reliable comparator data.

In addition, in many randomized trials patients receiving the standard therapy may be "crossed over" to the new experimental therapy when their tumor worsens. This may make the standard therapy arm look better than it really is and results in an underestimation of the true relative value of the new therapy.[708] Crossover introduces substantial uncertainty into the value of QALYs gained.[874,875] CADTH generally does a good job of evaluating efficacy of a new therapy, but its approval based on a price of less than $50,000 per QALY gained makes it difficult to take them seriously.

CADTH's recommendation to approve a drug "conditional on adjustment of the price to $50,000 per QALY" is sent to the Pan-Canadian Pharmaceutical Alliance (PCPA). PCPA then conducts confidential price negotiations with the drug company on behalf of the Canadian provinces and territories.[clxxvii] The target time to complete this is 6.5 months.[clxxviii] [876] While the price negotiated is confidential—so that a special deal for Canada will not adversely impact prices a company can charge in other countries—the negotiations result in rebates. On average, these reduce the

---

[clxxvii] PCPA is not bound to the price recommended by CADTH, and probably generally exceeds it.

[clxxviii] Individual provinces or territories may take longer to finalize negotiations.

net price paid by government payors to about 70% of the official Canadian list price.[877]

In sum, the delays discussed above—in companies applying to Health Canada, the time taken for the Health Canada review process, and the CADTH/PCPA funding processes—mean a Canadian patient may not have access to an effective cancer medication until more than two years after an American patient. The result is that thousands of Canadian years of life may be lost.[706,707]

The Canadian situation may get substantially worse. In Canada, companies face a Kafkaesque situation with yet another government body also overseeing drug prices, and with targets that differ from those of CADTH. The Patented Medicine Prices Review Board (PMPRB) acts in parallel to CADTH and PCPA. PMPRB is a quasi-judicial body that sets maximum prices allowed for patented medicines. In their deliberations, PMPRB has previously used the drug's average list price in the US, Switzerland, the United Kingdom, France, Germany, Italy, and Sweden to calculate the maximum target Canadian price.

The PMPRB was recently charged with further reducing Canadian drug prices. In 2019, Canada had the 6th highest per capita pharmaceutical sales among 28 reporting OECD[clxxix] countries[878] and had the 4th highest average price for patented medicines.[879] The high average price paid by Canadians is the reason that there is only a ten month delay between company applications to the FDA for drug approval and applications to the much smaller Canadian market. The bad news is that Canadians pay high drug prices, but the good news is that these Canadian high prices gain them faster access to effective new drugs.

In the 2015 Canadian federal election, the Liberals promised voters that they would bring down drug prices. In 2017, they mandated PMPRB to do this.[880] In response, PMPRB proposed to eliminate the two countries with the highest drug prices (US and Switzerland) from Canada's list of comparators and to add five countries with lower prices (Australia, Belgium Japan, Norway, Spain, and the Netherlands).

[clxxix] Organization for Economic Co-operation and Development. Comparison was at the US$ exchange rate.

In their presentation to the Canadian public to justify the change, PMPRB noted that health spending was increasing as a percent of Canada's gross domestic product (GDP), and that from 2013 to 2017, the spending on patented medicines had increased from 6.3% to 7.5% of all healthcare spending.[867] They failed to point out that their own data[879] indicated that patented medicine sales in 2017 were about the same percent of GDP as they were in 2003. Patented medicine sales as a percent of GDP in 2013— the comparator year that they chose to justify new price controls—were the lowest they had been since 2001. Furthermore, patented medicine sales in 2019 were about the same percent of GDP as they had been in 2002.[879] In their presentation, did PMPRB deliberately attempt to mislead the Canadian public?

With the new comparator countries included, the maximum amount a company could charge for their drug in Canada would be reduced by about 20%. To make the pill more bitter for drug companies, the PMPRB devised a complex set of rules that would mandate substantial further price reductions as sales rise.[881] This would mean both marked uncertainty for companies as well as an actual price reduction much greater than the initial 20% target.[882] The overall impact on prices of patented drugs in Canada would be an estimated reduction of 50-80%.[882]

In their presentation to the Canadian public, the PMPRB stated that "Countries with lower patented drug prices than Canada may have greater availability of new medicines". [867] When Dr John-Peter Bradford (CEO of the Life Saving Therapies Network) and I met with them in early 2021 to express our concern, they reiterated their confidence that cutting drug prices in Canada would not have any meaningful impact on drug access. However, our analysis of PMPRB data[867] and other analyses[883-885] clearly indicate that this type of price regulation results in delays in launching new medications. Aggressively reducing drug prices can lead to withdrawal of medications that are already marketed.[884]

Many of us were concerned that the proposed PMPRB changes would mean further delays in access to cancer medications for Canadian patients.[886] Rather than facing the expense of applying to Health Canada, companies would instead use their resources to first concentrate on other countries with a higher potential for profit. They might even avoid marketing their drugs in Canada if Canada's new low prices threatened to drag down prices internationally.

Pharmaceutical companies have stated they will delay or avoid drug launches in Canada if the proposed PMPRB changes proceed.[887] Not only would these new drugs no longer be available for sale in Canada, they would also no longer be available through compassionate access programs or clinical trials.[887]

After the 2017 announcement of the planned new PMPRB approach, there was a rapid drop in the proportion of new drugs launched globally that were also launched in Canada.[888] Companies said they would also delay availability of COVID-19 vaccines to Canadians unless they were exempted from the planned PMPRB measures.[889] After an avoidable delay, the vaccines were exempted, so Canadian access was much faster than it would have been without the exemption—but there was still a delay.

PMPRB claimed that their proposed new rules would have no negative impact on drug availability in Canada. The data they presented did not support their position.[867] Assessing PMPRB data, a 20% decrease in price would be expected to be associated with reduced drug availability. The countries with little apparent impact of low prices on access were European nations that were afforded access solely as a function of their EMA affiliation.[clxxx][867]

Patient advocacy groups strongly opposed the PMPRB position since they recognized how it would impact access to effective new therapies.[clxxxi] This opposition and court challenges resulted in putting off planned activation of the policy until January 2022,[890] and in late December 2021,

---

[clxxx] See slide 43 of the PMPRB public webinar. Of non-EMA countries, all seven with lower drug prices than Canada's had access to fewer drugs. Of all EMA members with drug prices less than or equal to the PMPRB target of 80% of the current Canadian price, 70% had access to fewer drugs than Canada. Canada would not have EMA membership to help preserve access despite lower prices.

[clxxxi] Internal PMPRB memos came to light that referred to patient advocacy groups as being "in the pocket of industry" and involved in "disinformation campaigns". In emails to colleagues, PMPRB's director of policy and economic analysis stated that "industry has been sucking Canada for decades." (https://www.macdonaldlaurier.ca/time-pmprb-go/).

the Canadian government announced that there would be a further delay in implementation until July, 2022.

However, this isn't good enough. The PMPRB proposal needs to be permanently trashed. If it proceeds, it will become a case study in how government ineptitude costs lives. Not only may proposed PMPRB rule changes delay or decrease access to effective new therapies, but they may also result in reduced clinical research in Canada[887,891] and reduced access of Canadian patients to new agents on a free, compassionate release basis.[887]

Some authors have expressed concern about PMPRB's overall behavior and have called for it to be dissolved.[892] I strongly agree with this. Canada does not need both CADTH and PMPRB overseeing drug prices. I disagree with CADTH's $50,000 per QALY metric, but I have even greater concerns with PMPRB.

Conclusions: In summary, there is no simple fix for the innumerable speed bumps slowing progress on the road to developing effective therapies. "Every system is perfectly designed to obtain exactly the results that it gets."[clxxxii] If we agree that faster progress is essential, we need to redesign the system.

The speed bumps did not appear by chance. They were put there for a reason. We need systems that protect patients, protect data integrity, and ensure that new, approved drugs are safe and effective. However, we must also recognize that these speed bumps kill. They delay access to therapies that could save lives and alleviate suffering.

They are also a major driving force in the rapid escalation of the costs of cancer care.

We need the safeguards that these speed bumps provide but we need to reengineer them. They must provide safety and data integrity while permitting much faster progress. The experience with the development of COVID-19 vaccines is unequivocal proof that this is possible if we put our minds to it. The world went from initial recognition of the COVID-19 problem to access to effective new vaccines in under one year. If we can do this with COVID-19, we can also markedly shorten the time required to

---

[clxxxii] There is some uncertainty about who first said this: Arthur Jones, W. Edwards Deming, Paul Batalden, or Donald Berwick: https://deming.org/quotes/10141/ Accessed 2020/12/22.

make effective new therapies available to patients with metastatic cancers and other lethal diseases.

Cancer is a much tougher problem than COVID-19. It can be more difficult to identify key drug targets in cancer, and it can be difficult to find and recruit the unique subpopulations of patients required for the clinical trials of personalized cancer therapies. But these challenges are responsible for only a very small part of the delays we face. Governments are either being dishonest with us that safe, effective COVID-19 vaccines could be developed in under one year—and I do not think they are being dishonest—or else they would be dishonest in telling us that we cannot also markedly accelerate the development of new therapies for much more lethal diseases like cancer.

The twelfth reason that cancer still sucks is that we have been highly ineffective in tackling the issues that slow progress. We need much faster progress <u>now</u>.

As I said at the start of the chapter, this is a call to action. Our regulators are a highly dedicated group working to assure protection while at the same time enabling progress. But no matter how highly dedicated they are, they cannot fix this problem without our help. If you agree with me and believe that changes are needed, please talk to your government representative.

More information can also be found on the website for our Life-Saving Therapies Network at <u>https://lifesavingtherapies.com/</u>.

— 13 —

If I had a Million Dollars: The Explosion of Cancer Therapy Costs

There have been major advances in the treatment of cancer, but the thirteenth reason that cancer still sucks is that the prices of new therapies are rising explosively.

Short Primer

Causes of high drug prices: By 2012, the average price for a year of treatment with cancer drugs in the US rose to $100,000 from less than $10,000 in the 1990s. It has continued to rise rapidly since then. Rising drug prices have created major issues for governments, insurance companies, and patients. In addition, a growing and aging population means that the number of cancer patients is increasing. Furthermore, more effective therapies mean that patients who could not previously have been treated now can be. These new, effective therapies also mean that patients may be treated for a longer time.

There are several factors contributing to high drug prices. The cost of drug development is one very important one. It cost an estimated $230 million to take a drug from discovery to approval and to marketing in 1987. By 2013, the cost had risen to about $2.9 billion. As I discussed in Chapter 11, this escalation in drug development costs has been driven largely by the rapidly rising cost of conducting the clinical trials required for drug approval.

Once the drug is approved, the drug development costs must be recovered through profits from sales. Otherwise, the company would go bankrupt. Companies would stop investing in new drug development, and progress would come to a halt.

Another factor contributing to rising drug prices is the "personalization" of therapy. In the 1990s, a new lung cancer drug might be used in all patients with lung cancer. The drug development costs could then be recouped by sales to many patients. However, the drug might only help 10% of the patients treated. With personalization, oncologists can predict which patients would be most likely to benefit. Hence, the drug is sold for use in only the 10% of patients it would probably help. However, the drug price goes up because the development costs must be recovered from a far smaller number of patients.

Short patent protection—usually twenty years—also contributes. If it takes twelve to fifteen years to get a drug approved, then all drug development costs must be recouped in the five to eight years of remaining patent protection after approval.

Subverted market forces contribute. In an ideal capitalist system, one company might cut the price of their drug to compete with another company. Instead, what happens is that one company decides to test the waters by charging a much higher price than anyone has charged before. When they get away with it, other companies follow their example and do the same thing. People get used to rapidly rising drug prices and come to accept them as a reality of life.

For two drugs that work in different ways, one may be used first against an incurable cancer, with a switch to the other when the first one stops working. Hence, there are no market forces to bring down drug prices, as there would be if it was a simple choice of one drug vs. another. Even when drugs from different companies work in the same way, there is often little competition over price because there appears to be a subtle unspoken agreement not to lower drug prices.

The existence of a pharmaceutical company "oligopoly" also plays a role. If a small company develops a promising new drug, the small company very often will be taken over by a big established company. This means that there are only a few large companies running most of the show.

The US has the highest drug prices in the world. Legislation directly contributes to this. The 2003 Medicare Prescription Drug, Improvement and Modernization Act made it illegal for Medicare to negotiate prices with drug companies. A company unilaterally sets its price and Medicare pays it! Similarly, the US Patient-Centered Outcomes Research Institute cannot use a drug's cost-effectiveness to make recommendations on its use. I do not know what one might call this, but it certainly is not capitalism. In a capitalist system, a buyer would be permitted to negotiate a price with the seller.

How can laws like this exist? One reason is that the pharmaceutical industry spends more than one billion dollars a year on lobbying and US political contributions. Politicians gladly accept this money.

When I worked at Houston's MD Anderson Cancer Center, I was told that Medicare paid only $0.38 for every dollar that I billed.[clxxxiii] If Medicare can discount physician billings to this degree, think what it could do on drug pricing if it were only permitted to negotiate!

High drug prices in one country do not only affect that country. There is a ripple effect. Many countries use drug prices in other countries to help set their pricing. A very high price in any jurisdiction can help drag up prices worldwide. Some authors have claimed that by paying high prices, the US is subsidizing drugs for the rest of the world. On the flip side, one could say instead that since US prices are inflating worldwide prices, the rest of the world is helping subsidize pharmaceutical company contributions to US politicians. Countries that bring down their drug prices compared to other countries can experience increasing difficulties and delays in accessing effective new drugs, as I discussed in Chapter 12.

Once a drug's patent expires, generic copies can compete with it. Generics have the potential to bring down prices, but often do not. Some pharma companies pay generic companies to not sell a competing drug. Other companies use a process called "evergreening" to maintain and extend their patents. They patent a new route of drug administration, a new indication for the drug, or a minor variant of the drug.

[clxxxiii]When I was young, Chris Forrester, an old farmer I worked with at Rideau River Provincial Park, Kemptville, Ontario told me, "Quit complaining. You're being paid almost as much as a good man!"

Drug prices are impacted by who pays for them. Several years ago, in an op-ed in the Ottawa Citizen I stated that I felt that the basic problem in both Canada and the US is that someone else (insurance or government) pays for most of the healthcare. The result in Canada has been delays and rationing. The US experiences massive spending, with very high administrative costs aimed at limiting utilization.[clxxxiv]

Many years ago, my grandfather told me about his cousin, Dr. Campbell Keenan who was a surgeon at Montreal's Royal Victoria Hospital in the early part of the twentieth century. A man dressed in rags came to him and asked how much it would cost to have his hernia repaired. When Dr. Keenan replied, "How much do you have in your pocket?" the man pulled out $25. "Then that is how much it will cost you," said Dr. Keenan. What Dr. Keenan did not know was that the patient was a wealthy lumber baron who had come from Ottawa to Montreal in his own private railway car. He had realized long ago that if he looked rich, he would be expected to pay much more than if he looked poor.

When I went to our family physician, Dr. Wilbert Byers, before Canadian Medicare came into effect in 1968, Dr. Byers would turn to my father and ask, "How's the landscaping business, Archie?" If my father replied, "Not so good," Dr. Byers would answer, "We'll get you next time, Archie." If my father answered, "It's going well," Dr. Byers would respond, "That will be $10."

I suspect prices of new therapies would drop substantially if all patients, rather than insurance or healthcare systems, had to pay for treatments out of pocket. In a free market economy, when buyers avoid a product due to price, the price generally falls, with true product development and production costs limiting how low it goes. Some would remain unable to afford it no matter how much the price dropped. But others who were unable or unwilling to pay a high price would elect to use their personal resources to access the therapy if the price dropped low enough.

When insurance or healthcare systems pay for drugs, personal resources are no longer a limiting factor. An individual patient does not decide the value of the product. So, this market force is mitigated.

---

clxxxiv I expand on this in Chapter 14.

Patients, oncologists, and payors may also overestimate a drug's benefits. The "magic of statistical power" in a trial with a very large number of patients may make a drug look more impressive than it is. A high level of "statistical significance" may belie a survival gain of only a few weeks.[clxxxv]

Under some circumstances, US oncologists are paid more for administering expensive drugs than for administering cheaper drugs. This is seen by some as a conflict of interest, and I agree that it could be.

Drug prices are also driven up in other ways. One way is the cut that intermediaries take. Examples are distributors and pharmacies who buy from drug companies and then sell to hospitals and patients. Prices are also kept high by restrictions on importing drugs. In the late days of his presidency, Donald Trump signed an executive order enabling Americans to buy cheaper drugs from Canada, but early indications are that Canada would not be willing to sell directly to Americans. Selling drugs directly to Americans could create shortages in Canada and would probably drive-up Canadian drug prices.

Solutions: To date, there have been two main strategies suggested to reduce high drug costs: either some form of government price controls or addressing the restraints on market forces. Some have called for "value-based" government price controls. However, it is very difficult to define value. If you have cancer, you may give the drug a much higher value than what a government is willing to pay. In addition, government price controls rarely work since they typically cause more problems than they solve. They create shortages and end runs that drive-up prices rather than reducing them. Government price controls also frequently involve "exceptions," by which the politically powerful are spared.

I feel that the most important step is to reduce the cost of new drug development markedly. If drug prices are forced down without substantially reducing the cost of new drug development, then companies will be unable to recoup these costs. If companies cannot recoup these costs, then investment will decrease, and progress will halt.

[clxxxv] See Chapter 12 for a more extensive discussion of this.

High drug prices have been a challenge for payors. However, my frequent collaborator and former colleague Dr. Razelle Kurzrock[clxxxvi] has reminded me that these high prices stimulate investment. Hence, they have been key in driving the rapid development of effective new cancer therapies. The data available support this view: lower prices would translate into fewer new drugs developed.[893]

Permitting US Medicare to negotiate drug prices with pharmaceutical companies could also help.[clxxxvii] So could outlawing the practice of companies bribing generic companies to keep competitors out of the market. Rich contributions to politicians buy influence, and this must stop. The recipients are every bit as much at fault as the contributors.

We need more extensive comparisons of drugs that have the same mechanism of action. If the efficacy and toxicity are proven to be similar, then payors can make coverage decisions based on price.

Further Details and References

In this section, I will go into more detail, and add some documentation.

The rising costs of cancer care are not sustainable.[865,894-897] When he gave the Karnofsky lecture in 1976, the great cancer pioneer Emil J Freireich presented "Freireich's Laws." One law was "Always be prepared for success."[796] We face the current dilemma in part because we have not been prepared for success.

Prices of anticancer drugs have risen rapidly over the past several decades.[898] Before 2000, the average price of anticancer drugs for a year of therapy in the US was under $10,000. By 2012, it had risen to more than $100,000.[899] By 2014, the average price of treatment for a year with new oral cancer medicines was more than $135,000.[900] Examples of individual therapies are presented in Table 1.

[clxxxvi] Previously at MD Anderson Cancer Center in Houston, then at University of California San Diego for 9 years, now at Medical College of Wisconsin.
[clxxxvii] Although as I discuss in Chapter 12, it could also delay drug access for American patients.

Table 1. Average cost in the US for a year of illustrative therapies			
Drug	Cancer Treated*	Year Assessed	Average cost to treat one patient for a year**
Ipilimumab[865]	Melanoma	2012	$120,000***
Provenge[865]	Prostate	2012	$90,000****
Bevacizumab[865]	Various	2012	$90,000
Abraxane[865]	Breast	2012	$80,000
Lenalidomide[865]	Myeloma	2012	$90,000
Bortezomid[865]	Myeloma	2012	$60,000
Imatinib[865]	CML	2012	$70,000
Dasatinib[865]	CML	2012	$110,000
Brigatinib[901]	Lung	2016–17	$154,000
Cabozantinib[901]	Kidney	2016-17	$182,000
Enasidenib[901]	AML	2016-17	$302,000
Midostaurin[901]	AML	2016-17	$190,000
Nertatinib[901]	Breast	2016-17	$127,000
Niraparib[901]	Ovary	2016-17	$173,000
Ribociclib[901]	Breast	2016-17	$102,000
Rucaparib[901]	Ovary	2016-17	$242,000
Venetoclax[901]	CLL	2016-17	$90,000
CAR T-cells[900]	ALL	2017	$475,000*****
* CML: chronic myelogenous leukemia; AML: acute myelogenous leukemia; CLL: chronic lymphocytic leukemia; ALL: acute lymphoblastic leukemia (in children) **for 2016–17 assessments, monthly costs reported in the reference were converted to annual costs ***only four treatments are given ****only three treatments are given *****only one treatment is given			

Moreover, some expensive therapies are more effective when given in combination with other expensive therapies. This compounds the problem. In addition, since new, expensive therapies are more effective than older, cheaper therapies, they may be given for a longer time. This further increases costs.[902] Furthermore, costs of diagnostic tests, radiotherapy, surgery, and physician incomes are rising at the same time.[900] These combined factors are putting tremendous pressure on healthcare systems.

The rapidly escalating costs are also creating huge problems for individual patients. A cancer diagnosis brings with it a greatly increased risk of personal bankruptcy.[782] The proportion of personal bankruptcies in the US that were due to health care costs increased from 46% in 2001 to 69% in 2007.[903] The high costs of treatment also mean that patients may delay or skip treatments that they can no longer afford, and this may result in an increased probability of later hospitalization.[900]

Causes of rapidly escalating drug costs: As noted in Chapter 12, "Every system is perfectly designed to get exactly the results that it gets."[clxxxviii] That is certainly the case with exploding drug costs. Contributing factors include the following:

High costs of drug development: As discussed earlier, the average cost of bringing a drug from discovery to approval has risen from around $4 million in 1962 and $231 million in 1987[773] to around $2.9 billion in 2013.[756] The major factor driving this is the exorbitant cost of the required clinical trials.[756] The costs directly attributable to each individual drug assessed are much lower, but only 3–8% of cancer drugs that enter clinical trials are approved for marketing.[775-777] The cumulative costs associated with the 92–97% of drugs that don't make it are very high. The sales of the very few drugs that are approved also must cover the development costs of failed drugs. No sane investor would invest in drug development companies if the cost of failures could not be recouped.

Personalization of therapy: When the chemotherapy agent cisplatin was approved in the late 1970s, it could be used for most patients with

clxxxviii   Quotation source uncertain: Arthur Jones, W. Edwards Deming, Paul Batalden, or Donald Berwick: https://deming.org/quotes/10141/ Accessed 2020/12/22 advanced cancers of the lung, ovary, bladder, head and neck, testicle, esophagus, and others. Hence, the potential market was very large.

Personalization has changed this. As therapies become more personalized, the number of patients for whom a drug can be used is rapidly shrinking.[759,778] Personalized therapies generally work much better than non-personalized therapies.[759] They are, however, more expensive than non-personalized therapies. Here's why. A new drug developed for lung cancer patients with an ROS1 fusion gene (for example) may be useful in only the 1% of lung cancer patients who have this gene alteration. The high costs of developing this drug must be recouped from just 1% of patients with metastatic lung cancer. If that's the only use for the drug, it makes sense to price the treatment up to 100 times higher than the price charged for a drug that could be used in all lung cancer patients. The smaller the potential market, the higher the minimum required price for economic viability.

Short patent life after approval: Patenting a drug gives a company protection for twenty years. However, the requirement for large clinical trials means that it takes an average of twelve to fifteen years or more to bring a new drug from discovery to marketing.[754] Hence, high drug development costs must be recouped within the five to eight years of remaining patent protection. Cost recovery is unlikely to happen within five to eight years unless the drug price is set very high.[865]

"Comparative pricing" and subverted market forces: Pharmaceutical companies have argued that drug prices that may initially be set very high will eventually be corrected to a reasonable level by "market forces."[864,904] However, numerous factors subvert these market forces.

It would be illegal for two companies to get together and fix prices on a given commodity. But it is not illegal to use what I will call "comparative pricing." This is one of the many factors impairing market forces. The story of paclitaxel (Taxol) illustrates how this works: when paclitaxel was first marketed in 1994, Bristol-Myers Squibb (BMS) charged about $2,600 for a month of treatment.[898] At that time, this was much higher than the price charged for most other drugs. BMS justified this high price by the fact that the drug was very difficult to manufacture, and it had no remaining patent protection.[clxxxix]

Most other anticancer agents introduced up to that point cost only a few hundred dollars per month.[cxc][898] However, prices of a number of new agents released in 1996 suddenly were in the ballpark range of the high 1994 paclitaxel price, despite there being no underlying special justification. The prices rose because paclitaxel had set a new level.

So, in comparative pricing one company sends out price shock waves. Others follow closely as soon as the shock starts to settle. As pointed out by Carrera et al., "Unfortunately, high launch prices of cancer drugs are largely based on the prices of existing therapies (not necessarily competitors), rather than innovation or clinical effectiveness, such that patients may be paying exorbitant costs without the expectation of much benefit."[901]

The boiling frog: According to a fable, if you throw a frog into boiling water, it will jump out. However, if you put the frog into cool water, then heat it up gradually, the frog will not detect a problem. It will not jump out. In a similar manner, patients and payors have come to expect the rapid heating up of drug prices, and this helps enable these high prices.[901]

Oligopoly: There are only a few large companies successfully developing anticancer agents. They compete based on the relative merits of their agents, rather than competing on price. This permits the companies to set high prices without facing price competition.[864,865]

A major reason that there are only a few large companies competing is they typically buy small companies that have particularly promising agents. This way, the large company benefits from the efforts of the small firm and reduces its risks in bringing a new drug to market. Few small companies can successfully bring a drug to market since they cannot afford the high clinical research costs. Hence, high drug development costs decrease competition by favoring large companies with deep pockets.[745,757]

[clxxxix] Taxol was initially developed by the US National Cancer Institute (NCI) but was very difficult to manufacture. They handed it over to BMS. Despite it being past patent, BMS was given exclusive marketing rights for five years under the Hatch-Waxman Act in 1993.

[cxc] There were some exceptions. "Biologic" agents are expensive to manufacture. This justifies their high price.

Impaired competition between drugs: You might think that drugs with similar efficacy against a given cancer would compete based on price. However, these drugs are often used sequentially if they have different mechanisms of action. One is used for as long as it works, and this is followed by the other drug. Hence, if one drug were avoided as first-line therapy since it is more expensive, it would still get its chance as second-line therapy, without having to reduce its price.[865] This issue is particularly important in oncology as advanced cancers (unlike, for example, infections) can generally not be cured. So, first-line therapy will not make second-line therapy unnecessary.[865]

Frustratingly, there is little price competition even when drugs are <u>not</u> used sequentially and when there is minimal difference in how they work. For example, in Canada, the price for the EGFR inhibitors gefitinib and afatinib are very similar.[cxci][905]

The PD-1/PD-L1 inhibitor immunotherapy drugs pembrolizumab, nivolumab, atezolizumab, and durvalumab have similar mechanisms of action and toxicity. Some have "worked" in situations where others have not. However, any apparent differences in efficacy could be due to minor differences in study design or study patient selection.[430]

There are no studies that have directly compared one to the other, and hence no way of saying if one is truly better than another. Just like the EGFR inhibitors, these immunotherapy drugs have prices that are similar.[cxcii] Economic gaming theory explains the virtual price equivalence of agents with similar efficacy and mechanism of action: "subtle collusive behaviors that keep prices high for particular commodities for long periods of time, regardless of market forces."[904]

---

cxci   The EGFR inhibitor osimertinib works better and can work in patients who have failed one of these other drugs. It is also much more expensive, costing about $8,000 Canadian per month instead of the $2,000 per month for gefitinib or afatinib.

cxcii   As discussed earlier, personalization of therapy generally reduces potential patient numbers. That is not the case with PD-1/PD-L1 inhibitors. No biomarker has yet been discovered that permits reliable selection of patients for most malignancies. Consequently, as a class, they can be used broadly across large number of patients. Hence, personalization of therapy cannot be used as a justification for their high prices.

Legislation: It is no accident that the US has the highest drug prices in the world. The Medicare Prescription Drug, Improvement and Modernization Act of 2003 forbade US Medicare from negotiating drug prices with the manufacturer.[864-866] This means that US Medicare pays substantially more for drugs than other US government programs (Veterans Affairs, Department of Defense, Medicaid)[866] or healthcare systems in other countries.[906]

In American elections, it is customary for Republicans to attack Democrats by raising the specter of socialism.[cxciii] This implies that Republicans favor capitalism. However, in capitalist free market economies, the buyer and seller typically negotiate with each other to set the final sales price. How is it a free market economy if US Medicare cannot negotiate the purchase price? This anomaly creates an upward pressure that impacts the rest of the world. Many healthcare systems use the average price of a drug in a group of other countries to determine the price they will pay. The US price increases this average and sets a ceiling price for the globe.

The Patient-Centered Outcomes Research Institute (PCORI), created by the US Patient Protection and Affordable Care Act (PPACA) of 2010, is an independent, nonprofit organization. Its research uses cost-effectiveness analyses to inform healthcare decisions. However, in a Kafkaesque twist, the PPACA specifically prohibits PCORI or the Secretary of Health from using cost-effectiveness measures to determine coverage, reimbursement, or incentive programs.[865] As I will discuss later, cost-effectiveness calculations can be problematic, but refusing to consider them makes little sense.

Why has American legislation evolved in this way? What part was played by lobbying and campaign donations by pharmaceutical companies? On average the pharmaceutical industry and related groups spend $233 million annually lobbying the US federal government. An additional $414 million is contributed to presidential and congressional electoral candidates, national party committees, and outside spending groups. Another $877 million is contributed to state candidates and committees.[907] These contributions typically target senior congressional

---

cxciii Ideology-based behavior can have major negative consequences, as exemplified by the high Covid-19 death rates in the United States linked to the sentiment that wearing a mask and social distancing are anti-freedom.

legislators who draft healthcare laws and state committees supporting or opposing initiatives on drug pricing and regulation.[907]

Allowing Medicare to negotiate drug prices would save an estimated US $40 billion to $80 billion per year.[864] Hence, lobbying and campaign donations bring high returns to the pharmaceutical industry.

Some have claimed that Americans are subsidizing drugs for the world by paying the world's highest prices.[906,908] However, perhaps the main problem is that they are instead underwriting the billions of dollars in drug company profits. These are much higher than for other publicly traded companies,[909] and they support the rich contributions to the politicians who facilitate their high profits.[907]

Having said all of this, I must add that things are rarely simple. The US approach markedly escalates costs, but it improves accessibility. US Medicare will rapidly begin coverage of a new therapy as soon as the FDA approves it. In Canada, after a drug is approved it takes months or years of further evaluation and negotiations before provincial healthcare plans begin paying for the therapy. This delay is detrimental to Canadian patients.[707] In Europe, there are similar delays between approval and funding.

In addition, US Medicare would not necessarily "play fair" in any negotiations with drug companies, if it was permitted to negotiate. Many US healthcare providers have refused to care for Medicare patients due to low reimbursement levels that are imposed on them. If Medicare could negotiate, would it be "good faith," or would Medicare unilaterally impose a bargain basement price on the manufacturer?

It's a double-edged sword. Exorbitant prices hurt payors and patients, but low profit levels could impede development of effective new therapies.[893] Prices that are too low can also be a major factor contributing to shortages of existing drugs.[910]

Impaired impact of generics: When a drug's patent expires, cheaper generic versions may become available. You might think that this would promote competition and lower drug prices. However, generics have far less of an impact on cancer treatment costs than might be anticipated.[865] Several factors are at play here, including various strategies used by drug companies to delay generic competition.[864] For example, a company may pay its potential competitor not to launch the generic.[cxciv] [911]

A pharma company can also attempt to prolong a drug's patent through processes referred to as "evergreening," such as filing a patent on a different use or delivery system[cxcv] for the drug.[912] A new, potentially better, patent-protected version of the drug may be made available when the original patent expires. An example is Abraxane as a new version of paclitaxel.[865] In addition, by the time a drug's patent expires, it may have become "obsolete" due to the availability of a new, better alternative. If this happens, there is no longer much call for the generic.[865]

For "biologic" agents such as monoclonal antibodies (e.g., Herceptin), it is not possible to create a generic agent since one cannot create an exact copy of the original drug. The manufacturing processes to produce it are too complex to permit exact copies. A company may instead create a "biosimilar"—a new agent that is very similar to the original drug but is not identical to it. Because the biosimilar is not identical to the original agent, expensive clinical trials may be required to get the biosimilar approved,[913] and the price of the biosimilar needs to be high to recoup these clinical trial costs.[cxcvi]

[cxciv] Generics can help reduce drug prices, but there can be a downside. As noted in the previous section, drug prices that are too low <u>may</u> be a causative factor in drug shortages. Over the past several years, shortages of several drugs have developed. Most companies may not be interested in making a drug if its price is low relative to its cost. Hence, there may be no one to fill the void if a low-cost manufacturer runs into production difficulties for a drug. Also, the margin on the drug may be too low to permit the lone manufacturer to make a rapid recovery. Prices that are too high can be bad for healthcare systems, but so can prices that are too low.

[cxcv] For example, a company can extend patent protection by developing a new formulation administered by subcutaneous injection rather than by intravenous administration.

[cxcvi] This is unlike the situation with generics, where only very limited clinical trials are needed.

Overestimation of benefit of a new therapy: Most new drugs[cxcvii] are approved for marketing based on a "statistically significant" gain when compared to an older therapy, with a "p value less than 0.05."[cxcviii] However, the difference in outcome for the two therapies can be very small despite being statistically significant, particularly if the study assessing it was very large.[524,865] The new therapy may be heralded as an advance based solely on statistical significance.[864] The fact that it has gained regulatory approval may be interpreted as indicating it has high value,[914] but the actual size of the gain may be small. In any case, the correlation between the price set for new cancer drugs and the degree of benefit is very weak.[904,915]

Some insist that funding of new drugs must be linked to randomized clinical trials that show improved life expectancy or quality of life compared to older agents.[916] However, defining the true degree of benefit may be more difficult than it might seem. For example, a new therapy may exhibit a high degree of benefit in patients with an uncommon cancer. But it may not be feasible to conduct large randomized trials comparing the new therapy to older options since the cancer is uncommon.[759]

In addition, if "crossover" is permitted, a patient who received the older standard therapy might be switched to the new therapy when their cancer worsens. If the new therapy helped them when they were switched to it, then patients who started initially on the standard treatment might live as long as those who started first on the experimental treatment. Hence, crossover may preclude determining the new drug's true impact on survival.[708]

---

cxcvii An exception is made for drugs that result in tumor regression in a high proportion of patients who would not be expected to benefit from standard therapy. As discussed in Chapters 11 and 12, these drugs may be approved based on a breakthrough drug designation, without the requirement of a randomized comparison to other agents.

cxcviii A "p value less than 0.05" is based on a statistical calculation that indicates that the gain with the new drug is large enough compared to standard therapy that there is less than a 5% probability that its apparent superiority is due to chance alone. The more patients included in a trial, the higher the "statistical power." Because of the nuances of statistical power, a large study may have a p value less than 0.05 based on a very small gain in efficacy, while a small study will have a wider margin of uncertainty. The small study will need to demonstrate a much larger gain for the p value to be less than 0.05. The net result is that large studies can detect very small gains.

Some studies have avoided this problem by forbidding crossover. In my opinion, forbidding crossover is highly unethical: patients are then being denied the option of receiving a potentially effective new therapy simply because they agreed to participate in the clinical trial. Patients are then being used as experimental animals just to prove the benefit of the drug.[708] Clinical trials must first and foremost be about helping patients.[796] Not about satisfying misguided statisticians, regulators, clinicians, and payors.

Potential oncologist conflict of interest: There are financial incentives for American oncologists to prescribe more expensive treatments.[865] When they give chemotherapy in their offices, US Medicare reimburses them the "average sales price" for the drug plus a percentage add-on of this price to cover drug procurement costs, etc. The dollar value of this add-on reimbursement will be larger for more expensive therapies than for less expensive therapies. However, others have challenged the suggestion that the size of this add-on reimbursement influences prescribing.[917] Its true impact is uncertain.

The therapies are getting better: As noted above, since the efficacy of cancer therapies is increasing, a higher proportion of patients are candidates for treatment, and these treatments may be continued for much longer. This proportionately increases the amount spent on these treatments.

Price hikes by middlemen: Prices charged by pharmaceutical companies may be high, but these prices are then inflated further by intermediaries such as distributors, pharmacies, and hospitals.[864]

Post licensing price hikes: When drugs are first introduced, the price may be high. One might expect price to then fall over time. However, often they instead progressively rise post licensing.[918] For example, when Gleevec was first marketed for treatment of chronic myelogenous leukemia, it cost about $26,000 for a year of treatment, but by the time its patent expired, it cost $120,000 annually.[919] There are no new development costs to explain this. The company raises the price because it can.

Restrictions on drug imports: Since drugs may be cheaper in one country than in another, it might appear to make sense to import drugs from a cheaper country. However, in the US (and Canada), there have been marked restrictions on this, in part due to concerns about the quality of drugs from other countries.[920] In July 2020, US President Donald Trump signed an executive order allowing importation into the US of cheaper drugs from Canada.[921]

This might look like a good idea, but it could cause major problems for Canada. First, it could lead to drug shortages in Canada since the US population is ten times the size of the Canadian population. Canadian production of drugs might not keep up with demand from this much larger market.[922]

The second problem is that it might mean that a company would withdraw their drug from Canada, or else they would then charge the same high price as they do in the US. Canada's relatively small population accounts for about 10% of the North American market for medications. If the price of a drug were (for example) 20% less in Canada than in the US, then it would make more economic sense for a company to just abandon the small Canadian market than to see its drug prices drop by 20% in the large US market due to importation of the cheaper drug from Canada.

It is generally a positive thing if another country wants to import your goods, but unfettered importation of Canadian drugs into the US would ultimately accomplish little for the US, while being a disaster for Canada. Canada would experience either loss of access to the drug or else much higher prices. The US would effectively export their high drug prices to Canada. The Canadian government understands this, and has disallowed the export of key drugs.[922]

Solutions: The high cost of drugs is a difficult problem. There are no easy solutions and there are many contributing factors. Several authors have expressed an opinion on this, and they have essentially fallen into the two main themes of government controlling the price vs. reducing restraints on market forces.

Government price controls: Some authors have proposed that cancer drugs be viewed like food during a famine or lifesaving items during a catastrophe. They would then be assigned a "just price" or "fair price" that provides a reasonable profit to the producer but is affordable to the patient and to society.[864,904] Other authors have proposed "value-based pricing," where the price of the drug would be proportional to the amount of benefit it provided.[864,865,900,902,923] Another option has been for government investment in commercial research and development projects to permit the investing government leverage in ultimate pricing decisions.[924]

The pharmaceutical industry has argued (with what I feel is some justification) that government price regulation could stifle innovation.[864]

In addition, history tells us that government attempts to control prices ultimately do not go well, and often end up increasing prices rather than decreasing them.[925] The main problem is that price controls are attacking the symptom (high prices) rather than addressing the underlying causes.[925] For example, in the 1970s, the US, Canada, and various European countries brought in "wage and price controls" to try to rein in inflation. These failed in all jurisdictions.[925-927] Inflation with high wages and prices was the symptom, while the true underlying cause of rampant inflation was that governments were simply printing too much money to finance spending deficits that were too high. Trying to artificially limit prices does not make the problem go away.[925]

Attempts at wage and price controls resulted in extensive gaming of the system, cuts in quality, "add-on" costs, shortages, queues, and political pressures to grant exceptions to "special groups."[925] As discussed in Chapter 12, groups with political strength have been successful in having their therapies prioritized.[788] History tells us that this would probably also happen with government-imposed regulation of cancer drug costs. Loopholes would begin to emerge, just as they have in the United Kingdom National Health Service's foray into value-based pricing.[900] The games that are played to evade controls also generally result in true costs that are far higher than the costs one is trying to control.[925]

In Chapter 12, I discussed potential impact of government price controls on drugs in more detail in the discussion of the potential for CADTH and PMPRB rules to delay access to new drugs. Government price controls generally create shortages. Current analyses indicate that lowering drug prices too much would reduce investment in new drug discovery. So, there would be fewer effective new drugs available.[893] Available data also indicate that countries that set low drug prices have delayed access to new drugs as a direct consequence of the low prices, if they are able to access them at all.[867,883-885]

There are also numerous logistical issues with value-based pricing.[cxcix] For example, if a clinical trial of an effective new agent permits crossover, the gain in survival will be less than if crossover is forbidden. If the

---

[cxcix] In Chapter 12, I discussed the difficulty with evaluating cost-effectiveness. The same issues create difficulty in assessing drug "value."

drug's perceived "value" is based on survival improvement, then less would ultimately be paid for a drug if crossover were permitted. It would then be tempting to specifically forbid crossover to enhance the drug's ultimate monetary value. I feel this would be unethical. Furthermore, drugs that are so good that they are approved rapidly based on breakthrough drug designation will have no formal, reliable survival comparison to older therapies, so may be judged to have little value.

Those who advocate value-based pricing assume that the approach would decrease drug costs. But would it instead result in a marked escalation of prices of effective therapies that are currently inexpensive? For example, to make value-based pricing "fair and equitable," should the price of each polio vaccine injection be raised to $1 million?[928] Strategies that only consider one side of an equation are ultimately prone to yield nasty surprises.

The chemotherapy drug cisplatin is key to curing a high proportion of young patients with metastatic testicular cancer. Cisplatin will increase cure rates by about 5% if added to radiotherapy or surgery in older patients with potentially curable lung cancer, and it will improve survival by a few months as part of combination chemotherapy for widely metastatic lung cancer. How will its value be calculated? Will it be granted a different price for each of these indications? Would pharmacies be required to keep their supplies of highly valuable testicular cancer cisplatin separate from their lower value supplies of lung cancer cisplatin? If one uses a "blended" price, how does one value each of these different indications? How would the system deal with the fact that generic cisplatin is relatively inexpensive (at least in Canada, at about $350 per 21-day cycle[929]) and that any move to true value-based pricing would require a marked increase in price, no matter which of the relevant malignancies were considered?

If two drugs are combined and the combination is more effective than either drug alone, does the price of both drugs go up? Should a drug that is priced based on its effect in a metastatic cancer have its price boosted if it is later found to increase cure rates when added to radiotherapy or surgery for localized cancers? Will a drug be priced higher if it is more effective in younger patients than older patients? In response to political pressure,[788] will drugs for some cancers be judged to be more valuable than drugs for other cancers? No matter how one does this, special interests will almost certainly intercede.

If it costs the same very high amount to develop two drugs, and one doubles life expectancy in the 15% of lung cancer patients with an *EGFR* mutation and the other doubles life expectancy in the 1% of patients with a ROS-1 fusion gene, will the price for both be the same? If the price for both is low, then development costs could potentially be recouped for the larger market EGFR drug but not for the ROS-1 drug. Conversely, if the price for both is high, one may be paying too much for the EGFR drug. If the price is not the same for both, then it is not value-based pricing. Etc., etc., etc.

Whether embracing a "value-based price" or a "just price" philosophy in setting drug prices, trying to deal with questions like the above will turn the whole process into a politically hijacked, bureaucratic, pseudoscientific horror show. It will only be tackling the symptom rather than the underlying causes of high prices using a methodology (government price fixing) that usually ultimately fails.

Another approach has been performance-based reimbursement, with the government only paying for a patient's treatment if the patient benefits. However, this is generally of little value, since the cost of the guarantee ends up being incorporated into a very high list price for the drug.[900]

Governments (including the US government) also have the option of a "compulsory licensing" approach, giving a generic company permission to produce the drug even if it is still under patent.[cc] However, this measure is not commonly used,[865,900] and is risky. Unless used under only extraordinary circumstances, it could lead to international breakdown of the patent system. What would happen if Canada and similar countries with little domestic drug development decided they were going to solve their drug cost problem by simply ignoring patents and producing generic versions of any drug they wanted?

Additional actions: In keeping with the factors identified as contributing to high costs, approaches that have been advocated include allowing US Medicare to negotiate drug prices,[864,865,901] encouraging

---

[cc]   The US government recently threatened to do this with COVID-19 vaccines to speed up production.

development of generics,[901] eliminating strategies taken by companies to block generic competitors,[864] allowing the US Patient-Centered Outcomes Research Institute, etc., to consider cost-effectiveness in their recommendations,[864] encouraging countries to band together as buying units to strengthen their purchasing power,[900] developing cancer treatment pathways and guidelines that incorporate the cost-benefit of drugs in their recommendations,[864] allowing importation of drugs from other countries,[864] creation of patient-driven advocacy groups,[864] publicly ranking pharmaceutical companies on steps they take to improve product availability, affordability, and accessibility,[864] and intervention of philanthropic foundations to help reduce costs of generic drugs.[865]

My thoughts: From my perspective, government price controls, importing cheaper drugs from another country, and governments ignoring a company's patent will not work well, for the reasons that I have outlined above. If we want less expensive drugs while maintaining rapid progress, we need to tackle the underlying cause, and not just the symptom of high prices. To do that, the top priority must be restoration of a true free market economy, as opposed to the current mirage.

I will now list some possible solutions that we can start with.

Costs of new drug development must be slashed: As outlined in Chapter 12, numerous factors delay access to effective new drugs while markedly increasing drug development costs. Despite a fatalistic mind set, we can fix this if we try. The current system costs lives, and it increases suffering while also meaning that drug prices must be high. Sales of drugs that eventually reach the market must not only pay for the development of the drugs that succeed but also must pay for the 92% to 97% of drugs that will fail in clinical trials.[775-777] If the cost of failures cannot be recouped through the successes, then no sane investor would continue to invest in new drug development. Progress would come to a halt. Venture capitalists have been losing interest in investing in small pharmaceutical company start-ups due to the high probability of failure,[930] and this trend must be reversed.

If development costs come down, then companies can no longer use high development costs as an excuse for high prices. Faster drug approval will not only save lives, but also results in longer patent protection since less of the 20-year protection period will be used up prior to marketing.

Consequently, recoupment of development costs can be amortized over a longer period.

Very importantly, slashing drug development costs would also mean that more start-ups and small companies could afford to bring a drug to market. This would increase competition. Currently, drug development costs are so high that only large companies can succeed. These high drug development costs perpetuate the current oligopoly situation that seriously impedes competition. The rules and regulations that currently govern clinical trials are vital for patient safety and data integrity, but they currently come with far too high a price. This must be rectified, as I discussed in Chapters 11 and 12.

Large pharmaceutical companies are now a major source of investment for small start-ups.[930] While broadening the scope of the potential new agents they can assess, they reduce their risk by concentrating on companies with new agents that look promising.[930] At the same time, a major part of the business strategy for small start-ups is to aim to become an attractive target for takeover by a large company, rather than aspiring to bring a drug all the way to market by themselves.[930]

On balance, I see this investment strategy by large pharmaceutical companies as being positive, provided they use this approach to increase the number of potentially effective new drugs rather than using it to quash competition from smaller companies. From my perspective, it would be even better if clinical research costs were low enough that a small company could bring their drug all the way to market on their own without the necessity of being swallowed by the oligopoly.

Medicare drug purchases: It goes directly against the principles of a free market economy that US Medicare is forbidden by law from negotiating the price of the drugs it buys. This is a major contributor to America having the world's highest drug prices.[906] These high prices mean rapid access of American patients to effective new therapies, and they are an important force driving new drug development.[893] But they also indirectly put an upward pressure on prices internationally by setting a very high initial mark against which other payors must negotiate.

Generics: Companies must be stopped from decreasing post-patent competition by bribing generic companies to withhold their product.

Pharmaceutical company contributions to politicians: Pharmaceutical companies make huge indirect or direct financial contributions to the American political system.[907] Regulations about contributions differ across countries, but these companies also make major political donations in other countries such as Australia.[931] (For Canada, the rules changed in 2017, such that any one person or corporation can currently only make a total contribution of $3300 toward an election campaign.[932]) The impact of donations is arguably larger in the US than elsewhere, but high US drug prices have a ripple effect across the world. Despite attempts to reform American campaign contribution laws there are simply too many indirect ways that large donations can be made.[933]

The influence purchased in this way destroys the competitive environment essential to bringing down costs. A modest contribution to a political candidate may *support* the candidate's stance–but a large contribution may *dictate* their stance. This must stop.

Clarity on comparability of similar drugs: A new agent is typically compared to a "standard therapy" during a clinical trial. It is approved if it is shown to be superior. However, several similar drugs may all add benefit over a standard therapy, but there is often insufficient evidence on how they compare to each other. For example, alectinib, ceritinib, brigatinib, and lorlatinib are all superior to the older agent crizotinib against lung cancers with the *ALK* fusion gene. It is not clear whether any one of these is a better alternative than another. The same is true for PD-1/PD-L1 inhibiting immunotherapy agents such as nivolumab, pembrolizumab, atezolizumab, durvalumab, and others across a broad range of cancers.

Unless they are directly compared to each other, we cannot tell if one is better than another. When outcomes are compared across trials, any apparent differences between the drugs may be due entirely to minor nuances of clinical trial design and patient selection methods.[430]

Unless differences are very large (as in the case of drugs designated as breakthrough therapies), only a direct comparison of drugs from the same class can determine whether one has real advantages over another. However, if a company thinks its marketed drug would likely be found to be similar to another agent, it would not generally choose to compare its drug directly to its competitors. Direct comparison is only an attractive option if you think your drug is better. This lack of direct comparisons inhibits the ability of payors to negotiate based on price. If there is proof that two drugs have a similar mechanism of action also have similar efficacy, then a payor can more confidently buy the cheaper one.

For example, all the PD-1/PD-L1 immunotherapy drugs can be helpful in some lung cancer situations. In randomized trials comparing them to chemotherapy or placebo, some have been found to work in patients with untreated metastatic disease, some have been shown to improve survival when given after combined chemotherapy and radiation for localized disease, etc. If one of the PD-1/PD-L1 inhibitors yielded positive results when compared to chemotherapy or placebo and another one did not, this may mean that the one PD-1/PD-L1 inhibitor is truly better than the other in that situation, or the fact that one was successful and the other was not may have instead been due to differences in clinical trial design or patient selection. The only way to tell if one of these agents is truly better than another is to directly compare them to each other in a randomized trial. But this has not been done.

If the drugs were compared to each other and were found to give similar outcomes, then a payor could have the companies compete based on best price. It is difficult to do this if you don't know if any apparent differences between them are real or are instead just artifacts of trial design.

A new, simple trial methodology called REaCT[934] could be used to directly compare similar approved drugs over a range of treatment situations.[cci] REaCT trials can recruit large numbers of patients at minimal cost and without time-consuming, expensive documentation requirements.[935]

Conclusions: In summary, the thirteenth reason that cancer still sucks is that effective new therapies are extremely expensive. But this is correctable. The same changes that could reduce deaths and suffering by making effective new therapies available much faster could also directly and indirectly foster a marked reduction in therapy costs.

[cci] The REaCT methodology was devised by my Ottawa colleague, Dr. Mark Clemons.

— 14 —

Ours is Better Than Yours: The Battle of the Healthcare Systems

Healthcare delivery differs among countries. When I talk to physicians from around the world, many say that their system is probably better than a neighboring country's system but that their system, nevertheless, has problems.

Many Canadians wish they could access the rapid private care that Americans can. Many others are thankful for Canada's publicly funded care and regard the US as having an evil system that bankrupts patients and deprives healthcare to millions of uninsured Americans. Many Americans wish they could have a universal, publicly funded healthcare system like Canada's. Many others have no wish to have a "socialist" system, with long waiting lists and a perceived rationing of care. Views may be strongly influenced by where one stands on the economic ladder.

During my career, I have cared for patients in both the Canadian system (1974 to 1976, 1980 to 2003, and 2011 to the present) and the

American system (1976 to 1980 and 2003 to 2011).[ccii] The fourteenth reason that cancer still sucks is that the healthcare systems on the north and south sides of the Canada-US border both have deficiencies.

A Short Primer

The United States has the world's most expensive healthcare system. It spends about twice as much each year on every American as the Canadian system spends on Canadians. Per capita, the US spends far more than Canada on drugs each year. The US also has far more healthcare capacity, with more specialists, nurses, hospital beds, CAT scanners, MRI scanners, PET scanners, and radiotherapy treatment units per capita than Canada. This higher capacity can be useful, but it costs a lot of money.

The higher healthcare spending in the US is primarily due to a much higher price for every medical procedure. It isn't due to more procedures being performed in the US.

Having a single payor system in Canada makes the Canadian system much less expensive to run. In Canada, hospitals and physicians easily submit a single monthly bill electronically to the provincial government. In the US, physicians and hospitals cannot automate sending their bills to each of hundreds of insurers. Instead, it involves a huge amount of time and expensive paperwork. Consequently, healthcare administrative costs are

---

[ccii] I left Canada for MD Anderson Hospital in Houston, Texas in 1976 for medical oncology training and accepted a position on faculty at MD Anderson when I completed my training. I returned to Ottawa in 1980, went back to MD Anderson in 2003 to pursue research opportunities, but returned to Ottawa in 2011. In both 1980 and 2011, my return to Ottawa was prompted by it being "home". There is comfort in being surrounded by the familiar. In addition, there are no direct flights from Houston to Ottawa, so every trip back to visit family was an adventure, very frequently involving missed connecting flights and being stranded at airports overnight. Christmas presents in checked baggage strayed to Europe almost as often as they made it to Ottawa. Looking after patients was at the same time rewarding and challenging in both places. MD Anderson and Cancer Care Ontario paid me very close to the same amount. While some US physicians in community practice can make a fortune, incomes in academic positions are similar in Canada and the US.

almost five times higher in the US than Canada. In this case, government bureaucracy is surprisingly more cost-efficient than the private sector.

Billing insurance companies is not the only expensive part of the US system. An insurance company must approve any nonemergency tests, procedures, or treatments before they can be performed. This extra administrative burden is costly.

When we moved to Texas, my wife, who is Canadian, was incredulous at the number of people working in US doctors' offices. The average Canadian physician shares a receptionist and maybe a nurse with a few other physicians. Conversely, a doctor's office in Houston is overflowing with staff, most of them dealing with insurance companies.

The Canadian approach is much simpler: provincial governments[cciii] build limited capacity. Canadian patients use this limited capacity to the maximum extent that available resources allow. If there is insufficient capacity, healthcare providers must apply for permission to build more capacity. That takes time. The net result is that there is never quite enough capacity. This increases wait times, yet it is also very cost-efficient administratively. The bottlenecks effectively control utilization, with no need for daily calls to insurance companies.

Because American healthcare is so expensive, US companies must pay a lot for employee healthcare insurance. This drives up labor costs in the US, while paradoxically keeping wages low. It is, therefore, more expensive to produce something in the US than in Canada and elsewhere. This is one major driving force behind US jobs shifting to other countries.

US physicians have a feral fear of liability. An American physician is much more likely to be sued than a Canadian physician. This and other factors drive-up the cost of US malpractice insurance. American physicians also follow more expensive "defensive medicine" processes, ordering tests that might not be necessary medically but that reduce the risk of a successful lawsuit.

Costs of compliance with government regulation are probably also higher in the US. When I worked at MD Anderson, I received frequent emails from the Office of Compliance that I was obligated to either do

---

[cciii] In Canada, provincial governments oversee day-to-day aspects of healthcare.

or avoid various specific things. I might be fired and might face criminal prosecution if I ignored them. For example, it was illegal to fill out forms requesting a motorized wheelchair for a patient. Such forms could only be completed by very specific professionals. I would have faced stiff criminal penalties if I completed one since I was not authorized to do so. We were also told that if we tried to arrange free chemotherapy for underinsured patients, the government might charge us with using coercion to try to attract patients.

Such legal threats from government are substantially less commonplace in Canada.[cciv] A Canadian physician must maintain a high level of professional conduct, in keeping with the standards of provincial medical licensing bodies. However, there are not constant threats from government, and no need for an institution to have an Office of Compliance.

While living in Houston, I was struck that overall, the relationship between the American people and their government appeared to be a somewhat uncomfortable one. This is in keeping with the US imprisonment rate. It's the highest in the world (639 prisoners per 100,000 population, compared to 104 per 100,000 in Canada).[936] In the US, prisons may be highly profitable, privately-owned capitalist ventures in which politicians and others may invest.[ccv][937]

I suspect this US discomfort with government plays at least some role in the strong support for the Second Amendment. It has probably also played a role in the 2016 election of Donald Trump as a president who promised to "drain the Washington swamp." In the "Frozen North," Canadians may strongly disagree with their government. We may even despise it, but we generally do not fear it. In Canada, governments control healthcare spending largely through strategic, though potentially misguided budget constraints rather than by heavy-handed threats.

[cciv] Of interest, when Medical Assistance in Dying (MAID, or physician-assisted suicide) was introduced in Canada, it created a lot of controversy. Some physicians were opposed to it, but it was designated as being illegal (with stiff penalties) for a physician to fail to refer a patient to a MAID program if a patient requested referral.

[ccv] Dick Cheney's family were noted investors in private prisons during his tenure as US Vice President.

Life expectancy: Despite the huge amount spent on US healthcare, American men live an average of 4.5 years less than Canadian men. American women live 3 years less than Canadian women. In fact, the US ranks a lowly 46th in the world in average life expectancy.

Part of this is due to many young Americans being underinsured. A country's average life expectancy will drop if a lot of young people die prematurely because they don't have health insurance. When we moved to Texas in 2003, we hired a company to install a swimming pool at our new house. In talking to one of the young workers, my wife was concerned to find that he had unrelenting, disabling stomach pain. He told my wife that because he had no insurance, he could not afford medical care. This would not have been an issue in Canada. A Canadian could always see a doctor. They could go a to any walk-in clinic if they had trouble finding a family physician. They might have to wait a few days or weeks for an appointment if they had a family doctor, but lack of insurance would not prevent them from seeing one.

Uninsured young people are not the only reason for a comparatively short US life expectancy, but they are a major contributor.

Access to care in Canada: Canadians on average have better access to primary care physicians than do Americans, but they wait much too long for many other things. It takes longer for a Canadian to see a specialist than in many other countries. It also takes longer for a Canadian specialist to initiate appropriate investigations and treatment. It takes much too long for Canadians to undergo elective surgery such as a hip replacement or cataract surgery, or to have scans. For example, my wife once waited eight months for an MRI scan.

Emergencies in Canada are dealt with expeditiously. But wait times can be excessive if it is not an emergency or if the primary care physician cannot deal with it.

Canadian life expectancy has dropped from 7th in the world in 1995 to 16th in 2020. Canada still does much better than the US in life expectancy but inadequate investment in healthcare is having a negative impact.

American vs. Canadian cancer outcomes: Cancer causes 30% of all deaths in Canada compared to 21% in the US. Americans are more likely than Canadians to die young of something else, before they can develop a cancer. Consequently, cancer is more common in Canada. However, an American who does develop a cancer is more likely to survive it than is a Canadian. The relative 5-year survival after a cancer diagnosis is currently around 67% in the US compared to 63% in Canada. The poorest Canadians have a higher probability of surviving cancer than the poorest Americans, but for people in middle- and high-income groups, Americans do better.

On average, Canadians are less likely to survive cancer than Americans since the Canadian system often moves too slowly in assessing patients who develop symptoms. Delays in cancer diagnosis and treatment reduce the probability of surviving cancer, and it takes too long in Canada for patients to undergo required scans and biopsies. Unfortunately, in Canada, there are far too few specialists, CAT scanners, MRI scanners, PET scanners, and other diagnostic resources.

Once a cancer is diagnosed, surgery may take too long due to shortages of surgeons and operating room availability.[ccvi] While radiotherapy can usually be started within two weeks of completing all required scans, it often takes too long to get these scans. In addition, while Canada has excellent radiotherapy equipment, there is much less of it available than in many other countries. Canada is the only G8 country without a proton beam[ccvii] facility. Most patients would not benefit from proton beam treatment, but some would if it was available.

In Canada, systemic therapies can usually be started promptly once all available testing is done. However, delays in required scans and "molecular testing"[ccviii] can delay therapy initiation. For some

---

[ccvi] This has been a long-standing problem but became much worse during the Covid-19 pandemic. Nurses, ICU beds, and other resources that would normally have been used for surgery patients were instead required for care of Covid-19 patients.

[ccvii] Proton beam therapy produces positively charged radioactive particles for radiotherapy treatments. It has the advantage over other types of radiation therapy that the beam can be set up so that the particles stop right after going through the tumor. Consequently, there is less damage to normal tissues behind the tumor. The radiation is not passing through this tissue on its way out of the body. One of the administrative assistants in the radiation oncology department in Ottawa displays the slogan, "Think like a proton. Be positive!"

[ccviii] Molecular testing means analyzing a tumor to see if there is a mutation or other factor that would make it sensitive or resistant to a particular therapy.

therapies, the oncologist must also apply to the government through an "Exceptional Access Program" for permission to use it. The therapy is delayed (sometimes by weeks) while the application is being considered, despite it being virtually certain that the application will eventually be approved. Canadians also may not have access to the best new drugs since drug companies have not applied to sell them in Canada—or because provincial governments have not agreed to pay for them. This is frustrating for patients and doctors alike.

Attempting to make things better: Most Americans over the age of 65 have healthcare coverage through Medicare. With the Affordable Care Act (Obamacare), the proportion of younger patients who are uninsured has decreased from 17.8% to 10%, but the high costs of insurance continue to limit the proportion of the population that is covered.

In the Canadian system, I have participated in several exercises to try to make things work better. The major issue is that many of the new processes that are put in place eventually fall apart as key components become victims of budget cuts or other factors. The exercise must then start again, as new solutions are sought.

Further details and References

I will now discuss some aspects of this in more detail.

Costs: American healthcare is more expensive than Canadian healthcare. In 2019, 16.8% of gross domestic product (GDP) was spent on healthcare in the US vs. 10.5% in Canada, with an average of $10,966 spent on each American vs. $5,418 spent on each Canadian.[ccix][938] The US has by far the most expensive healthcare system in the world, while in 2019, Canada ranked 11th out of 36 OECD[ccx] countries.[938]

In Chapter 13, I discussed why the prices of cancer therapies are generally much higher in the US than in Canada. There are several factors contributing to why other areas of healthcare are more expensive. One reason is that the American healthcare system has a lot of extra capacity

---

[ccix] Amounts for both the US and Canada are in US dollars. US GDP per capita is higher than Canada's.

[ccx] OECD is the Organization for Economic Co-operation and Development.

built in. Table 2 shows the number of physicians, nurses, hospital beds, CAT scanners, MRI scanners, and PET scanners per million population in the two countries.[ccxi] This extra capacity in the US means that resources are standing idle for a greater percentage of the day in the US than in Canada. There is a cost to this idle time.

In general, the higher spending in the US is due to higher prices for most medical procedures compared to other countries, rather than being due to higher utilization rates.[939] Despite the number of physicians available, the US had fewer physician consultations per capita in 2015 than most comparable countries (3.9 vs. 7.6).[939]

Table 2. Healthcare resources in the United States vs. Canada (converted to number per million population)	United States	Canada
Primary care physicians (2018)[940]	312*	1,296
Specialist physicians (2018)[940]	2,288	1,404
Nurses (2018)[940]	17,400	11,700
Hospital beds (2019)[941]	2,900	2,500**
CAT scanners (2019)[942]	44.94	14.82**
MRI scanners (2019)[943]	40.4	10.3**
PET scanners (2019)[944]	5.5	1.5**
Radiotherapy units (2019)[945]	11.7	2.9**

*The relative paucity of primary care physicians compared to specialists in the US has been proposed as one reason that the US does not do very well in preventative medicine and maternal care.
**Compared to other OECD countries supplying data for 2019, Canada ranked 32nd of 38 in hospital beds per million population, 30th of 36 in CAT scanners, 27th of 35 in MRI scanners, 20th of 30 in PET scanners (2017), 30th of 33 in radiotherapy equipment, and 20th out of 20 in wait times to see a specialist. In many of these categories, Canada ranked behind Turkey and many Eastern European nations.

[ccxi] When I moved to Texas in 2003, MD Anderson Hospital alone had 5 PET scanners while there were only 3 PET scanners all of Canada.

Another factor that reduces relative costs in Canada is the single payor system. Each month, physicians and hospitals in Canada simply submit their billings electronically to a single payment source, the government. Since there is just a single payor, submissions can be automated, and the paperwork is easy and inexpensive.

In the American system on the other hand, administrivia runs rampant. There are more than nine hundred different companies offering healthcare insurance,[946] each with its own set of criteria and forms. Therefore, payments cannot be automated. It takes far too many people in an American hospital or physician's office just to submit the billings. American physicians and hospitals must also call insurance companies for preapproval prior to setting up tests or treatments for a nonemergency. It takes a lot of expensive administrative staff to make these calls.

In addition, insurance company overhead and profits must be covered. The net result is that in 2017 it cost an average of $2,497 per capita (34.2% of healthcare expenditures) to cover healthcare administrative costs in the US, compared to $551 per capita (17.0% of healthcare expenditures) in Canada.[947]

Canada and the US are significant trading partners, with an estimated $718.4 billion in goods and services traded in 2019.[948] Economies of scale give the US an advantage here: since the US has ten times the population of Canada, it can produce many things more efficiently than Canada can. For this and other reasons, the productivity of US workers is higher than Canadian workers.[949] However, it is very expensive for an American employer to pay for health insurance for its workers. For example, in 2004 it was estimated that employee health insurance costs made it $1,300 more expensive for an automobile company to produce a midsize car in the US than in Canada.[950] There has been a lot of discussion about manufacturing jobs moving from the US to cheaper overseas sites, but bringing in a single payor healthcare system potentially could reverse this trend by cutting US manufacturing costs.

Another issue that affects healthcare costs is litigation. Physicians in the US are more likely to be sued than in Canada. This means that physician billings must cover the high costs of malpractice insurance while the physician is practicing "defensive medicine" and ordering expensive tests that might be unnecessary. According to a 2016 survey, 34% of

American physicians had at some point faced a malpractice lawsuit.[951] With an average of 15,000 to 19,000 lawsuits filed each year,[952] an American physician is more than twice as likely to be sued in a given year than is a Canadian physician.[953]

There are several factors that contribute to the lower risk of lawsuits in Canada. For example, some studies show that Americans are at high risk of being bankrupted by healthcare costs,[ccxii 782,903] although other authors have disputed this.[954] Canadians may face a lower risk of personal bankruptcy due to healthcare expenses thanks to Canada's single payor healthcare system. This may reduce a Canadian's financial incentive to sue. In addition, Canadian lawsuits are vigorously defended by the Canadian Medical Protective Association (CMPA) that insures most Canadian physicians, and maximum payments for pain and suffering are capped. Punitive damages are uncommon, and awards against physicians are less common and generally for much lower amounts than in the US.[955]

Life expectancy: What does the high cost of US medical care buy? Some argue "not much." In 2020, the US ranked 46th in the world with respect to average life expectancy (81.7 years for females and 76.6 years for males in the US, compared to 84.6 and 81.2 years in Canada).[956] A disproportionately high death rate in younger Americans negatively impacts the overall life expectancy of the population. This high death rate in young Americans can be explained in part by inadequate access to healthcare.[957] A young American will not have access to Medicare and is also less likely than an older American to have a job with healthcare benefits. In 2019, 17.5% of Americans aged 18 to 34 were uninsured. Many others were underinsured.[958] In 2019, excessive costs made 36.5% of uninsured Americans delay or forgo needed healthcare.[959]

This relative lack of access to healthcare for young people probably plays a role in the much higher maternal mortality rate in the US (29.9 per 100,000 live births) than in comparable countries (Japan, Australia, Switzerland, Canada, Austria, Sweden, United Kingdom, Netherlands,

---

[ccxii] In Houston, many of my cancer patients who were too young for Medicare would struggle valiantly to keep on working until a very few weeks before their death since they would lose their health insurance if they stopped working. This does not happen in Canada.

Belgium, France, Germany, which together had a maternal mortality rate of 6.1 per 100,000 live births).[960] It also probably plays a role in the US's much higher infant mortality rate than in other developed nations.[957] Ditto for the high US rate of premature death from heart disease, hypertension, diabetes, and other obesity-related deaths.[957,961,962] A scarcity of primary care physicians[940] and underinsurance can impede access to the preventative healthcare that might reduce this high rate of premature death. Differences in preventative approaches have probably also played a role in differences in death rates from COVID-19 in the US (1,815 deaths per million population, as of 2021/05/23) vs. Canada (663 deaths per million population).[963]

Lost economic productivity declines sharply when workers can access care for medical problems that slow them down.[964]

There are also other factors besides inadequate access to healthcare that led to a high death rate in young Americans. For example, death rates per 100,000 population from road accidents are 12.2 in the US vs. 6.7 in Canada,[957] from opioid overdoses they are 13.3 vs. 3.2,[957] from homicide by firearms they are 4.5 vs. 0.5,[965] and from homicides by other means they are 1.6 vs. 1.0.[957,965] As noted previously, the imprisonment rate is more than 6 times higher in the US than in Canada,[936] and imprisonment reduces life expectancy by an average of five years.[966]

Access to care in Canada: Overall, Canadians have better access to primary care physicians than do Americans.[940] In some Canadian provinces, these primary care physicians receive incentive payments to prioritize preventative measures like mammograms, Pap smears, senior flu shots, and colorectal cancer screening.[967] However, the Canadian healthcare system also has major issues. Canadians on average must wait longer to get care than do patients in similar countries. In 2016, only 43% of Canadians who needed care were able to book a same day or next day appointment, compared to an average of 57% across other countries.[968]

Among twenty OECD countries, Canada tied with Norway as having the longest wait times to be seen by a specialist, with 61% of patients waiting more than a month.[969] From the time a Canadian primary care physician referred a patient to a specialist for a nonemergency in 2019, it took an average of 14.6 weeks to see an orthopedic surgeon, 12.3 weeks for an ophthalmologist, 15.7 for a neurosurgeon, 8.5 for a general surgeon,

4.9 for a cardiovascular surgeon, and 6.1 for a general internist.[970] From my own experience in referring patients, it can take a year or more to have a patient seen by a rheumatologist for arthritis.

Once the patient is seen by the specialist, it typically takes months longer to initiate treatment for nonemergency conditions (average of 24.5 weeks for orthopedic surgery, 16.1 weeks for ophthalmology, etc.).[970] It typically takes several weeks for a specialist to be able to set up a bronchoscopy, colonoscopy, or gastroscopy to investigate symptoms that may indicate an underlying malignancy.[970]

An International Health Policy Survey in 2016 reported that 18% of Canadians had to wait four months or longer for elective surgery, compared to 12% in the United Kingdom, 8% in Australia, 4% in the United States, 2% in France, and almost none in Germany.[971] In 2019, only 72% of Canadians underwent hip or knee replacement or cataract surgery within the recommended wait time target of six months.[971,972] At that, six months is a long time to wait for a procedure that could alleviate pain and improve productivity and quality of life.

Emergencies in Canada are generally not an issue. For example, hip fractures[972] and emergency cardiovascular surgery[970] and other emergencies are generally dealt with immediately. Urgent but nonemergency cardiovascular surgery is generally done within one to two weeks.[970] However, anything deemed to be not urgent takes too long.

Wait times for scans are also too long. Again, urgent CT and MRI scans can be done rapidly, but nonurgent scans can often take weeks or months.[973] Some patients, whose situation does not initially seem to be urgent, may deteriorate rapidly while awaiting scans. Since diagnostic scans are needed before proper treatment can be initiated, scan delays also can potentially prolong avoidable suffering and anxiety.

Over the past fifty years, healthcare spending as a percent of GDP has gradually increased in both Canada and other OECD countries. Apart from minor short-term variations, Canada consistently has ranked around 6th or 7th in this metric. However, from 1995 to 2019 Canada dropped from an average rank of 5th or 6th in total per capita healthcare spending to 9th to 12th.[974] From 1985 to 1995, Canada ranked 7th in the world in average life expectancy.[975] By 1995–2005, Canada had dropped to 10th place, by 2005–2015 it was 13th,[975] and by 2020 it was 16th.[956] Canadian life expectancy is projected to drop to 27th in the world by 2040.[976] It is little cause for celebration that Canada is still much better off than the US (which, as noted earlier, currently ranks 46th in the world[956]). It is not clear that the gradual drop in per capita healthcare spending in Canada relative to other OECD countries has contributed to this reduced relative life expectancy, but I think it probably has.

American vs. Canadian cancer outcomes: When it comes to cancer, available data suggest that the American system is better. Cancer only accounts for 21% of all deaths in the US,[977] compared to 30% in Canada.[2] This is due in part to the unfortunate fact that Americans are more likely than Canadians to die young of something else, before they can develop cancer. However, in the US, the probability of surviving cancer is higher. Across the full spectrum of malignancies, the relative 5-year survival[ccxiii] is 67% in the US vs. 63% in Canada.[977,978]

While a difference this small could be due to chance if it were seen in only one time period, differences of a similar magnitude have been seen for most of the last several years. Table 3 shows Canadian vs. US survival for different cancer types.

[ccxiii] "Relative 5-year survival" means the probability of surviving 5-years compared to other people in the same age group. For example, if an average 70-year-old had an 80% probability of surviving 5 years, but a 70-year-old with cancer had a 54% probability of surviving 5 years, the relative 5-year survival of the patient with cancer would be 54/80= 0.675, or 67.5%.

Table 3. 5-Year Relative Survival for Selected Cancers in the United States vs. Canada

Malignancy	5-year relative survival	
	United States[977]	Canada[978]
Overall	67	63
Lung	21	19
Breast	90	88
Colorectal	65	65
Prostate	98	93
Pancreas	10	8
Esophagus	20	15
Kidney	75	71
Bladder	77	75
Melanoma	93	88
Non-Hodgkin lymphoma	73	68
Liver	20	19
Uterus	81	83
Ovary	49	45
Oral cavity and pharynx	66	64

Dividing the total number of patients dying from cancer each year by the number diagnosed gives very similar results. There are 32% as many deaths as diagnoses in the US vs. 37% in Canada.[ccxiv] In 2020, Canada (with a population of 38,000,000 people) had about 225,800 new cases of cancer, and about 83,300 deaths from it. The good news is that

---

[ccxiv] A caveat: if the incidence of the cancer is rising or falling rapidly over time, then this might be somewhat misleading, since some of the patients dying this year might have been diagnosed a few years ago. In addition, the way that death is attributed to a cancer might differ between countries. If a patient has a rapidly worsening cancer and dies of a myocardial infarct, the death might be attributed to the cancer in one country but to heart disease in another country.

approximately 142,500 Canadians who were diagnosed with cancer in 2020 will survive it. The bad news is that about 11,000 more would have survived their cancer if Canada's survival rates were the same as in the US. An estimated one out of every eight cancer deaths in Canada could be avoided if Canadian and American survival rates were the same.

In the US, the 5-year relative survival with cancer is lower for African Americans (63%) than for Caucasians (68%).[977] Canadians with cancer have the same survival probability as do African Americans, who often have inadequate access to healthcare. The poorest Canadians have a higher probability of surviving cancer than do the poorest Americans.[979] Hooray for Canada! However, survival is better in the US than in Canada for the 80% of patients in middle- and upper-income brackets.[979]

Some defenders of the Canadian system have argued that the better relative cancer survival in the US is an artifact. Explanations include differences in the way data are captured or an increased probability of finding a relatively benign cancer in an American. The US does have somewhat higher cancer screening rates than in Canada, due in part to differences between the countries in screening guidelines.[980] Despite this, cancer stage at diagnosis is on average somewhat higher in Americans than in Canadians (Table 4). This indicates that it is likely to be more advanced at the time of diagnosis in the US compared to Canada.

Table 4. Cancer stage at diagnosis for Canada vs. US

Cancer type	Stage: Canada 2011–2015* [981]				Stage: US 2010–2016[977]			
	Localized	Regional	Distant	Unknown	Localized	Regional	Distant	Unknown
Lung	29%	20%	50%	1%	17%	22%	57%	4%
Colorectal	47%	29%	20%	4%	38%	35%	22%	5%
Breast*	85%	12%	5%	1%	63%	30%	6%	1%
Prostate	74%	14%	9%	3%	76%	13%	6%	5%
Cervix*	68%	17%	12%	4%	44%	36%	16%	4%

*For Canada, Localized was reported as stage 1 and 2, Regional was reported as Stage 3, and Metastatic was reported as Stage 4 in the source publication. Canadian total for breast was greater than 100% in the source publication, while for cervix cancer it is 101% due to rounding of numbers.

As I explained in Chapter 2, only a low proportion of most cancers are diagnosed through screening. Far more are diagnosed after the onset of symptoms. Once symptoms develop, the Canadian system moves too slowly. In patients with early stage, potentially curable cancers of the breast, prostate, lung, colon/rectum, kidney, or pancreas, the risk of death increases by 1.2–3.2% per week of delay between cancer diagnosis and treatment initiation.[982] It would be reasonable to assume that there would be a similar increase in risk of death based on delays between the onset of cancer symptoms and establishment of the diagnosis. Similarly, the risk of a cancer recurring[983] and the probability of dying of a cancer[690,984] is increased by delays of just a few weeks of "adjuvant" radiotherapy or chemotherapy.

As I pointed out previously, Canada has fewer CAT scanners per million population than Turkey and many Eastern European nations..[942] It takes too long to get a patient in to see most specialists,[970] and it takes too long for these specialists to organize and execute the procedures required to establish a diagnosis.[970] After a biopsy is done, a pathologist must assess it to establish a cancer diagnosis. The number of pathologists per 100,000 population in Canada decreased by 43% from 1.4 in 1995 to 0.8 in 2019.[985] Once the diagnosis is established, it then takes too long to stage[ccxv] the disease due to shortages of CAT scanners,[942] PET scanners,[986] and MRI scanners.[940] It also takes too long to do molecular profiling.

After the excessively long time it may take for diagnosis and staging, it also takes too long for patients to undergo surgery. In Ontario, once the diagnostic and staging work-up has been completed, the patient is declared to be "ready to treat," and the clock starts ticking. By 2019, average wait time in Canada for cancer surgery from being declared ready to treat was 41 days for prostate cancer, 24 days for lung, 18 days for breast, 24 days for bladder, and 21 days for colorectal cancer.[972] It takes too long to make the determination that a patient is ready to treat, and it then takes too long to move on to surgery.

[ccxv] "Staging" the patient means using scans to determine whether the cancer is localized vs. widespread.

For many patients, waiting for diagnosis and treatment has no impact on probability of cure. Some will be cured despite even very long wait times, and some would not be cured even if things happened very rapidly.

But for some patients, waiting does make a difference.

Diagnostic and staging delays impact the initiation of therapy for both early, potentially curable cancers and for advanced, incurable cancers.[ccxvi] This has consequences. For example, while systemic treatment[ccxvii] costs are covered in Canada, fewer than 25% of patients with metastatic non-small cell lung cancer ever receive systemic treatment.[427] About 4% of remaining patients die during each week that initiation of therapy must be delayed,[424] and others deteriorate to the point that they are no longer strong enough to even consider therapy. Being eligible for "free" therapy is of little value to a patient who deteriorates to the point where they are no longer strong enough to receive this free therapy.

If a patient makes it to therapy, the best new drugs are often not an option for Canadian patients. After the US and Europe approve marketing of a new agent, it takes an average of an additional 1.3 years for it to be approved in Canada.[858] This Canadian delay does not occur because Health Canada is inefficient. It occurs because drug companies generally

---

[ccxvi] Despite there being substantially fewer oncologists in Canada than cardiologists, neurologists, gastroenterologists, respirologists, nephrologists, infectious disease specialists, general internists, dermatologists, or psychiatrists, the wait time to see a medical oncologist is shorter than for almost any other nonemergency specialty, at an average of about 2 weeks. This is not an accident. Cancer Care Ontario (CCO), the government cancer agency that until recently oversaw cancer treatment in Ontario, kept close track of wait times. The benchmark was that at least 80% of new patients were to be seen within two weeks of a consult request coming in, and your center's statistics were published each month on a public website, for all to see. Some of my colleagues complained about us being held to a much higher standard than other specialties, but I told them that I regarded this as being a fantastic problem to have. It meant that the government was at least interested, rather than trying to hide it from the public. Meeting this metric was a challenge, but overall, we have managed to do well. However, CCO also mandated that any systemic treatment should start within four weeks of the patient being seen by an oncologist, and much of the province fails on this metric. CCO doesn't control any of the radiology or pathology resources in the province, and it takes too long for us to access required staging scans and molecular testing.

[ccxvii] Systemic treatment means chemotherapy, immunotherapy, or targeted therapies.

only submit applications for approval in the relatively small, less profitable Canadian market long after their submissions to the large, highly profitable American and European markets.[858]

There is a further delay after this 1.3-year delay in Canadian approval for new anticancer drugs. As discussed in Chapter 12, provincial governments often take months or even years to agree to fund these new agents for their citizens. This delay in provincial drug funding results in thousands more years of life lost.[707] Chapters 11 and 12 discuss the consequences of the major worldwide delays in getting effective new agents to cancer patients. The situation is even worse in Canada than in the US or Europe.

Some patients have private insurance (typically covering 80–100% of drug costs) that will pay for drugs approved by Health Canada, but not yet funded by provincial health plans.[ccxviii] However, many patients do not. After Health Canada approval, some companies will provide their drug free of charge on a compassionate basis while funding negotiations are underway. Others do not.

Canadian provinces also vary in what they will fund. The largest province, Ontario, will typically pay for drugs that are administered intravenously, but will only pay for oral medications for patients over the age of 65. Under some circumstances, patients under the age of 65 will qualify for Ontario's "Trillium" program, which will pay for these approved medications. However, coverage by Trillium is not a given, and there can be life-threatening delays in approval after an application to Trillium is submitted.

Attempting to make things better: The US does a good job of covering its elderly with healthcare. Fewer than 1% of Americans over the age of 65 are uninsured.[987] However, the US does not do well with younger patients. The Affordable Care Act (Obamacare) attempted to

---

[ccxviii] In a Kafkaesque twist, Canadian politicians and federal government employees are typically provided with life-long private insurance that provides them and their dependents with 80–100% coverage for many medications not covered by provincial government programs. This drug insurance is paid for by the public through their taxes. Hence, government employees and politicians can access expensive new therapies through insurance paid for by the public, but the public who are paying for this insurance may not themselves be able to access these therapies.

make American healthcare more accessible to the young and the poor. With its introduction, the number of uninsured Americans under age 65 decreased from more than 46.5 million in 2010 (17.8% of those under 65) to fewer than 26.7 million in 2016 (10% of those under 65).[987] However, by 2019, it had climbed back up to 28.9 million (10.9%), with high costs being cited as the major reason that so many people still lacked insurance.[987] There is no question that a single payor system could make American healthcare much less expensive because it would slash the huge burden of administrative costs.[947] But would it be possible for the US to move to a single payor system without eventually creating the access delays that plague Canada?

Working within the Canadian system, I have participated in several exercises to improve access by enhancing efficiency. One recent example was The Ottawa Hospital's Lung Cancer Transformation exercise. It aimed to reduce substantially the time between patient referral and initiation of treatment. This exercise involved a large group consisting of medical oncologists, radiation oncologists, thoracic surgeons, respirologists, radiologists, nuclear medicine physicians, pathologists, nurses, information technology specialists, process experts, data analysts, and patient family members. They met once per week for more than a year, with support from business process technology software. This dedicated team identified constraints in 12 major processes occurring between referral and therapy initiation. They resolved 270 constraints to support 57 workflow process changes.[694] They managed to reduce the median time from referral to initiation of treatment by 48% (from 92 days down to 47 days).[694]

Our Lung Cancer Transformation exercise demonstrated that access timelines can be substantially improved in the Canadian system. However, there are problems with exercises like this. Over the past four decades, I have participated in several of them. There is always initial success, but things then deteriorate as key components in the solution are hamstrung, either by budget cuts or by restructuring as part of a different exercise. So far, Lung Cancer Transformation has helped, but wait times are starting to creep up again as diagnostic imaging and others are subjected to further cutbacks.

When I worked at MD Anderson in Houston, a joke among some physicians was that the institution's usual way to tackle any problem was to appoint another assistant vice president. They would hire a whole new department to write new policies and procedures to deal with the issue. These would make it more difficult for physicians to do their job. New support staff would be hired to provide physicians with a work-around for the new workload burden. Any added costs were simply passed on to patients. There was little attempt to address the underlying issue. Philosophically, I prefer the Canadian approach of trying to reduce inefficiencies, but it would be nice if we could make solutions stick.

Conclusions: So, which is better? The Canadian system or the American system? I love them both and I hate them both. Canadians only get what they pay for, and Americans pay far too much for what they get. The American system particularly fails underinsured younger patients, the poor, and minorities while the Canadian system particularly fails older patients, who suffer and die as they wait too long for "free" but hard-to-access therapies that could help them. The fourteenth reason that cancer still sucks is that neither system has got things quite right.

— 15 —

Say Not the Struggle Naught Availeth: The Future of Cancer Care

So far, I have discussed fourteen reasons cancer still sucks. But there <u>has</u> been rapid progress. And because of the efforts of researchers working together worldwide, progress <u>will</u> continue. In this chapter, I will touch on just a few of the many things that might be coming in the not-too-distant future. The reality probably will far exceed what I discuss here.

Short Primer

"Say Not the Struggle Naught Availeth" by Arthur Hugh Clough has been my favorite poem since I discovered it in my early teens. I have a framed copy on the wall of my office. To me, it's our battle cry for the fight against cancer.

Science leads the way in this fight, even though there are always skeptics. When I was young, my grandfather told me about his older cousin, David Keenan. Keenan was almost fired from his job as an elementary school teacher in the early 1880s. His heretical message to students that through science, houses in the distant future would all be lit by electricity and one-day people would fly was too much for the local school board.

In 1976, when I began my medical oncology fellowship in the Department of Developmental Therapeutics at Houston's MD Anderson Hospital, department head Dr. Emil J Freireich taught us that the only people who came close to predicting the future are the science fiction writers, and they usually under-predict.

When he said this, none of us thought we would soon be walking around with phones in our pockets, or that we would no longer need a map since a computer in our car would be receiving directions from a satellite. Yet many things that we now take for granted were predicted by science fiction writers long before these wonderous advances came into existence.[988] The difficulty is conceptualizing a new possibility. If we can imagine it, then we may eventually be able to do it.

Below are some of my expectations.

Here, I will reverse the order with which I discussed things earlier in the book and will start with how I think our overall systems will change. I predict a major change in regulatory approaches to new drug development. Regulation is essential. But it currently takes far too long and costs far too much to take a new therapy form discovery to marketing. This translates into huge numbers of years of life lost- lives that could have been saved if the new therapy was available sooner. It translates into suffering that could have been relieved. The regulatory morass is a major factor directly and indirectly driving the high drug prices that threaten to bankrupt healthcare systems.[ccxix]

There also will be major changes in the way cancer care is delivered. The number of cancer patients is increasing as the population grows and ages. On average, patients with advanced cancers now are living much longer than they used to and are receiving active anticancer care for a much longer time. There are not enough oncologists to deliver this care. We need new approaches that are both safe and effective.

There is no easy, effective prevention strategy on the horizon, but gene therapy should eventually be an option for patients with *BRCA* mutations or other inherited cancer predispositions.[ccxx] For the rest of the population, we may have "synthetic lethality" approaches that are selectively toxic to cells with cancer-causing mutations. There also may be effective "senotherapy" that eradicates damaged senescent cells that cannot

---

[ccxix] See Chapters 11, 12 and 13 for a further discussion on this.

[ccxx] See Chapter 1 for a discussion on inherited predisposition to cancers.

divide, but that eventually might regain their ability to divide and progress to a malignancy.[ccxxi]

Within 10–20 years we probably will have sensitive, more effective blood tests for detecting early cancers.[ccxxii]

In 10–15 years, we won't be offering "adjuvant" systemic therapies to patients who have undergone surgical removal of a cancer.[ccxxiii] Instead, patients with persistent tumor DNA detected by sensitive blood tests will be given therapy to treat "minimal residual disease." Those with a negative blood test will have a low risk of recurrence and will be observed carefully, with no further therapy unless recurrence is discovered.

Provided we continue to invest in them, there will be an explosion of new therapies for patients with advanced, metastatic cancers. This will include synthetic lethality approaches, new targeted therapies, new immunotherapies, and new antibody drug conjugates. Not only will there be new drugs—there will also be new methods to rapidly and inexpensively manufacture highly personalized versions of these drugs that are specific for an individual patient.

For example, bispecific T-cell engagers (BITEs) are monoclonal antibodies with one arm that will attach to a tumor cell and another arm that will attach to an immune system T-cell.[ccxxiv] This brings the T-cell close to the tumor cell and facilitates immune destruction of the tumor cell. Within 10–20 years, I anticipate that we will be able to take a standard precursor antibody that already has one arm specific for T-cells and immediately synthesize a "hook" for the other arm that will be specific for targets in an individual patient's tumor. The "hook" for each patient will be changed as needed, as the characteristics of surviving tumor cells evolve. Technologies like this will rapidly displace much more expensive, toxic CAR T-cell approaches.

We also will become better at reversing "epigenetic" changes in tumor cells. This will allow us to turn back on tumor suppressor genes that have

---

[ccxxi] See Chapters 5 and 8 for further discussion of senescent cells.

[ccxxii] See Chapter 2 for a discussion of screening.

[ccxxiii] See Chapters 4,5, 6, 8, and 9 for further discussions on adjuvant therapies.

[ccxxiv] See Chapter 7 for a further discussion of BITEs and other immunotherapy approaches.

been silenced.[ccxxv] There will also come a breakthrough moment when we eventually understand how tumor stem cells differ from normal stem cell cells and begin to uncover their vulnerabilities. A single new development in this area could prove dramatically effective against not just one or two, but almost all common cancers.

By 10–30 years from now, babies will undergo a full assessment of their genotype at birth. Through their life, this assessment will guide therapy choices for them. This will maximize the probability of benefit and minimize toxicity. This will have huge therapeutic and economic benefits. It will reduce useless spending on therapies that prove ineffective or excessively toxic. Since therapies will be more effective and less toxic, it will also reduce illness-related lost productivity.

We also will soon be analyzing clinical trial data in a more effective way. In addition to current analyses, we will be using "population kinetics" assessments to gain new insights into therapy biology and efficacy.

Overall, we will continue to make rapid progress. By a very few decades from now, the reality will have markedly surpassed the timid predictions I make in this chapter.

Further Details and References

As in the earlier chapters, I will now delve more deeply into the nerdy details.

The world of cancer treatment has changed rapidly over the past several decades. Rapid changes will continue in regulatory approaches, healthcare delivery systems, our understanding of cancer, and cancer management options.

Regulatory approaches: I anticipate that soon we will reach a tipping point in our regulatory approaches to new drug development. The paradigms will shift. Tipping points have occurred repeatedly over the course of history. Governance structures evolve in response to societal needs, politics, pressures, and economic forces, bounded by prevailing ideological constraints. But eventually it becomes obvious that the system or processes are no longer tenable. Two examples are the collapse of communism in the Soviet Union and the rapid evolution of China

---

[ccxxv] See Chapter 1 for a further discussion on epigenetic changes.

into a major capitalist force that remains excessively authoritarian but is communist in name only.

The same must happen to the regulatory structures that govern new drug development. However, history also tells us that when tipping points come, the tip can be in either a positive or a negative direction. The Nazi rise in Germany in the 1930s and the dark threats that overshadowed Washington DC from November 2020 to January 6, 2021, are negative examples. History teaches us that societies that tip towards greater freedom of thought and action tend to prosper, while those that tip towards excessive rigidity tend to stagnate.[989]

Development of effective new therapies takes far too long and costs far too much. I hope that when we do reach a regulatory tipping point, it is in a positive direction. But this cannot be guaranteed.

With new drug development, current regulatory systems have focused sharply on the unequivocal need for safety, informed consent, patient privacy, and data integrity. But as these regulatory systems have evolved, they have inadequately considered the undesirable consequences of this focus: long delays, inefficiency, and exorbitant costs—dollar costs and years of life lost due to delays accessing new therapies.

We pay a price for the highly ordered nature of current regulatory systems. A chaotic system with no rules would be even worse.

We need another approach. A third possibility is a "complexity" approach with boundaries but with substantial freedom within the confines of those boundaries.[990] This approach has major advantages. It is the paradigm that fosters flourishing societies and the approach that permits Germany's Autobahn to function safely without speed limits.

An analogy from industry might be the design of a drainage pipe. One might specify the material requirements for the system (e.g., whether it is constructed from PVC vs steel, the pipe diameter, and the wall thickness and strength, etc.), or one might instead only specify the performance requirements, leaving the material criteria unspecified. The important consideration is whether it gets the job done, and not the details of the components.[ccxxvi]

[ccxxvi] I thank Bill Allison for this analogy.

The development of AIDS treatments and COVID-19 vaccines has demonstrated clearly that it is possible to accelerate regulatory processes safely. The way we protect safety, informed consent, privacy, and data integrity must change, but without sacrificing these essential objectives. This shift may happen very quickly once the world finally wakes up and recognizes the unnecessarily huge costs we are paying. So far, the world remains asleep.

Many people have told me that the required changes are not possible. But they are wrong. *"Say Not the Struggle Naught Availeth."* Of paramount importance, we must first recognize that these changes are needed—urgently. I do not know precisely what our destination will look like, but as outlined in Chapter 12, there are several things that can and must change.

Sir Winston Churchill once said, "Sometimes doing your best is not good enough. Sometimes you must do what is required."

Healthcare Delivery: In his 1976 Karnofsky Lecture delivered to the American Society for Clinical Oncology, Dr. Emil Freireich presented his seven laws.[796] One law was, "Always be prepared for success!" We are not prepared.

Cancer is becoming more common as the population grows and ages. As treatments rapidly improve, patients with metastatic cancer are living longer. A higher proportion of cancer patients are candidates for treatment. On average, these treatments carry on for a longer time since they are more effective than older treatments.

With a population of 333 million people, the USA has about 12,400 physicians practicing oncology and closely related specialties.[991] The projected growth of numbers of US medical oncologists is insufficient to meet anticipated American needs.[992]

Canada has a population of about 38 million people. At some point in their lives, 40–50% of Canadians will develop cancer, and cancer accounts for 30% of all deaths in Canada.[978] However, Canada only has about 625 medical oncologists.[ccxxvii][993] Despite the size of the cancer problem, there

---

[ccxxvii] There are also about 400 hematologists in Canada. In most centers, hematologists are responsible for treatment of leukemias, lymphomas, and multiple myeloma, which constitute about 10% of malignancies. Most Canadian hematologists do not care for patients with other types of malignancies. Many primarily practice "benign hematology" involving patients with disorders such as anemia and blood clots.

are fewer medical oncologists in Canada than cardiologists, nephrologists, gastroenterologists, respirologists, neurologists, dermatologists, infectious disease specialists, or general internists. In Ontario, the number of medical oncologists is rigidly limited by government.[ccxxviii]

In keeping with the shortage of Canadian specialists that I discussed in Chapter 14, average clinical workload is substantially higher for medical oncologists in Canada than in other high-income countries.[994]

Primary care practitioners can manage many aspects of a patient's care when it comes to problems involving the heart, lungs, gastrointestinal tract, infections, etc., but most do not have the training needed to help oversee systemic therapies for metastatic cancers. We need new models of care to help manage the anticipated growth in the cancer load.

Delays in access to oncology care can be deadly. For example, about 4% of patients with metastatic non-small cell lung cancer will die during each week that initiation of therapy is delayed.[424] In recognition of this, many jurisdictions have transferred the responsibility for day-to-day care of hospitalized cancer patients to hospitalists[ccxxix] so that oncologists are freed up to see new cancer patients in the clinic as rapidly as possible.

The growing specialty of palliative care has taken on an increasing share of complex symptom management and care of patients at the end of life. Survivorship programs are evolving to oversee ongoing care of patients who have completed curative cancer therapies.[995] In some Canadian centers, primary care physicians may be given extra training as "general practitioners in oncology" (GPOs) to work with oncologists in the delivery of cancer care.[996] Various other countries also have programs to educate primary care physicians further on management of cancer patients.[996] Advanced practice nurses can also play an important role in cancer care delivery. By necessity, we can expect to see rapid further evolution in the delivery of cancer care.

[ccxxviii]   This is not the case for most other specialties, where availability is primarily limited by the number trained rather than by the number permitted to practice.

[ccxxix] A hospitalist is a physician who specializes in caring for patients admitted to hospital. Oncology has evolved from an inpatient specialty to predominantly an outpatient specialty over the past few decades.

Prevention: Prevention as an alternative to treatment is a very appealing idea. But a pill to prevent cancer is unlikely to become available for at least several decades.

Prevention is often difficult to execute and can be very expensive.[997] It is also generally no more cost-effective than treatment, despite a strong bias that it is.[998] But it is better for the people who avoid getting sick.

We <u>have</u> made progress. For example, in response to pressure and education campaigns, the proportion of the Canadian adult population who smoke has dropped from 50% in 1965 to about 15% currently.[999] Smoking rates have been falling in males since at least 1965. In females, they began to fall in the late 1970s.[999] With this, the rate of lung cancer began to decrease in Canadian males in 1990 and in Canadian females in 2013.[1000]

Chapter 2 outlines how screening has also helped reduce risk. For some cancer types, screening improves outcome by finding early cancers. Some colorectal and cervical cancers can be prevented by finding and eradicating premalignant changes.

As we tackle climate change, air pollution should improve, with at least a small positive effect on lung cancer rates.[1001] Wide adoption of vaccines against hepatitis B and human papillomavirus will also help reduce the rates of malignancies caused by these viruses.[201,204] Smoking-related cancers may be reduced further by new strategies to decrease nicotine addiction.[1002] New approaches[1003] to reduce other factors—like obesity and inflammation—that increase cancer risk[ccxxx] also could potentially prove beneficial.

As discussed in Chapter 1, some individuals inherit an increased susceptibility to cancer. In many cases this is due to a dysfunctional version of a DNA repair gene[1004] such as *TP53*,[1005] *BRCA1*,[211] or *BRCA2*.[211]

"Gene therapy" (replacing or correcting abnormal genes) using various technologies such as "CRISPR-Cas gene editing"[1006] and "base editing"[1007] has great promise for diseases that are due to genetic aberrations.[1006] However, gene therapy faces many challenges and progress has been slow.[1006,1008]

[ccxxx] See Chapter 1.

In addition to major technical challenges, gene therapy clinical trials have faced substantial regulatory impediments.[1009] Early in the days of gene therapy clinical research, a small number of unexpected patient deaths spawned hyper-cautious oversight approaches that have markedly slowed progress.

My personal feeling is that extreme caution is justifiable for many diseases, but not for lethal diseases like metastatic cancers.

In our gene therapy study assessing *TUSC2*-bearing nanoparticles in patients with advanced, incurable cancers, we were forced to disregard valuable data and interrupt the trial for close to a year simply because some patients needed routine antihistamines and corticosteroids to reduce inflammatory symptoms with the therapy.[1010]

This unfortunate regulatory hyper-vigilance has finally begun to change.[1011] Numerous studies are now underway.[1009] I hope they will face fewer impediments and delays.

In addition to the technical and regulatory challenges confronting gene therapy, there will also be clinical research challenges determining whether it reduces the risk of cancer development in patients with an inherited predisposition. With some diseases, rapid symptom improvement might demonstrate the benefit of gene therapy. On the other hand, many years of observation of large numbers of patients would be required to determine whether gene therapy for inherited DNA repair defects reduced the risk of developing cancers.

An inherited mutation is present in all cells in the body and is called a "germline" mutation. Individual cells can also develop non-inherited mutations at any point during a person's life. When an individual cell develops a non-inherited mutation, it is called a "somatic" mutation. Somatic mutations are much more common than germline mutations.[1012]

The exact mutation can be determined in germline mutations. Preventative gene therapy strategies potentially could be devised to counter that mutation. Preventative strategies are much more difficult with somatic mutations. Many different cells in the same person may have many different somatic mutations. Any one of these could lead to development of a malignant cell. It would be difficult or impossible to detect and counteract each of these many mutations as a preventative strategy.

On the other hand, if a cancer develops due to one of these mutations, all the cancer cells will likely have the same mutation. This mutation would be relatively easy to detect. Consequently, a gene therapy strategy targeting a somatic mutation that is present in all the cancer cells might be relatively "easy." But it would be difficult to simultaneously target many different mutations as a prevention strategy.

TP53 is important in repair of DNA mutations.[1013] If the DNA damage is too extensive to permit repair, *TP53* mechanisms instead induce the cell to undergo senescence[ccxxxi] or apoptosis,[ccxxxii] [1013] so that the cell cannot divide to produce daughter cells carrying these mutations. If *TP53* itself is damaged in a cell, then the cell can keep on dividing despite the presence of further mutations.

The *TP53* gene is mutated in a higher proportion of cancers than any other gene.[1014] A wide variety of different *TP53* mutations can be associated with cancer development.[1014] Because many different *TP53* mutations can contribute to cancer development, it would be very difficult to come up with an effective preventative gene therapy strategy that would target enough of them to make a difference.

Another more feasible approach would be a "synthetic lethality" strategy. With synthetic lethality, the presence of an abnormality in a cell may make it vulnerable to drugs that do not affect normal cells.[1015] An example of this is use of drugs called "PARP inhibitors" against cancers with a *BRCA* mutation. PARP inhibitors increase DNA strand breaks. In normal cells, the *BRCA* systems repair these DNA strand breaks, but if these repair systems are mutated, the PARP-induced damage can prove lethal.[1015,1016] Clinically, PARP inhibitors have proven effective against metastatic cancers with mutations in *BRCA* and similar DNA repair systems.[1017] PARP inhibitors can reduce development of cancers in *BRCA*-mutant animal models. There is clear potential for eventual clinical application as a preventative approach in patients with known germline *BRCA* mutations.

[ccxxxi] Senescence means that the cell remains alive but is no longer able to divide to form new cells.

[ccxxxii] Apoptosis is a system by which a cell is programmed to undergo a series of successive changes that ultimately result in cell death.

Various approaches may give synthetic lethality if *TP53* is mutated.[1015] Hence, a future cancer preventative strategy might involve periodic administration of agents conferring synthetic lethality on cells with a *TP53* mutations.

Detection: A major hunt is underway for a blood test that will detect early cancers. As discussed in Chapter 2, current screening methods save lives, but their impact is limited. I was hopeful that more sensitive blood tests to detect oncogene mutations might prove useful,[1018] but I now realize that there is a problem. Many of the oncogene mutations that drive cancer growth may also be present in benign conditions.[1019] In some cases, they can be even more common in benign conditions than in malignancies.[1019] For example, about 50% of malignant melanomas are driven by a *BRAF V600E* mutation,[1020] but this mutation is found in about 80% of benign skin moles.[1019]

Consequently, sensitive blood tests to detect oncogene mutations might not be very helpful, since there could be too many false positive tests.[1019] However, several other options are being assessed. For example, circulating cells with an abnormally high number of copies of a gene may be an indicator of an underlying malignancy. Detection of autoantibodies, tumor-associated antigens, and gene and protein expression patterns are all being explored as potential methods for early cancer detection.[1019] I anticipate that one or more of these methods will be more effective than current screening methods, and will be in widespread use within the next two to three decades.

Adjuvant treatment for cancers treated with curative intent: In Chapters 4 and 6, I discussed the use of "adjuvant" systemic therapies to increase the probability of cure for localized cancers in patients undergoing surgery. Most adjuvant approaches have a relatively modest effect. Many patients experience a recurrence of their cancer despite adjuvant therapy, and others would have been cured by the surgery alone without adjuvant therapy.

In Chapter 2, I also discussed some of the principles of Bayes' theorem. The impact of screening will be greater in a high-risk population than in a low-risk population. The same also generally holds true for adjuvant therapy.[ccxxxiii] High-risk populations will generally gain more from adjuvant treatment than will low-risk populations. Consequently, it is important that we be able to identify high-risk and low-risk populations.

The key question with adjuvant therapies is whether giving them after surgery is more effective than only offering therapy to those who relapse. On the one hand, some patients will have been cured by the surgery. We cannot tell who these patients are, but they will not benefit from the adjuvant therapy. The adjuvant therapy has direct and indirect costs, associated inconveniences and toxicities. Death from non-cancer causes may be increased significantly by adjuvant therapies,[1021] and they may be associated with premature aging.[621,659]

As noted in Chapter 8, up to 35% of patients may experience "chemo brain" that lasts months to years after the end of therapy,[301,621] and some patients will have MRI scan evidence of brain atrophy.[301] Assessment of patients who have survived childhood cancers tells us that, in the long term, cancer treatments can be associated with premature death from cardiovascular disease and other causes.[659]

On the other hand, if therapy is delayed until tumor recurrence, it may be less effective than adjuvant therapy. This is because tumor bulk will be higher at relapse than at the time of adjuvant therapy. In addition, patients who relapse may deteriorate and die so rapidly that they never make it to treatment for their recurrent cancer.

Early experience suggests that there is only a small risk of later cancer recurrence if sensitive blood testing methods do not show evidence of

---

ccxxxiii   As an example, let's say one group of patients has an 80% chance of being cured by surgery alone. Let's also say for this example that adjuvant treatment will cure 30% of the patients who would have relapsed after surgery alone. Therefore, the probability of cure would be 80% with surgery alone but would be 80% + (30% x 20%) = 86% with adjuvant therapy. Let's say in a different example that the chance of cure with surgery alone was 20%, but adjuvant therapy again cured 30% of patients who would have relapsed after surgery alone. In this case, the cure rate would be 20% with surgery alone but would be 20% + (30% x 80%) = 44% with adjuvant therapy. The 24% absolute gain in cure rate in the high-risk group would be much greater than the 6% increase in cure rate in the good risk group, despite the treatment saving 30% of patients destined to relapse in both situations.

tumor cell DNA after surgery or chemoradiation. Conversely, there is a high risk of recurrence if circulating tumor DNA remains detectable.[359,1022]

The methodology to detect minimal residual disease through assessments of circulating tumor DNA is becoming more sensitive.[1023] I predict that within less than ten years, we will no longer be considering adjuvant therapies. Instead, we will be offering treatment to patients identified by circulating tumor DNA as having "minimal residual disease" after surgery, while we will simply observe patients with no detectable circulating tumor DNA.

Systemic therapies for metastatic cancers: We can expect an explosion of new therapies for patients with metastatic cancers. We have been progressing very rapidly, and this progress is likely to continue, provided we continue investing.[ccxxxiv]

Gene therapies and synthetic lethality: The gene therapy and synthetic lethality approaches I discussed as possible methods of cancer prevention also could be effective against established cancers. For example, PARP inhibitors have already demonstrated efficacy as a synthetic lethality approach in patients with metastatic cancers harboring mutations in *BRCA* and other DNA repair systems. This includes cancers of the breast,[1024,1025] ovary,[1026] prostate,[1027] and pancreas.[1028]

Targeted therapies: As discussed in Chapter 6, many lung adenocarcinomas and some other malignancies have mutated growth factor receptors driving tumor cell growth. Inhibitors of these growth factor receptors can yield rapid tumor regression.

We have not identified exploitable tumor cell drivers for various other malignancies. For example, with both squamous[1029] and small cell lung cancers,[1030] several genetic alterations have been identified, but we cannot target them effectively.

However, progress tends to be stepwise. For example, researchers first discovered that the epidermal growth factor receptor gene was mutated in some lung adenocarcinomas and that blocking the mutated protein could lead to marked tumor regression.[456,457] This led to the rapid discovery of

---

[ccxxxiv]As discussed in Chapter 13, high drug prices yield high profits and this in turn stimulates the further investment needed for progress. Many different factors might severely limit future profits on drug sales, and this could slow progress.

multiple other mutated growth factor receptor genes and the development of matching targeted therapies.

The essential point is that a key new insight into one cancer will rapidly uncover vulnerabilities in a range of other cancers. Thus, the same kind of progress is likely to happen as research reveals how other changes drive growth in other malignancies.

Research is also rapidly revealing how cancers that are initially sensitive to a targeted therapy eventually develop resistance. In many cases, this is due to outgrowth of more slowly growing tumor cell clones that are resistant to the targeted therapy. Once these resistance mechanisms are discovered, they can be targeted. For example, the *EGFR T790M* mutation was found to be a frequent cause of acquired resistance to early EGFR inhibitors. Osimertinib is highly effective against cells with *T790M* mutations,[483] but osimertinib-resistant cells with mutations such as *C797S* eventually appear in patients receiving osimertinib.[1031] However, there are early indications that the monoclonal antibody amivantamab may be effective against lung cancers that are resistant to osimertinib due to *EGFR C797S* mutations.[1032]

As effective new targeted therapies emerge, acquired resistance also emerges. But strategies to tackle these acquired resistance mechanisms may in some cases be very close behind.

Immunotherapy: At least thirteen immune checkpoint systems help protect normal cells from attack by the body's immune system.[510] Cancer cells may escape destruction by immune system T-cells by hiding behind these checkpoints. As I discussed in Chapter 7, we now know that inhibiting two of these immune checkpoints—the PD-1/PDL-1 checkpoint and the CTLA4 checkpoint—can be effective in at least some patients with a wide range of malignancies. We also now know that it is feasible to combine the PD-1 inhibitor nivolumab with the CTLA-4 inhibitor ipilimumab.[1033-1035] At least in malignant melanoma, these two drugs together are more effective than either one alone, although toxicity is somewhat increased.[1036] However, this toxicity is usually (but not always) manageable.

I am optimistic that inhibition of some of the other eleven immune checkpoints may improve immunotherapy efficacy further. For example, early experience suggests that inhibitors of the immune checkpoint TIGIT may improve efficacy of PD-1/PDL-1 inhibitors in non-small cell lung cancer.[1037,1038]

In some patients, immunotherapy efficacy is limited by lack of expression of neoantigens that the immune system can target. This could be true particularly for cancers with defects in the antigen presenting machinery.[ccxxxv] [501] These defects could markedly limit the immune system's ability to detect tumor-associated neoantigens. Steps to increase the neoantigen load might increase the probability of detection by the immune system.

Various strategies have been proposed to increase the available tumor neoantigens that might be presented to the immune system. For example combining either chemotherapy,[569-573] radiotherapy,[1039,1040] or an "oncolytic virus"[1041] with immunotherapy may improve efficacy. Chemotherapy-, radiotherapy-, or virus-induced augmentation of neoantigen load is one potential mechanism by which this potentiation could occur.

There is also a growing appreciation that the microbiome—the bacteria that inhabit the body—may play a major role in modulating the immune system's reaction to a cancer.[1042] A better understanding of the mechanisms at play could provide us with new therapeutic tools. I anticipate that within a very few years, we will be manipulating the body's bacterial populations to enhance the efficacy of cancer immunotherapy.

Bispecific T-cell engagers (BITEs): Monoclonal antibodies have two arms. Both arms typically target the same antigen. A BITE is a monoclonal antibody in which one arm targets an antigen on the surface of a tumor cell and the other arm targets a T-cell.[1043] When the BITE attaches to both the tumor cell and the T-cell, it brings the T-cell in close enough to the tumor cell to destroy it.[1043] Early clinical studies indicate that BITEs can be effective.[1043,1044]

Currently, each BITE targets a specific pre-determined antigen on a tumor cell. Typically, this target is a normal protein that is overexpressed in the tumor. However, because it is a normal protein, some normal cells will also be impacted.

I anticipate that in the future, it will be possible to rapidly identify tumor neoantigens that are present only on tumor cells. This will enable inexpensive personalization of BITEs specific for that neoantigen. Immune

---

[ccxxxv] See Chapter 7 for a further discussion of this.

checkpoints might inactivate T-cells attached to BITEs. Thus, combining immune checkpoint inhibitors with a BITE may prove more effective than a BITE alone.

With CAR T-cells and similar strategies, a patient's T-cells are altered to increase the probability that they will attack a cancer.[1045] Large batches of these altered T-cells are grown in the laboratory before being infused into the patient. This process takes weeks, is very expensive, and has several limitations.[1045] Customized BITEs could bypass these problems by converting any of the patient's T-cells into potential anti-tumor T-cells, thereby avoiding the long, expensive process of growing extra T-cells in the laboratory.

Antibody drug conjugates (ADCs): ADCs are constructed by attaching a toxic drug to a monoclonal antibody. The monoclonal antibody is directed at an antigen that is overexpressed on tumor cells and will accumulate in the tumor. ADCs thereby have the potential to increase drug concentrations in tumor cells. There are now several ADCs that have proven therapeutically useful, including trastuzumab emtansine[494] and trastuzumab deruxtecan[1046] (targeting cells overexpressing HER2).

The ADC's target may be overexpressed on tumor cells, but also is expressed by some normal cells. These also may be damaged. Like BITEs, future ADCs also might be personalized against tumor neoantigens.

Another issue with ADCs is that "dose-response curves" flatten at higher drug doses for most malignancies.[ccxxxvi][436,437] That is probably why high dose chemotherapy with stem cell support does not work in these cancers.[436] In some cases, dose-response curve flattening may be due to limitations of drug uptake into tumor cells.[429] In this situation, ADCs

---

[ccxxxvi] "Dose-response curve" means the probability of response or benefit increases as the therapy dose increases. "Dose-response curve flattening" means that the degree of benefit stops increasing further as the therapy dose is increased. At high doses, dose-response curves typically remain steep for hematological malignancies like leukemias and lymphomas and for germ cell cancers like testicular carcinoma. Consequently, high dose chemotherapy can cure some patients with these malignancies. Conversely, dose-response curves typically begin to flatten at higher doses for the most common malignancies like adenocarcinomas, squamous cell carcinomas, and neuroendocrine carcinomas. Even very high dose chemotherapy cannot cure these common cancers with dose-response curves that flatten at higher doses. See Chapter 1 for a further discussion of this.

might increase drug efficacy by increasing drug uptake. However, several other factors may contribute to dose-response curve flattening.[429] These may limit the ultimate benefit of ADCs, and may result in BITEs being a better option than ADCs.[ccxxxvii]

Epigenetics: Cancers may develop because tumor suppressor genes have been mutated or deleted. Or, they may be silenced through "epigenetic" changes.[1047]

Almost every cell in your body has the same genes, but epigenetic mechanisms selectively silence a high proportion of all genes in each cell. The ones that remain active determine the cell's structure and function. That is why your liver looks and works differently than your kidney, why your brain is different from your heart and why your nose is different from your toe.[1048]

Genes are epigenetically silenced by two major mechanisms. One is through the addition of "methyl" groups to specific DNA sites. The other is via changes in the way a protein called "chromatin" interacts with DNA.[1048,1049]

Reactivating the tumor suppressor gene by reversing the abnormal epigenetic changes might cause tumor regression in cancers that develop because epigenetic mechanisms have silenced a tumor suppressor gene. Drugs called DNA methyltransferase inhibitors[1047,1049] and histone deacetylase inhibitors[1050] potentially can reverse epigenetic changes by altering DNA methylation and chromatin. These drugs have proven beneficial in only a limited number of malignancies. However, there is a potential for marked further benefit in the future.

Stem cell approaches: Most of our tissues have "stem cells." They are a mother lode of reserve cells that divide to produce daughter cells to replace aging, dying normal cells.

Tumors also have stem cells.[1051] We will be able to eliminate a cancer only if we can eradicate or inactivate its stem cells. However, stem cells are notoriously resistant to treatment, and we don't know exactly why. If we knew, we might be more successful at destroying them. We also don't know if and how defense mechanisms differ in tumor and normal stem cells. If

---

ccxxxvii   This, of course, remains conjecture.

they are identical, we may be unable to develop anti-stem cell strategies that selectively target tumors. However, if there is anything fundamentally different in tumor stem cells compared to normal stem cells, then a single new therapy could prove effective against a broad range of malignancies. There is an ongoing search for new agents that will target cancer stem cells.[1052-1054]

Patient selection/prediction of therapy benefit: As discussed in Chapter 6, targeted therapies may be highly effective at inducing regression of tumors expressing their target. With chemotherapy, numerous factors may be associated with resistance,[429] but we do not have any effective way to predict who will benefit. As discussed in Chapter 7, PD-1/PD-L1-inhibiting immunotherapy drugs are more likely to be effective against tumors with high expression of PD-L1 or with high tumor mutation burden. But these are relatively poor predictors of therapy benefit. Immunotherapies behave as if there is a factor that determines whether they will be effective. It is either there or not. This is the same way that presence—vs.—absence of a target determines whether a targeted therapy will be effective.[425] While we have not been successful yet in identifying this reliable predictor of immunotherapy benefit, I anticipate that researchers will find it within five to ten years.

Pharmacogenetics: As I discussed in Chapters 1 and 8, there may be several different "normal" versions of a given gene. These versions typically differ from each other based on variations in a single DNA nucleotide in the gene. These different normal versions are referred to as "single nucleotide polymorphisms" or SNPs.

While gene variants with different SNPs may be regarded as being normal, they may confer an altered level of function for the protein coded by the gene. When it comes to genes for proteins involved in drug metabolism, a person with one SNP may break down or activate a drug more effectively than a person with a different SNP. This means that one person might tolerate a high dose of a drug and might require a high dose for it to be effective against their cancer. Another person might tolerate or require only a low dose.

For example, different SNPs for various genes are associated with efficacy of the chemotherapy drug cisplatin in patients with advanced lung cancer.[668-670,1055] SNPs related to inflammation are associated with the risk of lung toxicity from radiotherapy.[422] Other SNPs are associated with the risk of severe toxicity with the chemotherapy drugs irinotecan[1056] and 5-fluorouracil.[1057]

Despite its huge potential, progress in the field of pharmacogenetics has been much slower than I anticipated. I am cautiously optimistic that this will change as genotyping becomes progressively cheaper. I envision a future where everyone is genotyped at birth, to define their full range of SNPs across a wide range of genes. This information will then become a standard factor in the individualized selection of therapy types and doses.

Current popular genetic testing platforms like 23andMe and Ancestry. com are an early step toward this. But at this stage, the information they provide generally is not reliable enough[1058] or extensive enough to be useful in the management of most patients.

Currently, if a patient requires treatment with a particular drug, they may be genotyped for SNPs directly relevant to that drug. For example, in some jurisdictions, a patient who requires treatment for colon cancer may undergo testing for SNPs in the gene dihydropyrimidine dehydrogenase that impact their ability to metabolize the chemotherapy drug 5-fluorouracil.[1057]

Patients with some SNPs will have little problem with standard doses of the drug, but patients with a different SNP will experience marked toxicity. These people either would not be treated with this drug or would be treated with a very low dose. I anticipate that in the future, genotyping related to a single drug will no longer be required, since we will already have extensive patient-specific genotyping data at our disposal.

Drug production: The future will also see a revolution in how drugs are manufactured. Currently, making some drugs is straight forward, but producing others is complicated and expensive. We can predict that future advances in drug manufacturing using approaches like click chemistry and the equivalent of 3D printing of drug molecules will make production much faster and cheaper.[1018] This will be a key component of approaches such as the personalization of BITEs and ADCs that I discussed earlier.

Conclusions: In this chapter, I have touched on just a few of the many strategies and forces that could impact future progress. There will be many more changes than the ones I have discussed, including new approaches to analyzing and interpreting clinical research data that will yield further insights.[424,425,432,577,708,1059]

Unequivocally, we are making progress despite the many obstacles we face. At times, progress seems far too slow, but what appears to be just one small step forward can have profound unforeseen consequences. The last verse of *"Say Not the Struggle Naught Availeth"* captures this perspective well:

> *And not by eastern windows only,*
> *When daylight comes, comes in the light,*
> *In front the sun climbs slow, how slowly,*
> *But westward, look, the land is bright.*

Acknowledgments

Many people helped in the creation of this book. I would particularly like to thank Dr. John-Peter Bradford (a frequent collaborator and CEO of the Life-Saving Therapies Network) who made numerous valuable suggestions about the structure and focus of the book. It was his suggestion to separate each chapter into a "short primer" vs. "further details" sections. He also spent a huge amount of time tackling an endless litany of split infinitives and a jungle of passive-voice statements. The book would have been less informative and much more difficult to read without his incredible input. John-Peter is a passionate, articulate, relentless advocate for cancer patients and their families.

Dr. William K. (Bill) Evans also reviewed the manuscript in detail and provided highly valued feedback and comments. Bill has had decades of experience as a medical oncologist and is one of the hardest working people that I know. He has held several senior positions in Ontario. At different times, he headed the cancer centers in Ottawa and Hamilton, and he very ably led Cancer Care Ontario's provincial Systemic Therapy Program from 2000 to 2005. He has unique insights into the issues that we face and has played a key role in enabling access to effective new anticancer agents for Ontario patients and their physicians.

My wife, Lesley, extensively reviewed the book from a lay perspective, and helped me word things in a way that would be more easily understood by patients and their families. My stepdaughter Jenika Alvarez did the important early footwork in figuring out how to get the book published and convinced me that we could actually do this. She also provided highly valued, detailed feedback on the book's contents.

I would like to thank my sons Andrew and Adam Stewart, my daughter Megan Stewart, and my stepson Grayson Adolph for their feedback and support. I am indebted to my entire family for putting up with me for the three years during which I was consumed with writing this book. Andrew also played a key role in the research required for our publication on number of years of life lost for every year that approval of an effective new drug is delayed.

Friend and talented artist Joni McCollam kindly provided the first draft of the cartoon crab for the book cover.

My cousin, aquatic ecologist Bill Allison, provided very valuable insights and a critical assessment of the book, from his perspective as a scientist.

Ottawa Hospital CEO Cameron Love gave greatly appreciated feedback on Chapter 14 (as well as doing a superb job leading The Ottawa Hospital through the COVID-19 pandemic).

Dr. Razelle Kurzrock provided several insightful comments. In particular, she drove home the fact that, while high drug prices are painful, these high prices are a key factor behind the investment that is fueling rapid progress. Razelle is a fellow Canadian, a frequent collaborator and a former colleague at MD Anderson Hospital in Houston. After MD Anderson, she was at the University of California in San Diego for 9 years. She is now with the Medical College of Wisconsin and is the Chief Medical Officer of the Worldwide Innovative Network (WIN) for Personalized Cancer Therapy. She has a better insight into both the biology of cancer and the complexity of new drug development than any other clinician I know.

Colleague Dr Paul Wheatley-Price has been a highly effective patient advocate through his work at Lung Cancer Canada. He gave me helpful advice on the structure of the book, as did several of my former medical school classmates, including Drs. Bob Reid, Gord Francis, Lee Ford-Jones, Ed Kostashuk, Mike Wright and Jim Brown.

I would also like to acknowledge the tremendous effort of my medical oncology colleagues and all those involved in cancer care at The Ottawa Hospital. It is an incredible team effort as they together do their best to ensure that our patients are well cared for despite the serious constraints and challenges we face. The team includes nursing staff, pharmacists, clerks, general practitioners in oncology, radiologists, pathologists, the molecular laboratory, hospital administrators and operations staff, emergency room teams, radiation oncologists, surgeons, general internists, internal medicine subspecialists and so many more at the Ottawa Hospital, the Queensway Carleton Hospital and all the smaller centers and chemotherapy administration clinics across the region.

I would like to recognize Paula Doering for her vision, leadership, support, and drive in the years she spent as Vice President for Cancer at The Ottawa Hospital. I also would like to recognize Dr. Michael Fung-Kee-Fung for his relentless, innovative, uniquely insightful efforts to streamline and transform cancer care in the Ottawa region, and Dr. Neil Reaume for the superb job that he has done leading the Division of Medical Oncology in Ottawa over the past 3 years.

In some of my earlier publications on the need for regulatory reform in new drug development, some reviewers have interpreted my concerns as a criticism of regulators in general and the FDA in particular. I once more want to stress that the regulators did not cause our current problems, and they cannot fix the problems on their own. We all have a role to play. I greatly appreciate the work done by Dr. Richard Pazdur and his colleagues at the FDA and the regulators at Health Canada, the EMA and other groups as they have worked to enable patient access to effective new therapies. We must work together to make things better.

My thanks to the team at Tellwell Talent, Ltd. who made publication of this book so much easier than I had anticipated. This includes Project Manager Sem Delima, Publishing Consultant Angelo Abadia, Cover Designer Gerardo Basilio Faelnar, Copy Editor Jodi McGuffin, Layout Designer Jine Mosquera and all the other members of their team.

Finally, I would like to thank our patients and their families for their patience and understanding as it takes us far too long to move things forward for them.

About the Author

David J. Stewart, MD, FRCPC is a medical oncologist at the Ottawa Hospital and a professor of medicine at the University of Ottawa in Ottawa, Ontario, Canada.

He grew up on a farm one mile west of Dalmeny, Ontario in a farmhouse built by his grandfather more than four decades before his birth. He attended elementary school at S.S. #23 Osgoode—a one-room schoolhouse a quarter of a mile down a gravel road from Dalmeny. Each year from grade one to grade eight, he and his classmates moved over one row of seats to the right. His early-life observations there sensitized him to the injustices often suffered by disadvantaged individuals. After S.S. #23, he then spent one and a half hours per day riding a school bus down back country roads to attend Osgoode Township High School in Metcalfe, Ontario.

In January 1968, he was one of nineteen Canadian delegates to an international students' conference in London, England, organized by the Council for Education in World Citizenship as part of UNESCO's International Year of Human Rights.

From 1968 to 1974 he attended premedicine and medical school at Queen's University in Kingston, Ontario. He then completed a residency in Internal Medicine at McGill University/Royal Victoria Hospital in Montreal, Quebec. In 1976, he began a medical oncology fellowship in the Department of Developmental Therapeutics at University of Texas MD Anderson Hospital and Tumor Institute in Houston. Following its completion, he accepted a staff position at MD Anderson.

In 1980, he moved back to Ottawa as a medical oncologist at The Ontario Cancer Treatment and Research Foundation Ottawa Regional Cancer Centre. From 1989 to 1999, he served as Head of Medical Oncology at the Ottawa Civic Hospital. In 2003, he moved back to MD Anderson Hospital in Houston to pursue research opportunities but returned again to Ottawa in 2011 as a Professor of Medicine and Head of the Division of Medical Oncology at the University of Ottawa and The Ottawa Hospital.

He completed his term as Division Head in 2019 and continues to practice and teach oncology in Ottawa.

He has 340 peer-reviewed publications, 13 invited publications, 11 published academic editorials, 26 book chapters, and 413 published scientific abstracts on presentations at national and international meetings. He coedited the book *Nausea and Vomiting: Recent Research and Clinical Advances* (CRC Press, Inc; 1991) and edited the book *Lung Cancer: Prevention, Management and Emerging Therapies* (Humana Press; 2010). He has also published thirteen letters to the editor in scientific journals and has published several op-eds in the lay press, primarily related to issues impacting patient access to care. His current research is on new methods to interpret clinical trial data and on impact of therapy delays.

In 2016, he and his collaborator Dr. Razelle Kurzrock were presented with the annual Federa Award by the Federation Foundation of Dutch Medical Scientific Societies. It recognized their call for reforms to clinical research methods so that access to effective new drugs could be accelerated.

Two books profoundly impacted his approach to life. One was Victor Hugo's *Les Misérables*—which taught him the importance of compassion and that all people have the potential for good. Ayn Rand's *Atlas Shrugged* taught him that you should never apologize for trying to achieve. His guiding principle is, "It is not what happens to you in life that is important. It's what you do about it."

Index

A

E

N

Endnotes

1. Siegel R. et al. Cancer statistics, 2019. CA Cancer J Clin 2019;69:7.

2. Cancer statisitcs at a glance. Canadian Cancer Society https://www.cancer.ca/en/cancer-information/cancer-101/cancer-statistics-at-a-glance/?region=on; accessed 04/28/2019.

3. Cunningham R. et al. The Major Causes of Death in Children and Adolescents in the United States. N Engl J Med 2018;379:2468.

4. How many years of life are potentially lost due to cancer? Canadian Cancer Statisitics 2013. Canadian Cancer Society accessed 2019/04/28 https://www.cancer.ca/~/media/cancer.ca/CW/cancer%20information/cancer%20101/Canadian%20cancer%20statistics%20supplementary%20information/2011-2015/2013-PYLL-EN.pdf?la=en.

5. Taksler G. at al. Assessing Years of Life Lost Versus Number of Deaths in the United States, 1995-2015. Am J Public Health 2017;107:1653.

6. Life Expectancy of the World Population. Worldometer https://www.worldometers.info/demographics/life-expectancy/. accessed 2021/05/23.

7. Ezkurdia I. et al. Multiple evidence strands suggest that there may be as few as 19,000 human protein-coding genes. Hum Mol Genet 2014;23:5866.

8. Piovesan A. et al. GeneBase 1.1: a tool to summarize data from NCBI gene datasets and its application to an update of human gene statistics. Database (Oxford) 2016;2016.

9. Roy A. et al. Toward mapping the human body at a cellular resolution. Mol Biol Cell 2018;29:1779.

10. The Molecular Biology of Cancer: A Bridge from Bench to Bedside. 2nd ed: Wiley and Sons; 2013.

11. McCulloch S. et al. The fidelity of DNA synthesis by eukaryotic replicative and translesion synthesis polymerases. Cell Res 2008;18:148.

12. Qian Y. et al. Senescence regulation by the p53 protein family. Methods Mol Biol 2013;965:37.

13. Epigenetics Wikipedia https://en.wikipedia.org/wiki/Epigenetics; accessed 04/30/2019.

14. Armitage P. et al. The age distribution of cancer*. Br J Cancer 1954;8:1.

15. Di Gregorio A. et al. Cell Competition and Its Role in the Regulation of Cell Fitness from Development to Cancer. Dev Cell 2016;38:621.

16. Talhout R. et al. Hazardous compounds in tobacco*. Int J Environ Res Public Health 2011;8:613.

17. Hecht S. Tobacco carcinogens, their biomarkers & cancer*. Nat Rev Cancer 2003;3:733.

* Throughout this section, "*" means I truncated the title to save space.

18. Warren G. Tobacco and lung cancer*. Am Soc Clin Oncol Educ Book 2013:359.

19. Kuper H. et al. Tobacco use and cancer causation*. J Intern Med 2002;252:206.

20. Liang P. et al. Cigarette smoking and colorectal cancer incidence & mortality: systematic review and meta-analysis. Int J Cancer 2009;124:2406.

21. Colamesta V. et al. Do the smoking intensity and duration, the years since quitting, the methodological quality and the year of publication of the studies affect the results of the meta-analysis on cigarette smoking and AML in adults?*. Crit Rev Oncol Hematol 2016;99:376.

22. Lugo A. et al. Strong excess risk of pancreatic cancer for low frequency and duration of cigarette smoking: A comprehensive review and meta-analysis. Eur J Cancer 2018;104:117.

23. Abdel-Rahman O. et al. Cigarette smoking as a risk factor for the development of and mortality from hepatocellular carcinoma*. J Evid Based Med 2017;10:245.

24. Sasco A. et al. Tobacco smoking and cancer*. Lung Cancer 2004;45 Suppl 2:S3.

25. Macacu A. et al. Active and passive smoking and risk of breast cancer: a meta-analysis. Breast Cancer Res Treat 2015;154:213.

26. Benowitz N. et al. Nicotine chemistry, metabolism, kinetics and biomarkers. Handb Exp Pharmacol 2009:29.

27. Malhotra J. et al. Association between Cigar or Pipe Smoking and Cancer Risk in Men: A Pooled Analysis of Five Cohort Studies. Cancer Prev Res (Phila) 2017;10:704.

28. Chang C. et al. Systematic review of cigar smoking and all cause and smoking related mortality. BMC Public Health 2015;15:390.

29. Tredaniel J. et al. Exposure to environmental tobacco smoke and risk of lung cancer: the epidemiological evidence. Eur Respir J 1994;7:1877.

30. Yan H. et al. Secondhand smoking increases bladder cancer risk in nonsmoking population: a meta-analysis. Cancer Manag Res 2018;10:3781.

31. Yang C. et al. Passive Smoking and Risk of Colorectal Cancer*. Asia Pac J Public Health 2016;28:394.

32. Lee P. et al. Epidemiological evidence on environmental tobacco smoke and cancers other than lung or breast. Regul Toxicol Pharmacol 2016;80:134.

33. Tindle H. et al. Lifetime Smoking History and Risk of Lung Cancer: Results From the Framingham Heart Study. J Natl Cancer Inst 2018;110:1201.

34. Coussens L. et al. Inflammation and cancer. Nature 2002;420:860.

35. Durham A. et al The relationship between COPD and lung cancer. Lung Cancer 2015;90:121.

36. Koshiol J. et al. Chronic obstructive pulmonary disease and altered risk of lung cancer in a population-based case-control study. PLoS One 2009;4:e7380.

37. Browman G. et al. Influence of cigarette smoking on the efficacy of radiation therapy in head and neck cancer. N Engl J Med 1993;328:159.

38. Warren G. et al. Smoking cessation after a cancer diagnosis and survival in cancer patients. J Clin Oncol 2018;36:1561.

39. Gritz E. et al. Smoking, the missing drug interaction in clinical trials: ignoring the obvious. Cancer Epidemiol Biomarkers Prev 2005;14:2287.

40. Walter V. et al. Smoking and survival of colorectal cancer patients*. Ann Oncol 2014;25:1517.

41. Duan W. et al. Smoking and survival of breast cancer patients*. Breast 2017;33:117.

42. O'Malley M. et al. Effects of cigarette smoking on metabolism and effectiveness of systemic therapy for lung cancer. J Thorac Oncol 2014;9:917.

43. Day G. et al. Second cancers following oral and pharyngeal cancers: role of tobacco and alcohol. J Natl Cancer Inst 1994;86:131.

44. Canistro D. et al. E-cigarettes induce toxicological effects that can raise the cancer risk. Sci Rep 2017;7:2028.

45. Rowell T. et al. Will chronic e-cigarette use cause lung disease? Am J Physiol Lung Cell Mol Physiol 2015;309:L1398.

46. Leventhal A. et al. Association of Electronic Cigarette Use With Initiation of Combustible Tobacco Product Smoking in Early Adolescence. JAMA 2015;314:700.

47. Aldington S. et al. Cannabis use and risk of lung cancer*. Eur Respir J 2008;31:280.

48. Cannabis and cannabinoids: Cancer risk and use to manage cancer symptoms. Ontario Health-Cancer Care Ontario 2020/07/24. https://www.cancercareontario.ca/sites/ccocancercare/files/guidelines/summary/OntarioHealthEvidenceSummary-Cannabis-2020-08-07 Clean.pdf. Accessed 2021/11/02.

49. Sledzinski P. et al. The current state and future perspectives of cannabinoids in cancer biology. Cancer Med 2018;7:765.

50. Narayanan D. et al. Ultraviolet radiation and skin cancer. Int J Dermatol 2010;49:978.

51. Savoye I. et al. Patterns of Ultraviolet Radiation Exposure and Skin Cancer Risk: the E3N-SunExp Study. J Epidemiol 2018;28:27.

52. van der Pols J. et al. Prolonged prevention of squamous cell carcinoma of the skin by regular sunscreen use. Cancer Epidemiol Biomarkers Prev 2006;15:2546.

53. Xie F. et al. Analysis of association between sunscreens use and risk of malignant melanoma. Int J Clin Exp Med 2015;8:2378.

54. Green A. et al. Point: sunscreen use is a safe and effective approach to skin cancer prevention. Cancer Epidemiol Biomarkers Prev 2007;16:1921.

55. Autier P. et al. Sunscreen use and increased duration of intentional sun exposure: still a burning issue. Int J Cancer 2007;121:1.

56. Gallagher R. et al. Broad-spectrum sunscreen use and the development of new nevi in white children: A randomized controlled trial. JAMA 2000;283:2955.

57. Watts C. et al. Sunscreen Use and Melanoma Risk Among Young Australian Adults. JAMA Dermatol 2018;154:1001.

58. Stern R. et al. Risk reduction for nonmelanoma skin cancer with childhood sunscreen use. Arch Dermatol 1986;122:537.

59. Wolin K. et al. Obesity and cancer. Oncologist 2010;15:556.

60. Berrington de Gonzalez A. et al. Body-mass index & mortality*. N Engl J Med 2010;363:2211.

61. Cao Y. et al. Obesity and Prostate Cancer. Recent Results Cancer Res 2016;208:137.

62. Renehan A. et al. Body-mass index and incidence of cancer*. Lancet 2008;371:569.

63. Welti L. et al. Weight Fluctuation and Cancer Risk in Postmenopausal Women: The Women's Health Initiative. Cancer Epidemiol Biomarkers Prev 2017;26:779.

64. De Pergola G. et al. Obesity as a major risk factor for cancer. J Obes 2013;2013:291546.

65. Pallavi R. et al. Insights into the beneficial effect of caloric/ dietary restriction for a healthy and prolonged life. Front Physiol 2012;3:318.

66. Duan Y. et al. Inflammatory Links Between High Fat Diets and Diseases. Front Immunol 2018;9:2649.

67. Pischon T. et al. General and abdominal adiposity and risk of death in Europe. N Engl J Med 2008;359:2105.

68. Westphal S. Obesity, abdominal obesity, and insulin resistance. Clin Cornerstone 2008;9:23.

69. Festa A. et al. Chronic subclinical inflammation as part of the insulin resistance syndrome: the Insulin Resistance Atherosclerosis Study (IRAS). Circulation 2000;102:42.

70. Tabung F. et al. Development and Validation of an Empirical Dietary Inflammatory Index. J Nutr 2016;146:1560.

71. Cornier M. et al. The metabolic syndrome. Endocr Rev 2008;29:777.

72. Paley C. et al. Abdominal obesity and metabolic syndrome: exercise as medicine? BMC Sports Sci Med Rehabil 2018;10:7.

73. Vissers D. et al. The effect of exercise on visceral adipose tissue in overweight adults: a systematic review and meta-analysis. PLoS One 2013;8:e56415.

74. Friedenreich C. et al. State of the epidemiological evidence on physical activity and cancer prevention. Eur J Cancer 2010;46:2593.

75. Lugo D. et al. The effects of physical activity on cancer prevention, treatment and prognosis: A review of the literature. Complement Ther Med 2019;44:9.

76. McTiernan A. et al. Physical Activity in Cancer Prevention and Survival: A Systematic Review. Med Sci Sports Exerc 2019;51:1252.

77. Moore S. et al. Association of Leisure-Time Physical Activity With Risk of 26 Types of Cancer in 1.44 Million Adults. JAMA Intern Med 2016;176:816.

78. McTiernan A. Mechanisms linking physical activity with cancer. Nat Rev Cancer 2008;8:205.

79. Sjostrom L. et al. Effects of bariatric surgery on cancer incidence in obese patients in Sweden (Swedish Obese Subjects Study): a prospective, controlled intervention trial. Lancet Oncol 2009;10:653.

80. Zhang X. et al. Intentional weight loss, weight cycling, and endometrial cancer risk: a systematic review and meta-analysis. Int J Gynecol Cancer 2019;29:1361.

81. Evans J. et al. Metformin and reduced risk of cancer in diabetic patients. BMJ 2005;330:1304.

82. Stevens V. et al. Weight cycling and cancer incidence*. Am J Epidemiol 2015;182:394.

83. Thompson H. et al. Weight cycling and cancer*. Cancer Prev Res (Phila) 2011;4:1736.

84. Mehta T. et al. Impact of weight cycling on risk of morbidity & mortality. Obes Rev 2014;15:870.

85. Stevens V. et al. Weight cycling and risk of endometrial cancer. Cancer Epidemiol Biomarkers Prev 2012;21:747.

86. Rohan T. et al. Body fat and breast cancer risk in postmenopausal women: a longitudinal study. J Cancer Epidemiol 2013;2013:754815.

87. Wang F. et al. Distinct Effects of Body Mass Index and Waist/Hip Ratio on Risk of Breast Cancer by Joint Estrogen and Progestogen Receptor Status*. Oncologist 2017;22:1431.

88. Key T. et al. Body mass index, serum sex hormones, and breast cancer risk in postmenopausal women. J Natl Cancer Inst 2003;95:1218.

89. Clemons M. et al. Estrogen and the risk of breast cancer. N Engl J Med 2001;344:276.

90. Canada's food guide. Government of Canada 2021. https://food-guide.canada.ca/en/. accessed 2022/01/26.

91. Aune D. et al. Fruit and vegetable intake and the risk of cardiovascular disease, total cancer and all-cause mortality*. Int J Epidemiol 2017;46:1029.

92. Schwingshackl L. et al. Food groups and risk of colorectal cancer. Int J Cancer 2018;142:1748.

93. Vieira A. et al. Foods and beverages and colorectal cancer risk*. Ann Oncol 2017;28:1788.

94. Dandamudi A. et al. Dietary Patterns and Breast Cancer Risk*. Anticancer Res 2018;38:3209.

95. Alpha-Tocopherol, Beta Carotene Cancer Prevention Study Group. The effect of vitamin E and beta carotene on the incidence of lung cancer and other cancers in male smokers. N Engl J Med 1994;330:1029.

96. Omenn G. et al. Effects of a combination of beta carotene and vitamin A on lung cancer and cardiovascular disease. N Engl J Med 1996;334:1150.

97. Karp D. et al. Randomized, double-blind, placebo-controlled, phase III chemoprevention trial of selenium supplementation in patients with resected stage I NSCLC*. J Clin Oncol 2013;31:4179.

98. Lippman S. et al. Effect of selenium and vitamin E on risk of prostate cancer & other cancers: the Selenium and Vitamin E Cancer Prevention Trial (SELECT). JAMA 2009;301:39.

99. Wang L. et al. Vitamin E and C supplementation and risk of cancer in men: posttrial follow-up in the Physicians' Health Study II randomized trial. Am J Clin Nutr 2014;100:915.

100. Manson J. et al. Vitamin D Supplements and Prevention of Cancer and Cardiovascular Disease. N Engl J Med 2019;380:33.

101. Lappe J. et al. Effect of Vitamin D and Calcium Supplementation on Cancer Incidence in Older Women: A Randomized Clinical Trial. JAMA 2017;317:1234.

102. Minihane A. et al. Low-grade inflammation, diet composition & health*. Br J Nutr 2015;114:999.

103. Aune D. et al. Whole grain consumption and risk of cardiovascular disease, cancer, and all cause and cause specific mortality*. BMJ 2016;353:i2716.

104. Reynolds A. et al. Carbohydrate quality and human health*. Lancet 2019;393:434.

105. McRae M. The Benefits of Dietary Fiber Intake on Reducing the Risk of Cancer*. J Chiropr Med 2018;17:90.

106. Zheng B. et al. Dietary fiber intake and reduced risk of ovarian cancer*. Nutr J 2018;17:99.

107. Aune D. et al. Nut consumption and risk of cardiovascular disease, total cancer, all-cause and cause-specific mortality*. BMC Med 2016;14:207.

108. Wu J. et al. Dietary Protein Sources and Incidence of Breast Cancer*. Nutrients 2016;8.

109. Red Meat and Processed Meat: Internationational Agency for Research on Cancer; 2018. https://monographs.iarc.who.int/wp-content/uploads/2018/06/mono114.pdf. accessed 2021/11/28.

110. Lippi G. et al. Meat consumption and cancer risk*. Crit Rev Oncol Hematol 2016;97:1.

111. Carr P. et al. Meat subtypes & association with colorectal cancer*. Int J Cancer 2016;138:293.

112. Zhang Z. et al. Poultry and Fish Consumption in Relation to Total Cancer Mortality: A Meta-Analysis of Prospective Studies. Nutr Cancer 2018;70:204.

113. Lian W. et al. Fish intake and the risk of brain tumor*. Nutr J 2017;16:1.

114. Huang R. et al. Fish intake and risk of liver cancer: a meta-analysis. PLoS One 2015;10:e0096102.

115. Yu X. et al. Fish consumption and risk of gastrointestinal cancers: a meta-analysis of cohort studies. World J Gastroenterol 2014;20:15398.

116. Song J. et al. Fish consumption and lung cancer risk*. Nutr Cancer 2014;au66:539.

117. Jakszyn P. Nitrosamine and related food intake and gastric and oesophageal cancer risk: a systematic review of the epidemiological evidence. World J Gastroenterol 2006;12:4296.

118. Ren J. et al. Pickled food & risk of gastric cancer*. Cancer Epidemiol Biomarkers Prev 2012;21:905.

119. D'Alessandro A. et al. Mediterranean Diet and cancer risk*. Int J Food Sci Nutr 2016;67:593.

120. Trichopoulou A. et al. Definitions and potential health benefits of the Mediterranean diet: views from experts around the world. BMC Med 2014;12:112.

121. Thun M. et al. Alcohol consumption and mortality among middle-aged and elderly U.S. adults. N Engl J Med 1997;337:1705.

122. Chen J. et al. Dose-Dependent Associations between Wine Drinking and Breast Cancer Risk - Meta-Analysis Findings. Asian Pac J Cancer Prev 2016;17:1221.

123. Turati F. et al. A meta-analysis of alcohol drinking and oral and pharyngeal cancers: results from subgroup analyses. Alcohol Alcohol 2013;48:107.

124. Purdue M. et al. Type of alcoholic beverage and risk of head and neck cancer--a pooled analysis within the INHANCE Consortium. Am J Epidemiol 2009;169:132.

125. Zhang C. at al. Consumption of beer and colorectal cancer incidence: a meta-analysis of observational studies. Cancer Causes Control 2015;26:549.

126. Longnecker M. at al. A meta-analysis of alcoholic beverage consumption in relation to risk of colorectal cancer. Cancer Causes Control 1990;1:59.

127. Wang P. et al. Alcohol drinking and gastric cancer risk*. Oncotarget 2017;8:99013.

128. Fang X. et al. Landscape of dietary factors associated with risk of gastric cancer*. Eur J Cancer 2015;51:2820.

129. Chao C. Associations between beer, wine, and liquor consumption and lung cancer risk: a meta-analysis. Cancer Epidemiol Biomarkers Prev 2007;16:2436.

130. Dennis L. Meta-analysis for combining relative risks of alcohol consumption and prostate cancer. Prostate 2000;42:56.

131. Vartolomei M. et al. The impact of moderate wine consumption on the risk of developing prostate cancer. Clin Epidemiol 2018;10:431.

132. Sun Q. et al. Alcohol consumption and the risk of endometrial cancer: a meta-analysis. Asia Pac J Clin Nutr 2011;20:125.

133. Wu S. et al. Alcohol consumption and risk of cutaneous basal cell carcinoma in women and men: 3 prospective cohort studies. Am J Clin Nutr 2015;102:1158.

134. Rivera A. et al. Alcohol Intake and Risk of Incident Melanoma: A Pooled Analysis of Three Prospective Studies in the United States. Cancer Epidemiol Biomarkers Prev 2016;25:1550.

135. Siiskonen S. et al. Alcohol Intake is Associated with Increased Risk of Squamous Cell Carcinoma of the Skin: Three US Prospective Cohort Studies. Nutr Cancer 2016;68:545.

136. Galeone C. et al. A meta-analysis of alcohol consumption and the risk of brain tumours. Ann Oncol 2013;24:514.

137. Psaltopoulou T. et al. Alcohol consumption and risk of hematological malignancies: A meta-analysis of prospective studies. Int J Cancer 2018;143:486.

138. Xu X. et al. Does beer, wine or liquor consumption correlate with the risk of renal cell carcinoma?*. Oncotarget 2015;6:13347.

139. Mao Q. et al. A meta-analysis of alcohol intake and risk of bladder cancer. Cancer Causes Control 2010;21:1843.

140. Seiwert T. et al. Safety and clinical activity of pembrolizumab for treatment of recurrent or metastatic squamous cell carcinoma of the head and neck (KEYNOTE-012)*. Lancet Oncol 2016;17:956.

141. Downer M. et al. Alcohol Intake and Risk of Lethal Prostate Cancer in the Health Professionals Follow-Up Study. J Clin Oncol 2019:JCO1802462.

142. List of countries by alcohol consumption per capita. accessed 2021/07/01. Wikipedia https://en.wikipedia.org/wiki/List_of_countries_by_alcohol_consumption_per_capita.

143. Paffenbarger R. et al. Characteristics that predict risk of breast cancer before and after the menopause. Am J Epidemiol 1980;112:258.

144. Gong T. et al. Age at menarche and risk of ovarian cancer*. Int J Cancer 2013;132:2894.

145. Gong T. et al. Age at menarche and endometrial cancer risk*. Sci Rep 2015;5:14051.

146. Reproductive history and breast cancer. National Cancer Institure https://www.cancer.gov/about-cancer/causes-prevention/risk/hormones/reproductive-history-fact-sheet. accessed 2019/05/11

147. Sood R. et al. Prescribing menopausal hormone therapy*. Int J Womens Health 2014;6:47.

148. Grady D. et al. Hormone replacement therapy and endometrial cancer risk: a meta-analysis. Obstet Gynecol 1995;85:304.

149. Chlebowski R. et al. Association of Menopausal Hormone Therapy With Breast Cancer Incidence and Mortality*. JAMA 2020;324:369.

150. Collaborative Group On Epidemiological Studies Of Ovarian Cancer. Menopausal hormone use and ovarian cancer risk*. Lancet 2015;385:1835.

151. Oral contraceptives and cancer risk. National Cancer Institute https://www.cancer.gov/about-cancer/causes-prevention/risk/hormones/oral-contraceptives-fact-sheet. accessed 2019/05/10.

152. Dugger S. et al. Drug development in the era of precision medicine. Nat Rev Drug Discov 2018;17:183.

153. Thun M. et al. Aspirin use and reduced risk of fatal colon cancer. N Engl J Med 1991;325:1593.

154. Garcia-Albeniz X. et al. Aspirin for the prevention of colorectal cancer. Best Pract Res Clin Gastroenterol 2011;25:461.

155. Cook N. et al. Low-dose aspirin in the primary prevention of cancer*. JAMA 2005;294:47.

156. McNeil J. et al. Effect of Aspirin on All-Cause Mortality in the Healthy Elderly. N Engl J Med 2018;379:1519.

157. Noto H. et al. Cancer risk in diabetic patients treated with metformin: a systematic review and meta-analysis. PLoS One 2012;7:e33411.

158. Demierre M. et al. Statins and cancer prevention. Nat Rev Cancer 2005;5:930.

159. Wilt T. et al. Five-alpha-reductase Inhibitors for prostate cancer prevention. Cochrane Database Syst Rev 2008:CD007091.

160. Deng Y. et al. Oral bisphosphonates and incidence of cancers in patients with osteoporosis: a systematic review and meta-analysis. Arch Osteoporos 2018;14:1.

161. Sheil A. Patterns of malignancies following renal transplantation. Transplant Proc 1999;31:1263.

162. Sorenson E. et al. Evidence-based adverse effects of biologic agents in the treatment of moderate-to-severe psoriasis: Providing clarity to an opaque topic. J Dermatolog Treat 2015;26:493.

163. Pereira R. et al. Safety of Anti-TNF Therapies in Immune-Mediated Inflammatory Diseases: Focus on Infections and Malignancy. Drug Dev Res 2015;76:419.

164. Canete J. et al. Safety profile of biological therapies for treating rheumatoid arthritis. Expert Opin Biol Ther 2017;17:1089.

165. Dommasch E. et al. Is there truly a risk of lymphoma from biologic therapies? Dermatol Ther 2009;22:418.

166. Shivaji U. et al. Review article: managing the adverse events caused by anti-TNF therapy in inflammatory bowel disease. Aliment Pharmacol Ther 2019;49:664.

167. Shelton E. et al. Cancer Recurrence Following Immune-Suppressive Therapies in Patients With Immune-Mediated Diseases: A Systematic Review and Meta-analysis. Gastroenterology 2016;151:97.
168. Stanley F. et al. Comprehensive survey of household radon gas levels and risk factors in southern Alberta. CMAJ Open 2017;5:E255.
169. Myatt T. et al. Assessing exposure to granite countertops--Part 1: Radiation. J Expo Sci Environ Epidemiol 2010;20:273.
170. Berrington de Gonzalez A. et al. Risk of cancer from diagnostic X-rays: estimates for the UK and 14 other countries. Lancet 2004;363:345.
171. Berrington de Gonzalez A. et al. Projected cancer risks from computed tomographic scans performed in the United States in 2007. Arch Intern Med 2009;169:2071.
172. Linet M. Cancer risks associated with external radiation from diagnostic imaging procedures. CA Cancer J Clin 2012;62:75.
173. Swift M. et al. Incidence of cancer in 161 families affected by ataxia-telangiectasia. N Engl J Med 1991;325:1831.
174. Cardis E. et al. The Chernobyl accident--an epidemiological perspective. Clin Oncol (R Coll Radiol) 2011;23:251.
175. Bagshaw M. et al. Exposure to cosmic radiation of British Airways flying crew on ultralonghaul routes. Occup Environ Med 1996;53:495.
176. Azzam E. et al. Low-dose ionizing radiation decreases the frequency of neoplastic transformation to a level below the spontaneous rate in C3H 10T1/2 cells. Radiat Res 1996;146:369.
177. Sigurdson A. et al. Cosmic radiation exposure and cancer risk among flight crew. Cancer Invest 2004;22:743.
178. Roychoudhuri R. et al. Radiation-induced malignancies following radiotherapy for breast cancer. Br J Cancer 2004;91:868.
179. Amendola B. Radiation-associated sarcoma: a review of 23 patients with postradiation sarcoma over a 50-year period. Am J Clin Oncol 1989;12:411.
180. Paloyan E. et al. Thyroid neoplasms after radiation therapy for adolescent acne vulgaris. Arch Dermatol 1978;114:53.
181. Simon N. Breast cancer induced by radiation*. JAMA 1977;237:789.
182. Sadetzki S. et al. Risk of thyroid cancer after childhood exposure to ionizing radiation for tinea capitis. J Clin Endocrinol Metab 2006;91:4798.
183. Electromagentic fields and cancer. National Cancer Institute https://www.cancer.gov/about-cancer/causes-prevention/risk/radiation/electromagnetic-fields-fact-sheet#q2 accessed 2019/05/18.
184. Cell phones and cancer risk. National Cancer Institute https://www.cancer.gov/about-cancer/causes-prevention/risk/radiation/cell-phones-fact-sheet. accessed 2019/05/18.
185. Gan Y. et al. Association between shift work and risk of prostate cancer: a systematic review and meta-analysis of observational studies. Carcinogenesis 2018;39:87.

186. Wang F. et al. A meta-analysis on dose-response relationship between night shift work and the risk of breast cancer. Ann Oncol 2013;24:2724.

187. Gallicchio L. et al. Sleep duration and mortality*. J Sleep Res 2009;18:148.

188. Stone C. et al. The association between sleep duration and cancer-specific mortality*. Cancer Causes Control 2019;30:501.

189. Bahri N. et al. The relation between stressful life events and breast cancer: a systematic review and meta-analysis of cohort studies. Breast Cancer Res Treat 2019;176:53.

190. Heikkila K. et al. Work stress and risk of cancer*. BMJ 2013;346:f165.

191. Santos M. et al. Association between stress and breast cancer in women: a meta-analysis. Cad Saude Publica 2009;25 Suppl 3:S453.

192. Duijts S. et al. The association between stressful life events and breast cancer risk: a meta-analysis. Int J Cancer 2003;107:1023.

193. Yang T. et al. Work stress and the risk of cancer*. Int J Cancer 2019;144:2390.

194. Pohl C. et al. Chronic inflammatory bowel disease and cancer. Hepatogastroenterology 2000;47:57.

195. Correa P. Helicobacter pylori and gastric carcinogenesis. Am J Surg Pathol 1995;19 Suppl 1:S37.

196. Heintz N. et al. Asbestos, lung cancers, and mesotheliomas: from molecular approaches to targeting tumor survival pathways. Am J Respir Cell Mol Biol 2010;42:133.

197. Amandus H. et al. Silicosis and lung cancer among workers in North Carolina dusty trades. Scand J Work Environ Health 1995;21 Suppl 2:81.

198. Yu Y. et al. Lung cancer risk following detection of pulmonary scarring by chest radiography in the prostate, lung, colorectal, and ovarian cancer screening trial. Arch Intern Med 2008;168:2326.

199. Park J. et al. Lung cancer in patients with idiopathic pulmonary fibrosis. Eur Respir J 2001;17:1216.

200. Meyer M. et al. A review of the relationship between tooth loss, periodontal disease, and cancer. Cancer Causes Control 2008;19:895.

201. Sanyal A. et al. The etiology of hepatocellular carcinoma and consequences for treatment. Oncologist 2010;15 Suppl 4:14.

202. Chiesa Fuxench Z. et al. The Risk of Cancer in Patients With Psoriasis: A Population-Based Cohort Study in the Health Improvement Network. JAMA Dermatol 2016;152:282.

203. Zeineddine N. et al. Systemic Sclerosis and Malignancy*. J Clin Med Res 2016;8:625.

204. Bansal A. et al. Human papillomavirus-associated cancers*. Int J Appl Basic Med Res 2016;6:84.

205. Thompson M. et al. Epstein-Barr virus and cancer. Clin Cancer Res 2004;10:803.

206. Boshoff C. et al. AIDS-related malignancies. Nat Rev Cancer 2002;2:373.

207. Rahner N. et al. Hereditary cancer syndromes. Dtsch Arztebl Int 2008;105:706.

208. Kharazmi E. et al. Familial risk of early and late onset cancer*. BMJ 2012;345:e8076.

209. Dimaras H. et al. Retinoblastoma. Lancet 2012;379:1436.

210. Jasperson K. et al. Hereditary and familial colon cancer. Gastroenterology 2010;138:2044.

211. Liede A. et al. Cancer risks for male carriers of germline mutations in BRCA1 or BRCA2: a review of the literature. J Clin Oncol 2004;22:735.

212. Hereditary cancer syndromes. MD Anderson Cancer Center accessed 2019/05/19 https://www.mdanderson.org/prevention-screening/family-history/hereditary-cancer-syndromes.html.

213. An intuitive (and short) explanation of Bayes' theorem. Better Explained. accessed 2019/05/19 https://betterexplained.com/articles/an-intuitive-and-short-explanation-of-bayes-theorem/.

214. Hartmann L. et al. Efficacy of bilateral prophylactic mastectomy in BRCA1 and BRCA2 gene mutation carriers. J Natl Cancer Inst 2001;93:1633.

215. Rebbeck T. Prophylactic oophorectomy in carriers of BRCA1 or BRCA2 mutations. N Engl J Med 2002;346:1616.

216. Yang H. et al. ATM sequence variants associate with susceptibility to non-small cell lung cancer. Int J Cancer 2007;121:2254.

217. Xing J. et al. Deficient G2-M and S checkpoints are associated with increased lung cancer risk: a case-control analysis. Cancer Epidemiol Biomarkers Prev 2007;16:1517.

218. Wang W. et al. Genetic variants in cell cycle control pathway confer susceptibility to lung cancer. Clin Cancer Res 2007;13:5974.

219. Lin J. et al. Systematic evaluation of apoptotic pathway gene polymorphisms and lung cancer risk. Carcinogenesis 2012;33:1699.

220. Tyndale R. et al. Genetic variation in CYP2A6-mediated nicotine metabolism alters smoking behavior. Ther Drug Monit 2002;24:163.

221. Hamra G. et al. Outdoor particulate matter exposure and lung cancer: a systematic review and meta-analysis. Environ Health Perspect 2014;122:906.

222. Villeneuve P. et al. Occupational exposure to asbestos and lung cancer in men: evidence from a population-based case-control study in eight Canadian provinces. BMC Cancer 2012;12:595.

223. Carbone M. et al. Malignant mesothelioma: facts, myths, and hypotheses. J Cell Physiol 2012;227:44.

224. Pallis A. et al. Lung cancer in never smokers: disease characteristics and risk factors. Crit Rev Oncol Hematol 2013;88:494.

225. Kurmi O. et al. Lung cancer risk and solid fuel smoke exposure*. Eur Respir J 2012;40:1228.

226. Gangemi S. et al. Occupational exposure to pesticides as a possible risk factor for the development of chronic diseases in humans. Mol Med Rep 2016;14:4475.

227. Erren T. et al. Meta-analyses of published epidemiological studies, 1979-2006, point to open causal questions in silica-silicosis-lung cancer research. Med Lav 2011;102:321.

228. Honaryar M. et al. Welding fumes and lung cancer*. Occup Environ Med 2019;76:422.

229. Narod S. Disappearing breast cancers. Curr Oncol 2012;19:59.

230. Withers HR, Lee SP. Modeling growth kinetics and statistical distribution of oligometastases. Semin Radiat Oncol 2006;16:111-9.

231. Burz C, Pop VV, Buiga R, et al. Circulating tumor cells in clinical research and monitoring patients with colorectal cancer. Oncotarget 2018;9:24561-71.

232. MacDonald I. et al. Cancer spread and micrometastasis development*. Bioessays 2002;24:885.

233. Breast Cancer: Screening US Preventive Services Task Force 2016. https://www.uspreventiveservicestaskforce.org/Page/Document/UpdateSummaryFinal/breast-cancer-screening1. accessed 2019/06/18.

234. Tonelli M. et al. Recommendations on screening for breast cancer in average-risk women aged 40-74 years. CMAJ 2011;183:1991.

235. Berringeon de Gonzalez A, Mahesh M, Kim KP, et al. Projected cancer risks from computed tomographic scans performed in the United States in 2007. Arch Intern Med 2009;169:2071-7.

236. Ekpo E. et al. Errors in Mammography Cannot be Solved Through Technology Alone. Asian Pac J Cancer Prev 2018;19:291.

237. Bassett L. et al. Mammography and breast cancer screening. Surg Clin North Am 1990;70:775.

238. Engmann N. et al. Longitudinal changes in volumetric breast density in healthy women across the menopausal transition. Cancer Epidemiol Biomarkers Prev 2019;28:1324.

239. Gotzsche P. et al. Screening for breast cancer with mammography. Cochrane Database Syst Rev 2013:CD001877.

240. Engmann N. et al. Combined effect of volumetric breast density and body mass index on breast cancer risk. Breast Cancer Res Treat 2019;177:165.

241. Breast cancer screening for women at high risk Cancer Care Ontario https://www.cancercareontario.ca/en/guidelines-advice/cancer-continuum/screening/breast-cancer-high-risk-women. accessed 2019/06/18.

242. Warner E. et al. MRI surveillance for hereditary breast-cancer risk. Lancet 2005;365:1747.

243. Lehman C. et al. Screening women at high risk for breast cancer with mammography and magnetic resonance imaging. Cancer 2005;103:1898.

244. Lehman C. et al. Cancer yield of mammography, MR, and US in high-risk women: prospective multi-institution breast cancer screening study. Radiology 2007;244:381.

245. Kuhl C. et al. Mammography, breast ultrasound, and magnetic resonance imaging for surveillance of women at high familial risk for breast cancer. J Clin Oncol 2005;23:8469.

246. Bakker M. et al. Supplemental MRI Screening for Women with Extremely Dense Breast Tissue. N Engl J Med 2019;381:2091.

247. Saadatmand S. et al. MRI versus mammography for breast cancer screening in women with familial risk (FaMRIsc): a multicentre, randomised, controlled trial. Lancet Oncol 2019;20:1136.

248. Burkett B. et al. A Review of Supplemental Screening Ultrasound for Breast Cancer: Certain Populations of Women with Dense Breast Tissue May Benefit. Acad Radiol 2016;23:1604.

249. Ludwig K. et al. Risk reduction and survival benefit of prophylactic surgery in BRCA mutation carriers, a systematic review. Am J Surg 2016;212:660.

250. Bell K. et al. Prevalence of incidental prostate cancer*. Int J Cancer 2015;137:1749.

251. Canadian Cancer Statistics: a 2018 special report on cancer incidence by stage. Can Cancer Soc https://www.cancer.ca/~/media/cancer.ca/CW/cancer%20information/cancer%20101/Canadian%20cancer%20statistics/Canadian-Cancer-Statistics-2018-EN.pdf?la=en accessed 2019/06/12.

252. Schroder F. et al. Screening and prostate-cancer mortality*. N Engl J Med 2009;360:1320.

253. Schroder F. et al. Screening and prostate cancer mortality*. Lancet 2014;384:2027.

254. Ilic D. et al. Prostate cancer screening with prostate-specific antigen (PSA) test: a systematic review and meta-analysis. BMJ 2018;362:k3519.

255. Tikkinen K. et al. Prostate cancer screening with prostate-specific antigen (PSA) test: a clinical practice guideline. BMJ 2018;362:k3581.

256. Naji L. et al. Digital Rectal Examination for Prostate Cancer Screening in Primary Care: A Systematic Review and Meta-Analysis. Ann Fam Med 2018;16:149.

257. Catalona W. et al. Screening for prostate cancer in high risk populations. J Urol 2002;168:1980.

258. Gleicher S. et al. Implications of High Rates of Metastatic Prostate Cancer in BRCA2 Mutation Carriers. Prostate 2016;76:1135.

259. An intuitive (and short) explanation of Bayes' theorem. Better Explained. accessed 2019/05/19. https://betterexplained.com/articles/an-intuitive-and-short-explanation-of-bayes-theorem/.

260. Dobbs R. et al. Is prostate cancer stage migration continuing for black men in the PSA era? Prostate Cancer Prostatic Dis 2017;20:210.

261. Mahal B. et al. Racial disparities in prostate cancer outcome among prostate-specific antigen screening eligible populations in the United States. Ann Oncol 2017;28:1098.

262. Eklund M. et al. The Stockholm-3 (STHLM3) Model can Improve Prostate Cancer Diagnostics in Men Aged 50-69 yr Compared with Current Prostate Cancer Testing. Eur Urol Focus 2018;4:707.

263. Holme O. et al. Flexible sigmoidoscopy versus faecal occult blood testing for colorectal cancer screening in asymptomatic individuals. Cochrane Database Syst Rev 2013;Issue 9:CD009259.

264. Tinmouth J. et al. Colorectal Cancer Screening in Average Risk Populations: Evidence Summary. Can J Gastroenterol Hepatol 2016;2016:2878149.

265. Whitlock E. et al. Screening for colorectal cancer*. Ann Intern Med 2008;149:638.

266. Johnson C. et al. Meta-analyses of colorectal cancer risk factors. Cancer Causes Control 2013;24:1207.

267. Nguyen T. et al. Strategies for detecting colon cancer in patients with inflammatory bowel disease. Cochrane Database Syst Rev 2017;9:CD000279.

268. Harpaz N. et al. Precancerous lesions in inflammatory bowel disease. Best Pract Res Clin Gastroenterol 2013;27:257.

269. Rex D. et al. Colorectal Cancer Screening: Recommendations for Physicians and Patients from the U.S. Multi-Society Task Force on Colorectal Cancer. Am J Gastroenterol 2017;112:1016.

270. Boland P. et al. Recent progress in Lynch syndrome and other familial colorectal cancer syndromes. CA Cancer J Clin 2018;68:217.

271. Tudyka V. et al. Surgical treatment in familial adenomatous polyposis. Ann Gastroenterol 2012;25:201.

272. Dickinson J. et al. Reduced cervical cancer incidence and mortality in Canada: national data from 1932 to 2006. BMC Public Health 2012;12:992.

273. Bansal A, Singh MP, Rai B. Human papillomavirus-associated cancers: A growing global problem. Int J Appl Basic Med Res 2016;6:84-9.

274. Porras C. et al. Efficacy of the bivalent HPV vaccine against HPV 16/18-associated precancer: long-term follow-up results from the Costa Rica Vaccine Trial. Lancet Oncol 2020;21:1643.

275. Chan K. et al. Primary HPV testing with cytology versus cytology alone in cervical screening*. Int J Cancer 2020;147:1152.

276. Gilham C. et al. HPV testing compared with routine cytology in cervical screening: long-term follow-up of ARTISTIC RCT. Health Technol Assess 2019;23:1.

277. Usman Ali M. et al. Screening for lung cancer*. Prev Med 2016;89:301.

278. Hoffman R. et al. Lung Cancer Screening with Low-Dose CT*. J Gen Intern Med 2020;35:3015.

279. van den Beuken-van Everdingen M. et al. Update on Prevalence of Pain in Patients With Cancer: Systematic Review and Meta-Analysis. J Pain Symptom Manage 2016;51:1070.

280. Chwistek M. Recent advances in understanding & managing cancer pain. F1000Res 2017;6:945.

281. Regan J. et al. Neurophysiology of cancer pain. Cancer Control 2000;7:111.

282. Freynhagen R. et al. Efficacy of pregabalin in neuropathic pain*. Pain 2005;115:254.

283. Fallon M. Neuropathic pain in cancer. Br J Anaesth 2013;111:105.

284. Koeller J. Understanding cancer pain. Am J Hosp Pharm 1990;47:S3.

285. Hauser W. et al. Cannabinoids in Pain Management and Palliative Medicine. Dtsch Arztebl Int 2017;114:627.

286. Mucke M. et al. Systematic review and meta-analysis of cannabinoids in palliative medicine. J Cachexia Sarcopenia Muscle 2018;9:220.

287. Trachtenberg A. et al. Cost analysis of medical assistance in dying in Canada. CMAJ 2017;189:E101.

288. Pereira J. Legalizing euthanasia or assisted suicide*. Curr Oncol 2011;18:e38.

289. Fonseca G. et al. Cancer Cachexia & Related Metabolic Dysfunction. Int J Mol Sci 2020;21:2321.

290. Laviano A. et al. Assessing pathophysiology of cancer anorexia. Curr Opin Clin Nutr Metab Care 2017;20:340.

291. Childs D. et al. A hunger for hunger: a review of palliative therapies for cancer-associated anorexia. Ann Palliat Med 2019;8:50.

292. Stewart D. Nausea and vomiting in cancer patients. In: Kucharczyk J, Stewart D, Miller A, eds. Nausea and vomiting: recent research and clinical advances. Boca Raton, Florida: CRC Press; 1991:177.

293. Peixoto da Silva S. et al. Cancer cachexia and its pathophysiology: links with sarcopenia, anorexia and asthenia. J Cachexia Sarcopenia Muscle 2020;11:619.

294. DeBerardinis R. et al. We need to talk about Warburg effect. Nature Metabolism 2020;2:127.

295. Liberti M. et al. The Warburg Effect: How Does it Benefit Cancer Cells? Trends Biochem Sci 2016;41:211.

296. Epstein T. et al. The Warburg effect as an adaptation of cancer cells to rapid fluctuations in energy demand. PLoS One 2017;12:e0185085.

297. Devic S. Warburg Effect - a Consequence or the Cause of Carcinogenesis? J Cancer 2016;7:817.

298. Gatenby R. et al. Why do cancers have high aerobic glycolysis? Nat Rev Cancer 2004;4:891.

299. Stewart D. Mechanisms of resistance to cisplatin and carboplatin. Crit Rev Oncol Hematol 2007;63:12.

300. Argiles J. Cancer cachexia, a clinical challenge. Curr Opin Oncol 2019;31:286.

301. Janelsins M. et al. Prevalence, mechanisms, and management of cancer-related cognitive impairment. Int Rev Psychiatry 2014;26:102.

302. Ezeoke C. et al. Pathophysiology of anorexia in the cancer cachexia syndrome. J Cachexia Sarcopenia Muscle 2015;6:287.

303. Stepanski E. et al. The relation of trouble sleeping, depressed mood, pain, and fatigue in patients with cancer. J Clin Sleep Med 2009;5:132.

304. Klock J. et al. Hemolytic anemia and somatic cell dysfunction in severe hypophosphatemia. Arch Intern Med 1974;134:360.

305. Weiss G. et al. Anemia of chronic disease. N Engl J Med 2005;352:1011.

306. Ellis L. et al. Autoimmune Hemolytic Anemia and Cancer. JAMA 1965;193:962.

307. Morton J. et al. Microangiopathic Hemolytic Anemia and Thrombocytopenia in Patients With Cancer. J Oncol Pract 2016;12:523.

308. Mundy G. et al. The hypercalcemia of cancer*. N Engl J Med 1984;310:1718.

309. Zagzag J. et al. Hypercalcemia and cancer*. CA Cancer J Clin 2018;68:377.

310. Major P. et al. Zoledronic acid is superior to pamidronate in the treatment of hypercalcemia of malignancy: a pooled analysis of two randomized, controlled clinical trials. J Clin Oncol 2001;19:558.

311. Hermes A. et al. Hyponatremia as prognostic factor in small cell lung cancer--a retrospective single institution analysis. Respir Med 2012;106:900.

312. List A. et al. The syndrome of inappropriate secretion of antidiuretic hormone (SIADH) in small-cell lung cancer. J Clin Oncol 1986;4:1191.

313.   Fiordoliva I. et al. Managing hyponatremia in lung cancer*. Ther Adv Med Oncol 2017;9:711.

314.   Polanski J. et al. Quality of life of patients with lung cancer. Onco Targets Ther 2016;9:1023.

315.   Ripamonti C, Bruera E. Dyspnea: pathophysiology and assessment. J Pain Symptom Manage 1997;13:220-32.

316.   Ripamonti C. Management of dyspnea in advanced cancer patients. Support Care Cancer 1999;7:233.

317.   Chalhoub M. et al. Indwelling pleural catheters*. J Thorac Dis 2018;10:4659.

318.   Efthymiou C. et al. Malignant pleural effusion in the presence of trapped lung. Five-year experience of PleurX tunnelled catheters. Interact Cardiovasc Thorac Surg 2009;9:961.

319.   Svoboda M. Malignant Pericardial Effusion and Cardiac Tamponade (Cardiac and Pericardial Symptoms). In: Olver IN, ed. The MASCC Textbook of Cancer Supportive Care and Survivorship. Boston, MA: Springer US; 2011:83.

320.   Berliner D. et al. The Differential Diagnosis of Dyspnea. Dtsch Arztebl Int 2016;113:834.

321.   Abdol Razak N. et al. Cancer-Associated Thrombosis*. Cancers (Basel) 2018;10:380.

322.   Carrier M. et al. Treatment algorithm in cancer-associated thrombosis: Canadian expert consensus. Curr Oncol 2018;25:329.

323.   Gogakos A. et al. Heimlich valve and pneumothorax. Ann Transl Med 2015;3:54.

324.   Kvale P. Chronic cough due to lung tumors*. Chest 2006;129:147S.

325.   Karkhanis V. et al. Pleural effusion*. Open Access Emerg Med 2012;4:31.

326.   Harle A. et al. A cross sectional study to determine the prevalence of cough and its impact in patients with lung cancer: a patient unmet need. BMC Cancer 2020;20:9.

327.   Ryan N. et al. Gabapentin for refractory chronic cough*. Lancet 2012;380:1583.

328.   Shim J. et al. A systematic review of symptomatic diagnosis of lung cancer. Fam Pract 2014;31:137.

329.   Gershman E. et al. Management of hemoptysis in patients with lung cancer. Ann Transl Med 2019;7:358.

330.   Herbst R. et al. Pembrolizumab versus docetaxel for previously treated, PD-L1-positive, advanced non-small-cell lung cancer (KEYNOTE-010)*. Lancet 2016;387:1540.

331.   Panda A. et al. Bronchial artery embolization in hemoptysis*. Diagn Interv Radiol 2017;23:307.

332.   Rolston K. Infections in Cancer Patients with Solid Tumors: A Review. Infect Dis Ther 2017;6:69.

333.   Morris A. Cellulitis and erysipelas. BMJ Clin Evid 2008;2008:1708.

334.   Lorenz J. Management of Malignant Biliary Obstruction. Semin Intervent Radiol 2016;33:259.

335.   Hansson E. et al. Herpes zoster risk after 21 specific cancers*. Br J Cancer 2017;116:1643.

336. Johnstone C. Bleeding in cancer patients and its treatment*. Ann Palliat Med 2018;7:265.
337. Gerstner E. et al. VEGF inhibitors in the treatment of cerebral edema in patients with brain cancer. Nat Rev Clin Oncol 2009;6:229.
338. Dietrich J. et al. Corticosteroids in brain cancer patients*. Expert Rev Clin Pharmacol 2011;4:233.
339. Englot D. et al. Epilepsy and brain tumors. Handb Clin Neurol 2016;134:267.
340. Oberndorfer S. et al. [The frequency of seizures in patients with primary brain tumors or cerebral metastases*]. Wien Klin Wochenschr 2002;114:911-6.
341. Donato J. et al. Intracranial hemorrhage in patients with brain metastases treated with therapeutic enoxaparin: a matched cohort study. Blood 2015;126:494.
342. Zwicker J. et al. A meta-analysis of intracranial hemorrhage in patients with brain tumors receiving therapeutic anticoagulation. J Thromb Haemost 2016;14:1736.
343. Prasad D. et al. Malignant spinal-cord compression. Lancet Oncol 2005;6:15.
344. Schiff D. et al. Intramedullary spinal cord metastases*. Neurology 1996;47:906.
345. Honnorat J. et al. Paraneoplastic neurological syndromes. Orphanet J Rare Dis 2007;2:22.
346. Diskin C. et al. Towards an understanding of oedema. BMJ 1999;318:1610.
347. Gounden V. et al. Hypoalbuminemia. StatPearls. Treasure Island (FL)2021. https://www.ncbi.nlm.nih.gov/books/NBK526080/.
348. Soeters P. et al. Hypoalbuminemia*. JPEN J Parenter Enteral Nutr 2019;43:181.
349. Chojkier M. Inhibition of albumin synthesis in chronic diseases: molecular mechanisms. J Clin Gastroenterol 2005;39:S143.
350. Stick C. et al. On the edema-preventing effect of the calf muscle pump. Eur J Appl Physiol Occup Physiol 1989;59:39.
351. Saif M. et al. Management of ascites due to gastrointestinal malignancy. Ann Saudi Med 2009;29:369.
352. Narayanan G. et al. Safety and efficacy of the PleurX catheter for the treatment of malignant ascites. J Palliat Med 2014;17:906.
353. Shephard E. et al. Recognising laryngeal cancer in primary care: a large case-control study using electronic records. Br J Gen Pract 2019;69:e127.
354. Song S. et al. CT evaluation of vocal cord paralysis due to thoracic diseases: a 10-year retrospective study. Yonsei Med J 2011;52:831.
355. Stachler R. et al. Clinical Practice Guideline: Hoarseness (Dysphonia) (Update). Otolaryngol Head Neck Surg 2018;158:S1.
356. Alghonaim Y. et al. Evaluating the timing of injection laryngoplasty for vocal fold paralysis in an attempt to avoid future type 1 thyroplasty. J Otolaryngol Head Neck Surg 2013;42:24.
357. Burz C. et al. Circulating tumor cells in clinical research and monitoring patients with colorectal cancer. Oncotarget 2018;9:24561.
358. Foster J. et al. Liver Metastases. Curr Probl Surg 1981;18:157.
359. Chaudhuri A. et al. Early Detection of Molecular Residual Disease in Localized Lung Cancer by Circulating Tumor DNA Profiling. Cancer Discov 2017;7:1394.

360. Fujii T. et al. Effectiveness of an Adjuvant Chemotherapy Regimen for Early-Stage Breast Cancer: A Systematic Review and Network Meta-analysis. JAMA Oncol 2015;1:1311.

361. Petrelli F. et al. Adjuvant dose-dense chemotherapy in breast cancer: a systematic review and meta-analysis of randomized trials. Breast Cancer Res Treat 2015;151:251.

362. Wu X. et al. Postoperative adjuvant chemotherapy for stage II colorectal cancer: a systematic review of 12 randomized controlled trials. J Gastrointest Surg 2012;16:646.

363. Burdett S. et al. Adjuvant chemotherapy for resected early-stage non-small cell lung cancer. Cochrane Database Syst Rev 2015:CD011430.

364. Freedman O. et al. Adjuvant endocrine therapy for early breast cancer*. Curr Oncol 2015;22:S95.

365. Davari M. et al. Effectiveness of trastuzumab as adjuvant therapy in patients with early stage breast cancer*. Med J Islam Repub Iran 2017;31:88.

366. Wu Y. et al. Osimertinib in Resected EGFR-Mutated Non-Small-Cell Lung Cancer. N Engl J Med 2020;383:1711.

367. Laurie S. et al. Canadian consensus: oligoprogressive, pseudoprogressive, and oligometastatic non-small-cell lung cancer. Curr Oncol 2019;26:e81.

368. Withers H. et al. Modeling growth kinetics and statistical distribution of oligometastases. Semin Radiat Oncol 2006;16:111.

369. Putnam J. et al. Analysis of prognostic factors in patients undergoing resection of pulmonary metastases from soft tissue sarcomas. J Thorac Cardiovasc Surg 1984;87:260.

370. Kesler K. et al. Survival after resection for metastatic testicular nonseminomatous germ cell cancer to the lung or mediastinum. Ann Thorac Surg 2011;91:1085.

371. Riggs S. et al. Postchemotherapy surgery for germ cell tumors*. Oncologist 2014;19:498.

372. Jedi. Wikipedia https://en.wikipedia.org/wiki/Jedi. Accessed 2022/01/18.

373. Reisz J. et al. Effects of ionizing radiation on biological molecules*. Antioxid Redox Signal 2014;21:260.

374. Hubenak J. et al. Mechanisms of injury to normal tissue after radiotherapy: a review. Plast Reconstr Surg 2014;133:49e.

375. Jordan B. et al. Targeting tumor perfusion and oxygenation to improve the outcome of anticancer therapy. Front Pharmacol 2012;3:94.

376. Free radical damage to DNA. Wikipedia https://en.wikipedia.org/wiki/Free_radical_damage_to_DNA#:~:text=Hydrogen%20abstraction%20causes%20radical%20formation,from%20both%20strands%20of%20DNA. accessed 2022/02/01.

377. Youssef A. et al. Hypofractionation Radiotherapy vs. Conventional Fractionation for Breast Cancer: A Comparative Review of Toxicity. Cureus 2018;10:e3516.

378. Lippitz B. et al. Stereotactic radiosurgery in the treatment of brain metastases: the current evidence. Cancer Treat Rev 2014;40:48.

379. Gianfaldoni S. et al. An Overview on Radiotherapy: From Its History to Its Current Applications in Dermatology. Open Access Maced J Med Sci 2017;5:521.

380. Gossage L. et al. Base excision repair factors are promising prognostic and predictive markers in cancer. Curr Mol Pharmacol 2012;5:115.

381. Curtin N. Therapeutic potential of drugs to modulate DNA repair in cancer. Expert Opin Ther Targets 2007;11:783.

382. Douple E. et al. A review of interactions between platinum coordination omplexes and ionizing radiation: implications for cancer therapy. In: Prestayko AW, Crooke ST, Carter SK, eds. Cisplatin: current status and new developments. New York: Academic; 1980:125.

383. Suit HD. Application of radiobiologic principles to radiation therapy. Cancer 1968;22:809.

384. Wike-Hooley J. et al. The relevance of tumour pH to the treatment of malignant disease. Radiother Oncol 1984;2:343.

385. Coleman C. Hypoxia in tumors*. J Natl Cancer Inst 1988;80:310.

386. Vaupel P. et al. Blood flow, oxygen and nutrient supply, and metabolic microenvironment of human tumors: a review. Cancer Res 1989;49:6449.

387. Norton L. A Gompertzian model of human breast cancer growth. Cancer Res 1988;48:7067.

388. Peters L. et al. Applying radiobiological principles to combined modality treatment of head and neck cancer--the time factor. Int J Radiat Oncol Biol Phys 1997;39:831.

389. Mitchell J. et al. The role of glutathione in radiation and drug induced cytotoxicity. Br J Cancer Suppl 1987;8:96.

390. Southorn P. et al. Free radicals in medicine. I. Chemical nature and biologic reactions. Mayo Clin Proc 1988;63:381.

391. Bump E. et al. Role of glutathione in the radiation response of mammalian cells in vitro and in vivo. Pharmacol Ther 1990;47:117.

392. Sies H. Relationship between free radicals & vitamins*. Int J Vitam Nutr Res Suppl 1989;30:215.

393. Chaudiere J. et al. Intracellular antioxidants*. Food Chem Toxicol 1999;37:949-62.

394. Rousseau E. et al. Protection by beta-carotene and related compounds against oxygen-mediated cytotoxicity and genotoxicity*. Free Radic Biol Med 1992;13:407.

395. Lawenda B. et al. Should supplemental antioxidant administration be avoided during chemotherapy and radiation therapy? J Natl Cancer Inst 2008;100:773.

396. Sayin V. et al. Antioxidants accelerate lung cancer progression in mice. Sci Transl Med 2014;6:221ra15.

397. Kaiser J. Antioxidants could spur tumors by acting on cancer gene*. Science 2014;343:477.

398. Le Gal K. et al. Antioxidants can increase melanoma metastasis in mice. Sci Transl Med 2015;7:308re8.

399. Chen C. et al. Enhanced tumorigenesis in p53 knockout mice exposed in utero to high-dose vitamin E. Carcinogenesis 2006;27:1358.

400. Diao Q. et al. Vitamin E promotes breast cancer cell proliferation by reducing ROS production and p53 expression. Eur Rev Med Pharmacol Sci 2016;20:2710.

401. Jacobs C. et al. Is there a role for oral or intravenous ascorbate (vitamin C) in treating patients with cancer? A systematic review. Oncologist 2015;20:210.

402. Meyn R. et al. Receptor signaling as a regulatory mechanism of DNA repair. Radiother Oncol 2009;92:316.

403. Douple E. et al. Carboplatin as a potentiator of radiation therapy. Cancer Treat Rev 1985;12 Suppl A:111-24.

404. Raghunand N. et al. pH and drug resistance in tumors. Drug Resist Updat 2000;3:39.

405. Stewart D. et al. Factors affecting platinum concentrations in human surgical tumour specimens after cisplatin. Br J Cancer 1995;71:598.

406. Stewart D. et al. Human tissue distribution of platinum after cis-diamminedichloroplatinum. Cancer Chemother Pharmacol 1982;10:51.

407. Richmond R. et al. Radiation sensitization of bacterial spores by cis-dichlorodiammineplatinum(II). Radiat Res 1976;68:251.

408. Stewart D. et al. Human central nervous system distribution of cis-diamminedichloroplatinum and use as a radiosensitizer in malignant brain tumors. Cancer Res 1982;42:2474.

409. Stewart D. et al. Weekly Cisplatin during cranial irradiation for malignant melanoma metastatic to brain. J Neurooncol 1983;1:49.

410. Rose P. et al. Concurrent cisplatin-based radiotherapy and chemotherapy for locally advanced cervical cancer. N Engl J Med 1999;340:1144.

411. Forastiere A. et al. Concurrent chemotherapy and radiotherapy for organ preservation in advanced laryngeal cancer. N Engl J Med 2003;349:2091.

412. Curran W. et al. Sequential vs. concurrent chemoradiation for stage III non-small cell lung cancer: randomized phase III trial RTOG 9410. J Natl Cancer Inst 2011;103:1452.

413. Murray N. et al. Importance of timing for thoracic irradiation in the combined modality treatment of limited-stage small-cell lung cancer*. J Clin Oncol 1993;11:336.

414. Byhardt R. et al. Response, toxicity, failure patterns & survival in 5 RTOG trials of sequential &/or concurrent chemotherapy & radiotherapy for locally advanced NSCLC*. Int J Radiat Oncol Biol Phys 1998;42:469.

415. Furuse K. et al. Phase III study of concurrent versus sequential thoracic radiotherapy in combination with mitomycin, vindesine & cisplatin in unresectable stage III NSCLC*. J Clin Oncol 1999;17:2692.

416. Herskovic A. et al. Combined chemotherapy and radiotherapy compared with radiotherapy alone in patients with cancer of the esophagus. N Engl J Med 1992;326:1593.

417. Coppin C. et al. Improved local control of invasive bladder cancer by concurrent cisplatin and preoperative or definitive radiation*. J Clin Oncol 1996;14:2901.

418. Cox J. Chemoradiotherapy for inoperable non-small cell lung cancer. In: Stewart DJ, ed. Lung Cancer: prevention, management and emerging therapies. New York: Humana Press; 2010:161.

419. Robertson J. et al. Breast conservation therapy. Severe breast fibrosis after radiation therapy in patients with collagen vascular disease. Cancer 1991;68:502.

420. Morris M. et al. Irradiation in the setting of collagen vascular disease: acute and late complications. J Clin Oncol 1997;15:2728.

421. Kim H. et al. Preliminary result of definitive radiotherapy in patients with non-small cell lung cancer who have underlying idiopathic pulmonary fibrosis*. Radiat Oncol 2019;14:19.

422. Hildebrandt M. et al. Genetic variants in inflammation-related genes are associated with radiation-induced toxicity following treatment for non-small cell lung cancer. PLoS One 2010;5:e12402.

423. Fernet M. et al. Genetic biomarkers of therapeutic radiation sensitivity. DNA Repair (Amst) 2004;3:1237.

424. Stewart D. et al. The need for speed in advanced non-small cell lung cancer*. Cancer Med 2021; Nov 11, https:/doi.org/10.1002/cam4.4411.

425. Stewart D. et al. Potential insights from population kinetic assessment of progression-free survival curves. Crit Rev Oncol Hematol 2020;153:103039.

426. Brule S. et al. Palliative systemic therapy for advanced non-small cell lung cancer: Investigating disparities between patients who are treated versus those who are not. Lung Cancer 2016;97:15.

427. Sacher A. et al. Real-world chemotherapy treatment patterns in metastatic non-small cell lung cancer: Are patients undertreated? Cancer 2015;121:2562.

428. Slevin M. et al. Attitudes to chemotherapy*. BMJ 1990;300:1458.

429. Stewart D. Tumor and host factors that may limit efficacy of chemotherapy in non-small cell and small cell lung cancer. Crit Rev Oncol Hematol 2010;75:173.

430. Stewart D. et al. Fool's gold, lost treasures, and randomized trials*. BMC Cancer 2013;13:193.

431. Kohno N. et al. Interactions of doxorubicin and cis-platin in squamous carcinoma cells in culture. Br J Cancer 1988;58:330.

432. Stewart D. et al. Optimal frequency of scans for patients on cancer therapies: A population kinetics assessment Cancer Medicine 2019;8:6871.

433. Gatenby R. et al. Adaptive therapy. Cancer Res 2009;69:4894.

434. Shen D. et al. Reduced expression of small GTPases and hypermethylation of the folate binding protein gene in cisplatin-resistant cells. Br J Cancer 2004;91:270.

435. Liang X. et al. A pleiotropic defect reducing drug accumulation in cisplatin-resistant cells. J Inorg Biochem 2004;98:1599.

436. Stewart D. et al. Extensive disease small cell lung cancer dose-response relationships: implications for resistance mechanisms. J Thorac Oncol 2010;5:1826.

437. Stewart D. et al. Chemotherapy dose--response relationships in non-small cell lung cancer and implied resistance mechanisms. Cancer Treat Rev 2007;33:101.

438. Stewart D. et al. Active vs. passive resistance, dose-response relationships, high dose chemotherapy, and resistance modulation: a hypothesis. Invest New Drugs 1996;14:115.

439. West H. et al. Performance Status in Patients With Cancer*. JAMA Oncol 2015;1:998.

440. Hickish T. et al. Clinical benefit from palliative chemotherapy in non-small-cell lung cancer extends to the elderly and those with poor prognostic factors. Br J Cancer 1998;78:28.

441. Chemotherapy in non-small cell lung cancer: a meta-analysis using updated data on individual patients from 52 randomised clinical trials. NSCLC Collaborative Group*. BMJ 1995;311:899.

442. Clemons M. et al. Tamoxifen ("Nolvadex"): a review. Cancer Treat Rev 2002;28:165.

443. Heinlein C. et al. Androgen receptor in prostate cancer. Endocr Rev 2004;25:276.

444. Pilepich M. et al. Phase III RTOG trial 86-10 of androgen deprivation adjuvant to definitive radiotherapy in locally advanced carcinoma of the prostate*. Int J Radiat Oncol Biol Phys 2001;50:1243.

445. Kumar S. et al. Neo-adjuvant and adjuvant hormone therapy for localised and locally advanced prostate cancer. Cochrane Database Syst Rev 2006:CD006019.

446. Hellerstedt B. The current state of hormonal therapy for prostate cancer. CA: a cancer jounral for clinicians 2008;52:154.

447. Osborne C. et al. Fulvestrant: an oestrogen receptor antagonist with a novel mechanism of action. Br J Cancer 2004;90 Suppl 1:S2.

448. Tancredi R. et al. Endocrine Therapy in Premenopausal Hormone Receptor Positive/Human Epidermal Growth Receptor 2 Negative Metastatic Breast Cancer*. Oncologist 2018;23:974.

449. Miller W. Aromatase inhibitors: mechanism of action and role in the treatment of breast cancer. Semin Oncol 2003;30:3.

450. Rugo H. et al. Palbociclib plus letrozole as first-line therapy in estrogen receptor-positive/human epidermal growth factor receptor 2-negative advanced breast cancer with extended follow-up. Breast Cancer Res Treat 2019;174:719.

451. Turner N. et al. Overall Survival with Palbociclib and Fulvestrant in Advanced Breast Cancer. N Engl J Med 2018;379:1926.

452. Denmeade S. et al. A history of prostate cancer treatment. Nat Rev Cancer 2002;2:389.

453. Kelly K. The benefits of achieving stable disease in advanced lung cancer. Oncology (Williston Park) 2003;17:957.

454. Ranson M. et al. ZD1839, a selective oral epidermal growth factor receptor-tyrosine kinase inhibitor, is well tolerated & active in patients with solid, malignant tumors*. J Clin Oncol 2002;20:2240.

455. Perez-Soler R. et al. Determinants of tumor response and survival with erlotinib in patients with non--small-cell lung cancer. J Clin Oncol 2004;22:3238.

456. Paez J. et al. EGFR mutations in lung cancer: correlation with clinical response to gefitinib therapy. Science 2004;304:1497.

457. Lynch T. et al. Activating mutations in the epidermal growth factor receptor underlying responsiveness of non-small-cell lung cancer to gefitinib. N Engl J Med 2004;350:2129.

458. Cancer Genome Atlas Research Network. Comprehensive molecular profiling of lung adenocarcinoma. Nature 2014;511:543.

459. Gazdar A. et al. Mutations and addiction to EGFR: the Achilles 'heal' of lung cancers? Trends Mol Med 2004;10:481.

460. Zhai X. et al. Insight into the Therapeutic Selectivity of the Irreversible EGFR Tyrosine Kinase Inhibitor Osimertinib through Enzyme Kinetic Studies. Biochemistry 2020;59:1428.

461. Druker B. et al. Efficacy and safety of a specific inhibitor of the BCR-ABL tyrosine kinase in chronic myeloid leukemia. N Engl J Med 2001;344:1031.

462. Baselga J. et al. Phase II study of weekly intravenous trastuzumab (Herceptin) in patients with HER2/neu-overexpressing metastatic breast cancer. Semin Oncol 1999;26:78.

463. Tubbs R. et al. Discrepancies in clinical laboratory testing of eligibility for trastuzumab therapy: apparent immunohistochemical false-positives do not get the message. J Clin Oncol 2001;19:2714.

464. Soria J. et al. Osimertinib in Untreated EGFR-Mutated Advanced Non-Small-Cell Lung Cancer. N Engl J Med 2018;378:113.

465. Shaw A. et al. Crizotinib versus chemotherapy in advanced ALK-positive lung cancer. N Engl J Med 2013;368:2385.

466. Shaw A. et al. Crizotinib in ROS1-rearranged NSCLC*. N Engl J Med 2014;371:1963.

467. Hong D. et al. KRAS(G12C) Inhibition with Sotorasib in Advanced Solid Tumors. N Engl J Med 2020;383:1207.

468. Drilon A. et al. Efficacy of Selpercatinib in RET Fusion-Positive Non-Small-Cell Lung Cancer. N Engl J Med 2020;383:813.

469. Wolf J. et al. Capmatinib in MET Exon 14-Mutated or MET-Amplified NSCLC Non-Small-Cell Lung Cancer. N Engl J Med 2020;383:944.

470. Haratake N. et al. NTRK Fusion-positive Non-small-cell Lung Cancer: The Diagnosis and Targeted Therapy. Clin Lung Cancer 2021;22:1.

471. Mazieres J. et al. Lung cancer that harbors an HER2 mutation: epidemiologic characteristics and therapeutic perspectives. J Clin Oncol 2013;31:1997.

472. Robert C. et al. Improved overall survival in melanoma with combined dabrafenib and trametinib. N Engl J Med 2015;372:30.

473. Planchard D. et al. Dabrafenib plus trametinib in patients with previously untreated BRAF(V600E)-mutant metastatic non-small-cell lung cancer*. Lancet Oncol 2017;18:1307.

474. Hyman D. et al. Vemurafenib in Multiple Nonmelanoma Cancers with BRAF V600 Mutations. N Engl J Med 2015;373:726.

475. Kelly C. et al. The management of metastatic GIST*. J Hematol Oncol 2021;14:2.

476. Robson M. et al. Olaparib for Metastatic Breast Cancer in Patients with a Germline BRCA Mutation. N Engl J Med 2017;377:523.

477. Kaufman B. et al. Olaparib monotherapy in patients with advanced cancer and a germline BRCA1/2 mutation. J Clin Oncol 2015;33:244.

478. Sekulic A. et al. Efficacy and safety of vismodegib in advanced basal-cell carcinoma. N Engl J Med 2012;366:2171.

479. Bang Y. et al. Trastuzumab in combination with chemotherapy versus chemotherapy alone for treatment of HER2-positive advanced gastric or gastro-oesophageal junction cancer (ToGA): a phase 3, open-label, randomised controlled trial. Lancet 2010;376:687.

480. Mok T. et al. Gefitinib or carboplatin-paclitaxel in pulmonary adenocarcinoma. N Engl J Med 2009;361:947.

481. Chapman P. et al. Improved survival with vemurafenib in melanoma with BRAF V600E mutation. N Engl J Med 2011;364:2507.

482. McCormick F. KRAS as a Therapeutic Target. Clin Cancer Res 2015;21:1797.

483. Khozin S. et al. Osimertinib for the Treatment of Metastatic EGFR T790M Mutation-Positive Non-Small Cell Lung Cancer. Clin Cancer Res 2017;23:2131.

484. Oxnard G. et al. Assessment of Resistance Mechanisms and Clinical Implications in Patients With EGFR T790M-Positive Lung Cancer and Acquired Resistance to Osimertinib. JAMA Oncol 2018;4:1527.

485. Baselga J. et al. Everolimus in postmenopausal hormone-receptor-positive advanced breast cancer. N Engl J Med 2012;366:520.

486. Moynahan M. et al. Correlation between PIK3CA mutations in cell-free DNA and everolimus efficacy in HR(+), HER2(-) advanced breast cancer: results from BOLERO-2. Br J Cancer 2017;116:726.

487. Motzer R. et al. Efficacy of everolimus in advanced renal cell carcinoma: a double-blind, randomised, placebo-controlled phase III trial. Lancet 2008;372:449.

488. Escudier B. et al. Sorafenib in advanced clear-cell renal-cell carcinoma. N Engl J Med 2007;356:125.

489. Motzer R. et al. Sunitinib in patients with metastatic renal cell carcinoma. JAMA 2006;295:2516.

490. Jonker D. et al. Cetuximab for the treatment of colorectal cancer. N Engl J Med 2007;357:2040.

491. Bonner J. et al. Radiotherapy plus cetuximab for squamous-cell carcinoma of the head and neck. N Engl J Med 2006;354:567.

492. Cheson B. Rituximab: clinical development and future directions. Expert Opin Biol Ther 2002;2:97.

493. Kurzrock R. et al. Exploring the Benefit/Risk Associated with Antiangiogenic Agents for the Treatment of Non-Small Cell Lung Cancer Patients. Clin Cancer Res 2017;23:1137.

494. Verma S. et al. Trastuzumab emtansine for HER2-positive advanced breast cancer. N Engl J Med 2012;367:1783.

495. Yun J. et al. Antitumor Activity of Amivantamab (JNJ-61186372), an EGFR-MET Bispecific Antibody, in Diverse Models of EGFR Exon 20 Insertion-Driven NSCLC. Cancer Discov 2020;10:1194.

496. Keast D. Immunosurveillance and cancer. Lancet 1970;2:710.

497. Crispin J. et al. Cancer immunosurveillance by CD8 T cells. F1000Res 2020;9.

498. Sharma P. et al. Natural Killer Cells - Their Role in Tumour Immunosurveillance. J Clin Diagn Res 2017;11:BE01.

499. Hoong B. et al. cGAS-STING pathway in oncogenesis and cancer therapeutics. Oncotarget 2020;11:2930.

500. Chang R. et al. The interplay between innate and adaptive immunity in cancer shapes the productivity of cancer immunosurveillance. J Leukoc Biol 2020;108:363.

501. Sabbatino F. et al. Role of Human Leukocyte Antigen System as A Predictive Biomarker for Checkpoint-Based Immunotherapy in Cancer Patients. Int J Mol Sci 2020;21:7295.

502. Wosen J. et al. Epithelial MHC Class II Expression and Its Role in Antigen Presentation in the Gastrointestinal and Respiratory Tracts. Front Immunol 2018;9:2144.

503. Johnson A. et al. Cancer Cell-Specific Major Histocompatibility Complex II Expression as a Determinant of the Immune Infiltrate Organization and Function in the NSCLC Tumor Microenvironment. J Thorac Oncol 2021;16:1694.

504. Korman A. et al. Checkpoint blockade in cancer immunotherapy. Adv Immunol 2006;90:297.

505. Leach D. et al. Enhancement of antitumor immunity by CTLA-4 blockade. Science 1996;271:1734.

506. Ishida Y. et al. Induced expression of PD-1, a novel member of the immunoglobulin gene superfamily, upon programmed cell death. EMBO J 1992;11:3887.

507. Freeman G. et al. Engagement of the PD-1 immunoinhibitory receptor by a novel B7 family member leads to negative regulation of lymphocyte activation. J Exp Med 2000;192:1027.

508. Iwai Y. et al. Involvement of PD-L1 on tumor cells in the escape from host immune system and tumor immunotherapy by PD-L1 blockade. Proc Natl Acad Sci U S A 2002;99:12293.

509. Iwai Y. et al. PD-1 blockade inhibits hematogenous spread of poorly immunogenic tumor cells by enhanced recruitment of effector T cells. Int Immunol 2005;17:133.

510. Ai M. et al. Immune checkpoint combinations from mouse to man. Cancer Immunol Immunother 2015;64:885.

511. Mathe G. et al. The immunological approach to cancer treatment. J R Coll Physicians Lond 1970;5:62.

512. Baraniskin A. et al. Efficacy of bevacizumab in first-line treatment of metastatic colorectal cancer: A systematic review and meta-analysis. Eur J Cancer 2019;106:37.

513. Hernberg M. et al. Regimens with or without interferon-alpha as treatment for metastatic melanoma and renal cell carcinoma: an overview of randomized trials. J Immunother 1999;22:145.

514. Interferon alfa versus chemotherapy for chronic myeloid leukemia: CMLTrialists' Collaborative Group*. J Natl Cancer Inst 1997;89:1616.

515. Rosenberg S. et al. Durable complete responses in heavily pretreated patients with metastatic melanoma using T-cell transfer immunotherapy. Clin Cancer Res 2011;17:4550.

516. Rosenberg S. et al. Durability of complete responses in patients with metastatic cancer treated with high-dose interleukin-2*. Ann Surg 1998;228:307.

517. Rosenberg S. et al. Prospective randomized trial of high-dose interleukin-2 alone or in conjunction with lymphokine-activated killer cells for the treatment of patients with advanced cancer. J Natl Cancer Inst 1993;85:622.

518. Small E. et al. Placebo-controlled phase III trial of immunologic therapy with sipuleucel-T (APC8015) in patients with metastatic, asymptomatic hormone refractory prostate cancer. J Clin Oncol 2006;24:3089.

519. Ostrand-Rosenberg S. Animal models of tumor immunity, immunotherapy and cancer vaccines. Curr Opin Immunol 2004;16:143.

520: Long J. et al. Drug discovery oncology in a mouse: concepts, models and limitations. Future Science OA 2021;7:https://doi.org/10.2144/fsoa-021-0019.

521. Kim S. et al. Tumor Burden and Immunotherapy: Impact on Immune Infiltration and Therapeutic Outcomes. Front Immunol 2020;11:629722.

522. Slavin S. et al. Immunotherapy of minimal residual disease by immunocompetent lymphocytes and their activation by cytokines. Cancer Invest 1992;10:221.

523. Mathe G. et al. Clinical trials of the treatment of minimal residual tumours. Drugs Exp Clin Res 1986;12:83.

524. Stewart D. et al. Cancer: the road to Amiens. J Clin Oncol 2009;27:328.

525. Downey S. et al. Prognostic factors related to clinical response in patients with metastatic melanoma treated by CTL-associated antigen-4 blockade. Clin Cancer Res 2007;13:6681.

526. Hodi F. et al. Improved survival with ipilimumab in patients with metastatic melanoma. N Engl J Med 2010;363:711.

527. Robert C. et al. Ipilimumab plus dacarbazine for previously untreated metastatic melanoma. N Engl J Med 2011;364:2517.

528. Yang J. et al. Ipilimumab (anti-CTLA4 antibody) causes regression of metastatic renal cell cancer associated with enteritis and hypophysitis. J Immunother 2007;30:825.

529. Topalian S. et al. Survival, durable tumor remission, and long-term safety in patients with advanced melanoma receiving nivolumab. J Clin Oncol 2014;32:1020.

530. Robert C. et al. Anti-programmed-death-receptor-1 treatment with pembrolizumab in ipilimumab-refractory advanced melanoma*. Lancet 2014;384:1109.

531. Weber J. et al. Nivolumab versus chemotherapy in patients with advanced melanoma who progressed after anti-CTLA-4 treatment (CheckMate 037)*. Lancet Oncol 2015;16:375.

532. Robert C. et al. Pembrolizumab versus Ipilimumab in Advanced Melanoma. N Engl J Med 2015;372:2521.

533. Ribas A. et al. Association of Pembrolizumab With Tumor Response and Survival Among Patients With Advanced Melanoma. JAMA 2016;315:1600.

534. Rizvi N. et al. Activity and safety of nivolumab, an anti-PD-1 immune checkpoint inhibitor, for patients with advanced, refractory squamous NSCLC (CheckMate 063)*. Lancet Oncol 2015;16:257.

535. Garon E. et al. Pembrolizumab for the treatment of NSCLC*. N Engl J Med 2015;372:2018.

536. Brahmer J. et al. Nivolumab versus Docetaxel in Advanced Squamous-Cell Non-Small-Cell Lung Cancer. N Engl J Med 2015;373:123.

537. Borghaei H. et al. Nivolumab versus Docetaxel in Advanced Nonsquamous Non-Small-Cell Lung Cancer. N Engl J Med 2015;373:1627.

538. Antonia S. et al. Nivolumab alone and nivolumab plus ipilimumab in recurrent small-cell lung cancer (CheckMate 032): a multicentre, open-label, phase 1/2 trial. Lancet Oncol 2016;17:883.

539. Reck M. et al. Pembrolizumab versus Chemotherapy for PD-L1-Positive Non-Small-Cell Lung Cancer. N Engl J Med 2016;375:1823.

540. Ott P. et al. Pembrolizumab in Patients With Extensive-Stage Small-Cell Lung Cancer: Results From the Phase Ib KEYNOTE-028 Study. J Clin Oncol 2017;35:3823.

541. Motzer R. et al. Nivolumab for Metastatic Renal Cell Carcinoma: Results of a Randomized Phase II Trial. J Clin Oncol 2015;33:1430.

542. McDermott D. et al. Survival, Durable Response, and Long-Term Safety in Patients With Previously Treated Advanced Renal Cell Carcinoma Receiving Nivolumab. J Clin Oncol 2015;33:2013.

543. Motzer R. et al. Nivolumab versus Everolimus in Advanced Renal-Cell Carcinoma. N Engl J Med 2015;373:1803.

544. Rosenberg J. et al. Atezolizumab in patients with locally advanced and metastatic urothelial carcinoma who have progressed following treatment with platinum-based chemotherapy: a single-arm, multicentre, phase 2 trial. Lancet 2016;387:1909.

545. Massard C. et al. Safety and Efficacy of Durvalumab (MEDI4736), an Anti-Programmed Cell Death Ligand-1 Immune Checkpoint Inhibitor, in Patients With Advanced Urothelial Bladder Cancer. J Clin Oncol 2016;34:3119.

546. Ferris R. et al. Nivolumab for Recurrent Squamous-Cell Carcinoma of the Head and Neck. N Engl J Med 2016;375:1856.

547. Ansell S. et al. PD-1 blockade with nivolumab in relapsed or refractory Hodgkin's lymphoma. N Engl J Med 2015;372:311.

548. Le D. et al. PD-1 Blockade in Tumors with Mismatch-Repair Deficiency. N Engl J Med 2015;372:2509.

549. Hamanishi J. et al. Safety and Antitumor Activity of Anti-PD-1 Antibody, Nivolumab, in Patients With Platinum-Resistant Ovarian Cancer. J Clin Oncol 2015;33:4015.

550. Nghiem P. et al. PD-1 Blockade with Pembrolizumab in Advanced Merkel-Cell Carcinoma. N Engl J Med 2016;374:2542.

551. Nanda R. et al. Pembrolizumab in Patients With Advanced Triple-Negative Breast Cancer: Phase Ib KEYNOTE-012 Study. J Clin Oncol 2016;34:2460.

552. Muro K. et al. Pembrolizumab for patients with PD-L1-positive advanced gastric cancer (KEYNOTE-012): a multicentre, open-label, phase 1b trial. Lancet Oncol 2016;17:717.

553. Armand P. et al. Programmed Death-1 Blockade With Pembrolizumab in Patients With Classical Hodgkin Lymphoma After Brentuximab Vedotin Failure. J Clin Oncol 2016;34:3733.

554. Younes A. et al. Nivolumab for classical Hodgkin's lymphoma after failure of both autologous stem-cell transplantation and brentuximab vedotin*. Lancet Oncol 2016;17:1283.

555. Morris V. et al. Nivolumab for previously treated unresectable metastatic anal cancer (NCI9673): a multicentre, single-arm, phase 2 study. Lancet Oncol 2017;18:446.

556. Alley E. et al. Clinical safety and activity of pembrolizumab in patients with malignant pleural mesothelioma (KEYNOTE-028)*. Lancet Oncol 2017;18:623.

557. Kudo T. et al. Nivolumab treatment for oesophageal squamous-cell carcinoma: an open-label, multicentre, phase 2 trial. Lancet Oncol 2017;18:631.

558. El-Khoueiry A. et al. Nivolumab in patients with advanced hepatocellular carcinoma (CheckMate 040)*. Lancet 2017;389:2492.

559. Ott P. et al. Safety and Antitumor Activity of Pembrolizumab in Advanced Programmed Death Ligand 1-Positive Endometrial Cancer*. J Clin Oncol 2017;35:2535.

560. Zinzani P. et al. Safety and tolerability of pembrolizumab in patients with relapsed/refractory primary mediastinal large B-cell lymphoma. Blood 2017;130:267.

561. Hsu C. et al. Safety and Antitumor Activity of Pembrolizumab in Patients With Programmed Death-Ligand 1-Positive Nasopharyngeal Carcinoma*. J Clin Oncol 2017;35:4050.

562. Frenel J. et al. Safety and Efficacy of Pembrolizumab in Advanced, Programmed Death Ligand 1-Positive Cervical Cancer: Results From the Phase Ib KEYNOTE-028 Trial. J Clin Oncol 2017;35:4035.

563. Tawbi H. et al. Pembrolizumab in advanced soft-tissue sarcoma and bone sarcoma (SARC028): a multicentre, two-cohort, single-arm, open-label, phase 2 trial. Lancet Oncol 2017;18:1493.

564. Giaccone G. et al. Pembrolizumab in patients with thymic carcinoma: a single-arm, single-centre, phase 2 study. Lancet Oncol 2018;19:347.

565. Fehrenbacher L. et al. Atezolizumab versus docetaxel for patients with previously treated non-small-cell lung cancer (POPLAR)*. Lancet 2016;387:1837.

566. Rittmeyer A. et al. Atezolizumab versus docetaxel in patients with previously treated non-small-cell lung cancer (OAK)*. Lancet 2017;389:255.

567. Cohen E. et al. Pembrolizumab versus methotrexate, docetaxel, or cetuximab for recurrent or metastatic head-and-neck squamous cell carcinoma (KEYNOTE-040)*. Lancet 2019;393:156.

568. Mok T. et al. Pembrolizumab versus chemotherapy for previously untreated, PD-L1-expressing, locally advanced or metastatic non-small-cell lung cancer (KEYNOTE-042)*. Lancet 2019;393:1819.

569. Gandhi L. et al. Pembrolizumab plus Chemotherapy in Metastatic Non-Small-Cell Lung Cancer. N Engl J Med 2018;378:2078.

570. Socinski M. et al. Atezolizumab for First-Line Treatment of Metastatic Nonsquamous NSCLC. N Engl J Med 2018;378:2288.

571. Paz-Ares L. et al. Pembrolizumab plus Chemotherapy for Squamous Non-Small-Cell Lung Cancer. N Engl J Med 2018;379:2040.

572. Horn L. et al. First-Line Atezolizumab plus Chemotherapy in Extensive-Stage Small-Cell Lung Cancer. N Engl J Med 2018;379:2220.

573. Antonia S. et al. Overall Survival with Durvalumab after Chemoradiotherapy in Stage III NSCLC. N Engl J Med 2018;379:2342.

574. Arora S. et al. Existing and Emerging Biomarkers for Immune Checkpoint Immunotherapy in Solid Tumors. Adv Ther 2019;36:2638.

575. Nowicki T. et al. Mechanisms of Resistance to PD-1 and PD-L1 Blockade. Cancer J 2018;24:47.

576. Fuentes-Antras J. et al. Hyperprogression as a distinct outcome after immunotherapy. Cancer Treat Rev 2018;70:16.

577. Stewart D. et al. Exponential decay nonlinear regression analysis of patient survival curves: preliminary assessment in non-small cell lung cancer. Lung Cancer 2011;71:217.

578. Petrelli F. et al. Association of Steroids use with Survival in Patients Treated with Immune Checkpoint Inhibitors: A Systematic Review and Meta-Analysis. Cancers (Basel) 2020;12:546.

579. Huang X. et al. Antibiotic use and the efficacy of immune checkpoint inhibitors in cancer patients: a pooled analysis of 2740 cancer patients. Oncoimmunology 2019;8:e1665973.

580. Lee K. et al. The impact of smoking on the effectiveness of immune checkpoint inhibitors - a systematic review and meta-analysis. Acta Oncol 2020;59:96.

581. Wang X. et al. Association between Smoking History and Tumor Mutation Burden in Advanced Non-Small Cell Lung Cancer. Cancer Res 2021;81:2566.

582. Cavanna L. et al. Immune checkpoint inhibitors in EGFR-mutation positive TKI-treated patients with advanced non-small-cell lung cancer network meta-analysis. Oncotarget 2019;10:209.

583. Oya Y. et al. Efficacy of Immune Checkpoint Inhibitor Monotherapy for Advanced Non-Small-Cell Lung Cancer with ALK Rearrangement. Int J Mol Sci 2020;21:2623.

584. Hastings K. et al. EGFR mutation subtypes and response to immune checkpoint blockade treatment in non-small-cell lung cancer. Ann Oncol 2019;30:1311.

585. Fruh M. et al. EGFR mutation subtype impacts efficacy of immune checkpoint inhibitors in non-small-cell lung cancer. Ann Oncol 2019;30:1190.

586. Nagahashi M. et al. Common driver mutations and smoking history affect tumor mutation burden in lung adenocarcinoma. J Surg Res 2018;230:181.

587. Shire N. et al. STK11 (LKB1) mutations in metastatic NSCLC: Prognostic value in the real world. PLoS One 2020;15:e0238358.

588. Haanen J. et al. Autoimmune diseases and immune-checkpoint inhibitors for cancer therapy: review of the literature and personalized risk-based prevention strategy. Ann Oncol 2020;31:724.

589. Kittai A. et al. Immune Checkpoint Inhibitors in Organ Transplant Patients. J Immunother 2017;40:277.

590. Friedlaender A. et al. Rethinking the Optimal Duration of Immune Checkpoint Inhibitors in Non-small Cell Lung Cancer Throughout the COVID-19 Pandemic. Front Oncol 2020;10:862.

591. Park J. et al. Long-Term Follow-up of CD19 CAR Therapy in Acute Lymphoblastic Leukemia. N Engl J Med 2018;378:449.

592. Locke F. et al. Phase 1 Results of ZUMA-1: A Multicenter Study of KTE-C19 Anti-CD19 CAR T Cell Therapy in Refractory Aggressive Lymphoma. Mol Ther 2017;25:285.

593. Stewart D. et al. Human autopsy tissue concentrations of mitoxantrone. Cancer Treat Rep 1986;70:1255.

594. Stewart D. et al. Concentrations of doxorubicin and its metabolites in human autopsy heart and other tissues. Anticancer Res 1993;13:1945.

595. Stewart D. Concentration of vinblastine in human intracerebral tumor and other tissues. J Neurooncol 1983;1:139.

596. Stewart D. et al. Human autopsy tissue distribution of the epipodophyllotoxins etoposide and teniposide. Cancer Chemother Pharmacol 1993;32:368.

597. Gregg R. et al. Cisplatin neurotoxicity: the relationship between dosage, time, and platinum concentration in neurologic tissues, and morphologic evidence of toxicity. J Clin Oncol 1992;10:795.

598. Burotto M. et al. Gefitinib and erlotinib in metastatic non-small cell lung cancer: a meta-analysis of toxicity and efficacy of randomized clinical trials. Oncologist 2015;20:400.

599. Sakai K. et al. In-frame deletion in the EGF receptor alters kinase inhibition by gefitinib. Biochem J 2006;397:537.

600. Fakih M. et al. Adverse events associated with anti-EGFR therapies for the treatment of metastatic colorectal cancer. Curr Oncol 2010;17 Suppl 1:S18.

601. Marin-Acevedo J. et al. Immune Checkpoint Inhibitor Toxicities. Mayo Clin Proc 2019;94:1321.

602. Spain L. et al. Management of toxicities of immune checkpoint inhibitors. Cancer Treat Rev 2016;44:51.

603. Kennedy L. et al. Preexisting Autoimmune Disease: Implications for Immune Checkpoint Inhibitor Therapy in Solid Tumors. J Natl Compr Canc Netw 2019;17:750.

604. Mundt A. et al. Principles of Radiation Oncology. In: Kufe DW, Bast Jr RC, Hait WN, al. E, eds. Cancer Medicine. 7 ed. Hamilton, ON: BC Decker, Inc; 2006:530.

605. McTyre E. et al. Whole brain radiotherapy for brain metastasis. Surg Neurol Int 2013;4:S236.

606. Tsao A. et al. Collateral damage associated with chemotherapy. In: Yeung S-CJ, Escalante CP, Gagel RF, eds. Medical Care of Cancer Patients. Shelton, CT: BC Decker Inc; 2009:18.

607. Stewart D. Cancer therapy, vomiting, and antiemetics. Can J Physiol Pharmacol 1990;68:304.

608. Stewart D. et al. Costs of treating and preventing nausea and vomiting in patients receiving chemotherapy. J Clin Oncol 1999;17:344.

609. Hesketh P. et al. The oral neurokinin-1 antagonist aprepitant for the prevention of chemotherapy-induced nausea and vomiting: a multinational, randomized, double-blind, placebo-controlled trial in patients receiving high-dose cisplatin*. J Clin Oncol 2003;21:4112.

610. Morrow G. Anticipatory nausea. Cancer Invest 1988;6:327.

611. Bellono N. et al. Enterochromaffin Cells Are Gut Chemosensors that Couple to Sensory Neural Pathways. Cell 2017;170:185.

612. Darmani N. Mechanisms of Broad-Spectrum Antiemetic Efficacy of Cannabinoids against Chemotherapy-Induced Acute and Delayed Vomiting. Pharmaceuticals (Basel) 2010;3:2930.

613. Van Ryckeghem F. Corticosteroids, the oldest agent in the prevention of chemotherapy-induced nausea and vomiting: What about the guidelines? J Transl Int Med 2016;4:46.

614. Gilliam L. et al. Chemotherapy-induced weakness and fatigue in skeletal muscle: the role of oxidative stress. Antioxid Redox Signal 2011;15:2543.

615. Bower JE. Cancer-related fatigue*. Nat Rev Clin Oncol 2014;11:597.

616. Smets E. et al. Fatigue and radiotherapy: (B) experience in patients 9 months following treatment. Br J Cancer 1998;78:907.

617. Hsiao C. et al. The Etiology and management of radiotherapy-induced fatigue. Expert Rev Qual Life Cancer Care 2016;1:323.

618. Larkin J. et al. Fatigue in renal cell carcinoma: the hidden burden of current targeted therapies. Oncologist 2010;15:1135.

619. Abdel-Rahman O. et al. Treatment-associated Fatigue in Cancer Patients Treated with Immune Checkpoint Inhibitors; a Systematic Review and Meta-analysis. Clin Oncol (R Coll Radiol) 2016;28:e127.

620. Wang Y. et al. Treatment-Related Adverse Events of PD-1 and PD-L1 Inhibitors in Clinical Trials: A Systematic Review and Meta-analysis. JAMA Oncol 2019;5:1008.

621. Kovalchuk A. et al. Chemo brain: From discerning mechanisms to lifting the brain fog-An aging connection. Cell Cycle 2017;16:1345.

622. Janelsins M. et al. An update on cancer- and chemotherapy-related cognitive dysfunction: current status. Semin Oncol 2011;38:431.

623. Haslam I. et al. Chemotherapy-Induced Hair Loss: The Use of Biomarkers for Predicting Alopecic Severity and Treatment Efficacy. Biomark Insights 2019;14:1177271919842180.

624. Watanabe T. et al. A multicenter survey of temporal changes in chemotherapy-induced hair loss in breast cancer patients. PLoS One 2019;14:e0208118.

625. Lawenda B. et al. Permanent alopecia after cranial irradiation: dose-response relationship. Int J Radiat Oncol Biol Phys 2004;60:879.

626. Rugo H. et al. Scalp cooling with adjuvant/neoadjuvant chemotherapy for breast cancer and the risk of scalp metastases: systematic review and meta-analysis. Breast Cancer Res Treat 2017;163:199.

627. Furze R. et al. Neutrophil mobilization and clearance in the bone marrow. Immunology 2008;125:281.

628. Jie Z. et al. Large-scale ex vivo generation of human neutrophils from cord blood CD34+ cells. PLoS One 2017;12:e0180832.

629. Pizzo P. Granulocytopenia and cancer therapy. Past problems, current solutions, future challenges. Cancer 1984;54:2649.

630. Sender R. et al. Revised Estimates for the Number of Human and Bacteria Cells in the Body. PLoS Biol 2016;14:e1002533.

631. Crawford J. et al. Chemotherapy-induced neutropenia: risks, consequences, and new directions for its management. Cancer 2004;100:228.

632. van der Velden W. et al. Mucosal barrier injury, fever and infection in neutropenic patients with cancer: introducing the paradigm febrile mucositis. Br J Haematol 2014;167:441.

633. Scott S. Identification of cancer patients at high risk of febrile neutropenia. Am J Health Syst Pharm 2002;59:S16.

634. Bodey G. et al. Quantitative relationships between circulating leukocytes and infection in patients with acute leukemia. Ann Intern Med 1966;64:328.

635. Skipper H. Kinetics of mammary tumor cell growth & implications for therapy. Cancer 1971;28:1479.

636. Lyman G. Impact of chemotherapy dose intensity on cancer patient outcomes. J Natl Compr Canc Netw 2009;7:99.

637. Lyman G. et al. The effect of filgrastim or pegfilgrastim on survival outcomes of patients with cancer receiving myelosuppressive chemotherapy. Ann Oncol 2015;26:1452.

638. Budman D. et al. Dose and dose intensity as determinants of outcome in the adjuvant treatment of breast cancer. The Cancer and Leukemia Group B. J Natl Cancer Inst 1998;90:1205.

639. Berry D. et al. High-dose chemotherapy with autologous hematopoietic stem-cell transplantation in metastatic breast cancer*. J Clin Oncol 2011;29:3224.

640. Larson R. et al. A randomized controlled trial of filgrastim during remission induction and consolidation chemotherapy for adults with acute lymphoblastic leukemia*. Blood 1998;92:1556.

641. Lyman G. et al. Cost-effectiveness of pegfilgrastim versus filgrastim primary prophylaxis in women with early-stage breast cancer receiving chemotherapy in the United States. Clin Ther 2009;31:1092.

642. Kubista E. et al. Bone pain associated with once-per-cycle pegfilgrastim is similar to daily filgrastim in patients with breast cancer. Clin Breast Cancer 2003;3:391.

643. Moon H. et al. Effects of granulocyte-colony stimulating factor and the expression of its receptor on various malignant cells. Korean J Hematol 2012;47:219.

644. Goldvaser H. et al. Influence of control group therapy on the benefit from dose-dense chemotherapy in early breast cancer*. Breast Cancer Res Treat 2018;169:413.

645. Wang Q. et al. Analyses of Risk, Racial Disparity, and Outcomes Among US Patients With Cancer and COVID-19 Infection. JAMA Oncol 2021;7:220.

646. Saini K. et al. Mortality in patients with cancer and coronavirus disease 2019: A systematic review and pooled analysis of 52 studies. Eur J Cancer 2020;139:43.

647. Liu C. et al. COVID-19 in cancer patients: risk, clinical features, and management. Cancer Biol Med 2020;17:519.

648. Sun L. et al. Immune Responses to SARS-CoV-2 Among Patients With Cancer: What Can Seropositivity Tell Us? JAMA Oncol 2021;7:1123.

649. Stanworth S. et al. Risk of bleeding and use of platelet transfusions in patients with hematologic malignancies: recurrent event analysis. Haematologica 2015;100:740.

650. Boral L. Platelet transfusion therapy. Laboratory Medicine 1985;16:221.

651. Bussel J. et al. A Review of Romiplostim Mechanism of Action and Clinical Applicability. Drug Des Devel Ther 2021;15:2243.

652. van Vliet M. et al. The role of intestinal microbiota in the development and severity of chemotherapy-induced mucositis. PLoS Pathog 2010;6:e1000879.

653. Stein A. et al. Chemotherapy-induced diarrhea: pathophysiology, frequency and guideline-based management. Ther Adv Med Oncol 2010;2:51.

654. Chaveli-Lopez B. et al. Treatment of oral mucositis due to chemotherapy. J Clin Exp Dent 2016;8:e201.

655. Wang D. et al. Incidence of immune checkpoint inhibitor-related colitis in solid tumor patients: A systematic review and meta-analysis. Oncoimmunology 2017;6:e1344805.

656. Collins M. et al. Management of Patients With Immune Checkpoint Inhibitor-Induced Enterocolitis: A Systematic Review. Clin Gastroenterol Hepatol 2020;18:1393.

657. Dodd M. et al. Randomized clinical trial of the effectiveness of 3 commonly used mouthwashes to treat chemotherapy-induced mucositis. Oral Surg Oral Med Oral Pathol Oral Radiol Endod 2000;90:39.

658. Krames E. The role of the dorsal root ganglion in the development of neuropathic pain. Pain Med 2014;15:1669.

659. Cupit-Link M. et al. Biology of premature ageing in survivors of cancer. ESMO Open 2017;2:e000250.

660. Sculier J. et al. Chemotherapy improves low performance status lung cancer patients. Eur Respir J 2007;30:1186.

661. Vansteenkiste J. et al. Influence of cisplatin-use, age, performance status and duration of chemotherapy on symptom control in advanced non-small cell lung cancer*. Lung Cancer 2003;40:191.

662. Dall'Olio F. et al. ECOG performance status >/=2 as a prognostic factor in patients with advanced non small cell lung cancer treated with immune checkpoint inhibitors*. Lung Cancer 2020;145:95.

663. Chen Y. et al. The impact of clinical parameters on progression-free survival of NSCLC patients harboring EGFR-mutations receiving first-line EGFR-tyrosine kinase inhibitors*. Lung Cancer 2016;93:47.

664. Salzberg S. Open questions: How many genes do we have? BMC Biol 2018;16:94.

665. Huang R. et al. Pharmacogenetics and pharmacogenomics of anticancer agents. CA Cancer J Clin 2009;59:42.

666. Ando Y. et al. Polymorphisms of UDP-glucuronosyltransferase gene and irinotecan toxicity: a pharmacogenetic analysis. Cancer Res 2000;60:6921.

667. Lee A. et al. Dihydropyrimidine dehydrogenase deficiency: impact of pharmacogenetics on 5-fluorouracil therapy. Clin Adv Hematol Oncol 2004;2:527.

668. Pu X. et al. PI3K/PTEN/AKT/mTOR pathway genetic variation predicts toxicity and distant progression in lung cancer patients receiving platinum-based chemotherapy. Lung Cancer 2011;71:82.

669. Wu X. et al. Genome-wide association study of survival in non-small cell lung cancer patients receiving platinum-based chemotherapy. J Natl Cancer Inst 2011;103:817.

670. Wu X. et al. Germline genetic variations in drug action pathways predict clinical outcomes in advanced lung cancer treated with platinum-based chemotherapy. Pharmacogenet Genomics 2008;18:955.

671. Jatau A. et al. Use and toxicity of complementary and alternative medicines among patients visiting emergency department: Systematic review. J Intercult Ethnopharmacol 2016;5:191.

672. Horneber M. et al. How many cancer patients use complementary and alternative medicine: a systematic review and metaanalysis. Integr Cancer Ther 2012;11:187.

673. Myers SP, Vigar V. The State of the Evidence for Whole-System, Multi-Modality Naturopathic Medicine: A Systematic Scoping Review. J Altern Complement Med 2019;25:141-68.

674. Pan S. et al. New Perspectives on How to Discover Drugs from Herbal Medicines*. Evid Based Complement Alternat Med 2013;2013:627375.

675. Vogelzang N. et al. VP-16-213 (etoposide): the mandrake root from Issyk-Kul. Am J Med 1982;72:136.

676. Cassileth B. et al. Complementary and alternative therapies for cancer. Oncologist 2004;9:80.

677. Ballotari P. et al. Diabetes and risk of cancer incidence: results from a population-based cohort study in northern Italy. BMC Cancer 2017;17:703.

678. Chowdhury T. Diabetes and cancer. QJM 2010;103:905.

679. Devries S. Does cancer have a sweet tooth?. Gaples Institute https://www.gaplesi nstitute.org/does-cancer-have-a-sweet-tooth/?gclid=EAIaIQob ChMIitWYgZ-c5AIVav jBx3DLg73EAAYASAAEgIM2PD BwE. Accessed 2019/08/24.

680. Porporato P. Understanding cachexia as a cancer metabolism syndrome. Oncogenesis 2016;5:e200.

681. Cameron E. et al. Supplemental ascorbate in the supportive treatment of cancer: reevaluation of prolongation of survival times in terminal human cancer. Proc Natl Acad Sci U S A 1978;75:4538.

682. Sacks H. et al. Randomized versus historical controls for clinical trials. Am J Med 1982;72:233.

683. Price D. et al. A comprehensive review of the placebo effect: recent advances and current thought. Annu Rev Psychol 2008;59:565.

684. Pauling L. et al. A proposition: megadoses of vitamin C are valuable in the treatment of cancer. Nutr Rev 1986;44:28.

685. Chaft J. et al. Phase II trial of neoadjuvant bevacizumab plus chemotherapy and adjuvant bevacizumab in patients with resectable nonsquamous NSCLC*. J Thorac Oncol 2013;8:1084.

686. Blumenthal G. et al. Drug Development, Trial Design, and Endpoints in Oncology: Adapting to Rapidly Changing Science. Clin Pharmacol Ther 2017;101:572.

687. Big Pharma conspiracy theory. Wikipedia 2019. https://en.wikipedia.org/wiki/Big_Pharma_conspiracy_theory. accessed 09/17/2019.

688. Tohme S, Simmons RL, Tsung A. Surgery for Cancer: A Trigger for Metastases. Cancer Res 2017;77:1548-52.

689. DeWys W. Studies correlating the growth rate of a tumor and its metastases and providing evidence for tumor-related systemic growth-retarding factors. Cancer Res 1972;32:374.

690. Biagi J. et al. Association between time to initiation of adjuvant chemotherapy and survival in colorectal cancer: a systematic review and meta-analysis. JAMA 2011;305:2335.

691. Zhan Q. et al. Survival and time to initiation of adjuvant chemotherapy among breast cancer patients: a systematic review and meta-analysis. Oncotarget 2018;9:2739.

692. Camphausen K. et al. Radiation therapy to a primary tumor accelerates metastatic growth in mice. Cancer Res 2001;61:2207.

693. Waiting your turn: Wait times for health care in Canada, 2018 report. Fraser Institute https://www.fraserinstitute.org/sites/default/files/waiting-your-turn-2018.pdf. Accessed 2019/09/18.

694. Fung-Kee-Fung M. et al. Regional process redesign of lung cancer care*. Curr Oncol 2018;25:59.

695. Naing A. et al. Chemotherapy resistance and retreatment: a dogma revisited. Clin Colorectal Cancer 2010;9:E1.

696. Seltzer V. et al. Recurrent ovarian carcinoma: retreatment utilizing combination chemotherapy including cis-diamminedichloroplatinum in patients previously responding to this agent. Gynecol Oncol 1985;21:167.

697. Kucukoztas N. et al. Response rates of taxane rechallenge in metastatic breast cancer patients previously treated with adjuvant taxanes. J BUON 2016;21:1076.

698. Petrelli F. et al. Platinum rechallenge in patients with advanced NSCLC: a pooled analysis. Lung Cancer 2013;81:337.

699. Postmus P. et al. Retreatment with the induction regimen in small cell lung cancer relapsing after an initial response to short term chemotherapy. Eur J Cancer Clin Oncol 1987;23:1409.

700. Hanovich E. et al. Rechallenge Strategy in Cancer Therapy. Oncology 2020;98:669.

701. Horning S. et al. Developing standards for breakthrough therapy designation in oncology. Clin Cancer Res 2013;19:4297.

702. Oxnard G. et al. Variability of lung tumor measurements on repeat computed tomography scans taken within 15 minutes. J Clin Oncol 2011;29:3114.

703. Eisenhauer E. et al. New response evaluation criteria in solid tumours: revised RECIST guideline (version 1.1). Eur J Cancer 2009;45:228.

704. Rothenberg M. FDA Grants Crizotinib Breakthrough Designation for MET+ NSCLC and ALK+ ALCL. 2018/05/29. https://www.onclive.com/view/fda-grants-crizotinib-breakthrough-designation-for-met-nsclc-and-alk-alcl. accessed 2021/12/04.

705. Drilon A. et al. Antitumor activity of crizotinib in lung cancers harboring a MET exon 14 alteration. Nat Med 2020;26:47.

706. Stewart D. et al. The importance of greater speed in drug development for advanced malignancies. Cancer Med 2018;7:1824.

707. Gotfrit J. et al. Potential Life-Years Lost: The Impact of the Cancer Drug Regulatory and Funding Process in Canada. Oncologist 2020;25:e130.

708. Stewart D. et al. A novel, more reliable approach to use of progression-free survival as a predictor of gain in overall survival*. Crit Rev Oncol Hematol 2020;148:102896.

709. Harmon A. New Drug Stirs Debate on Rules of Clinical Trials. New York Times 2010/09/19. https://www.nytimes.com/2010/09/19/health/research/19trial.html. accessed 2021/12/20.

710. Ziliak ST, McCloskey DN. The Cult of Statistical Significance: How the Standard Error Costs Us Jobs, Justice, and Lives. Ann Arbor: The University of Michigan Press; 2011.

711. Greenland S, Senn SJ, Rothman KJ, et al. Statistical tests, P values, confidence intervals, and power: a guide to misinterpretations. Eur J Epidemiol 2016;31:337-50.

712. Senn S. Contribution to the discussion of '"A critical evaluation of the current p-value controversy"'. Biom J 2017;59:892-4.

713. Gagnier JJ, Morgenstern H. Misconceptions, Misuses, and Misinterpretations of P Values and Significance Testing. J Bone Joint Surg Am 2017;99:1598-603.

714. Stang A, Poole C, Kuss O. The ongoing tyranny of statistical significance testing in biomedical research. Eur J Epidemiol 2010;25:225-30.

715. Sprowls S. et al. Improving CNS Delivery to Brain Metastases by Blood-Tumor Barrier Disruption. Trends Cancer 2019;5:495.

716. Deshpande K. et al. Clinical Perspectives in Brain Metastasis. Cold Spring Harb Perspect Med 2020;10:a037051.

717. Achrol A. et al. Brain metastases. Nat Rev Dis Primers 2019;5:5.

718. Ahluwalia M. et al. Epidermal Growth Factor Receptor Tyrosine Kinase Inhibitors for Central Nervous System Metastases from Non-Small Cell Lung Cancer. Oncologist 2018;23:1199.

719. Di Lorenzo R. et al. Targeted therapy of brain metastases*. Ther Adv Med Oncol 2017;9:781.

720. Venur V. et al. Systemic therapy for brain metastases. Handb Clin Neurol 2018;149:137.

721. Nam J. et al. Current chemotherapeutic regimens for brain metastases treatment. Clin Exp Metastasis 2017;34:391.

722. Dempke W. Brain Metastases in NSCLC*. Anticancer Res 2015;35:5797.

723. Blecharz K. et al. Control of the blood-brain barrier function in cancer cell metastasis. Biol Cell 2015;107:342.

724. Fokas E. et al. Biology of brain metastases and novel targeted therapies: time to translate the research. Biochim Biophys Acta 2013;1835:61.

725. Fortin D. The blood-brain barrier: its influence in the treatment of brain tumors metastases. Curr Cancer Drug Targets 2012;12:247.

726. Ewend M. et al. Brain metastases. Curr Treat Options Oncol 2001;2:537.

727. Haughton M. et al. Treatment of brain metastases of lung cancer in the era of precision medicine. Front Biosci (Elite Ed) 2016;8:219.

728. Bhatt V. et al. Epidermal Growth Factor Receptor Mutational Status and Brain Metastases in Non-Small-Cell Lung Cancer. J Glob Oncol 2017;3:208.

729. Stewart D. A critique of the role of the blood-brain barrier in the chemotherapy of human brain tumors. J Neurooncol 1994;20:121.

730. Oh Y. et al. Systemic therapy for lung cancer brain metastases: a rationale for clinical trials. Oncology (Williston Park) 2008;22:168.

731. Deeken J. et al. The blood-brain barrier and cancer*. Clin Cancer Res 2007;13:1663.

732. Peters S. et al. Alectinib versus Crizotinib in Untreated ALK-Positive Non-Small-Cell Lung Cancer. N Engl J Med 2017;377:829.

733. Solomon B. et al. Intracranial Efficacy of Crizotinib Versus Chemotherapy in Patients With Advanced ALK-Positive Non-Small-Cell Lung Cancer*. J Clin Oncol 2016;34:2858.

734. Ali A. et al. Survival of patients with non-small-cell lung cancer after a diagnosis of brain metastases. Curr Oncol 2013;20:e300.

735. Riihimaki M. et al. Metastatic sites and survival in lung cancer. Lung Cancer 2014;86:78.

736. Hsu F. et al. Patterns of spread and prognostic implications of lung cancer metastasis in an era of driver mutations. Curr Oncol 2017;24:228.

737. Xu Z. et al. Clinical associations and prognostic value of site-specific metastases in non-small cell lung cancer: A population-based study. Oncol Lett 2019;17:5590.

738. Ashour Badawy A. et al. Site of Metastases as Prognostic Factors in Unselected Population of Stage IV Non-Small Cell Lung Cancer. Asian Pac J Cancer Prev 2018;19:1907.

739. Van Schil P. et al. The 8(th) TNM edition for lung cancer*. Ann Transl Med 2018;6:87.

740. Sorensen J. et al. Brain metastases in adenocarcinoma of the lung: frequency, risk groups, and prognosis. J Clin Oncol 1988;6:1474.

741. Walbert T. et al. The role of chemotherapy in the treatment of patients with brain metastases from solid tumors. Int J Clin Oncol 2009;14:299.

742. Nieder C. et al. Integration of chemotherapy into current treatment strategies for brain metastases from solid tumors. Radiat Oncol 2006;1:19.

743. Fidler I. The Biology of Brain Metastasis: Challenges for Therapy. Cancer J 2015;21:284.

744. Bonomi P. et al. Making Lung Cancer Clinical Trials More Inclusive*. J Thorac Oncol 2018;13:748.

745. Stewart D. et al. The Urgent Need for Clinical Research Reform to Permit Faster, Less Expensive Access to New Therapies for Lethal Diseases. Clin Cancer Res 2015;21:4561.

746. Hardin G. The tragedy of the commons*. Science 1968;162:1243.

747. Berton P. Vimy. Toronto, ON: McClelland and Stewart, LTD; 1986.

748. Fanelli D. How many scientists fabricate and falsify research?*. PLoS One 2009;4:e5738.

749. Take a stand. Nature 2012;487:139-40.

750. Khamsi R. Painkiller verdict shows mistrust of Merck. Nature 2005;436:1070.

751. Azuine R. et al. Overcoming Challenges in Conducting Clinical Trials in Minority Populations: Identifying and Testing What Works. Int J MCH AIDS 2015;3:81.

752. Lowenstein P. et al. Challenges in the evaluation, consent, ethics and history of early clinical trials.* Curr Opin Mol Ther 2009;11:481.

753. Hilts PJ. Protecting America's Health: the FDA, Business, and One Hundred Years of Regulation. New York: Alfred A. Knopf; 2003.

754. Dickson M. et al. The cost of new drug discovery and development. Discov Med 2004;4:172.

755. Chen J. et al. Drug discovery and drug marketing*. Am J Transl Res 2018;10:4302.

756. DiMasi J. et al. Innovation in the pharmaceutical industry: R&D costs*. J Health Econ 2016;47:20.

757. Stewart D. et al. Equipoise lost: ethics, costs, and the regulation of cancer clinical research. J Clin Oncol 2010;28:2925.

758. Salas-Vega S. et al. Assessment of Overall Survival, Quality of Life, and Safety Benefits Associated With New Cancer Medicines. JAMA Oncol 2017;3:382.

759. Stewart D. et al. Redefining cancer: a new paradigm*. J Popul Ther Clin Pharmacol 2014;21:e56.

760. Stewart D. et al. To benefit from new cancer drugs, reform the regulatory regime. Globe and Mail 2015.

761. Antonov-Ovseyenko A. The time of Stalin: portrait of a tyranny: Harper & Row; 1981.

762. Darrow J. et al. New FDA breakthrough-drug category*. N Engl J Med 2014;370:1252.

763. Doyle C. Committed to Excellence: Oncology Drug Development Marches on Amid a Pandemic. The ASCO Post 2020: https://ascopost.com/issues/november-10-2020/oncology-drug-development-marches-on-amid-a-pandemic/ accessed 2020/12/16.

764. El-Maraghi R. et al. Review of phase II trial designs used in studies of molecular targeted agents: outcomes and predictors of success in phase III. J Clin Oncol 2008;26:1346.

765. Vidaurre T. et al. Stable disease is not preferentially observed with targeted therapies and as currently defined has limited value in drug development. Cancer J 2009;15:366.

766. Bruzzi P. et al. Objective response to chemotherapy as a potential surrogate end point of survival in metastatic breast cancer patients. J Clin Oncol 2005;23:5117.

767. Aitken M. Global Oncology Trend Report: a review of 2015 and outlook to 2020. 2016: https://morningconsult.com/wp-content/uploads/2016/06/IMS-Institute-Global-Oncology-Report-05.31.16.pdf; accessed 2020/10/31.

768. Hudson K. et al. The 21st Century Cures Act - A View from the NIH. N Engl J Med 2017;376:111.

769. Burris J. et al. Impact of Federal Regulatory Changes on Clinical Pharmacology and Drug Development: the Common Rule and the 21st Century Cures Act. J Clin Pharmacol 2018;58:281.

770. Winners and losers of the 21st Century Cures Act. STAT https://www.statnews.com/2016/12/05/21st-century-cures-act-winners-losers/ accessed 2017/01/12.

771. Biden J. Cancer Moonshot. Report to the President from the Vice President. 2016. https://obamawhitehouse.archives.gov/sites/default/files/docs/finalvp_exec_report_10-17-16final_3.pdf. accessed 2022/01/15.

772. Right-to-try law. Wikipedia https://en.wikipedia.org/wiki/Right-to-try_law. accessed 2020/10/28.

773. Tamimi N. et al. Drug development: from concept to..*. Nephron Clin Pract 2009;113:c125.

774. Ubel P. et al. In a survey, marked inconsistency in how oncologists judged value of high-cost cancer drugs in relation to gains in survival. Health Aff (Millwood) 2012;31:709.

775. Global Oncology Trends 2019. The IQVIA Institute https://www.iqvia.com/insights/the-iqvia-institute/reports/global-oncology-trends-2019. accessed 2020/11/1.

776. Kola I. et al. Can the pharmaceutical industry reduce attrition*. Nat Rev Drug Discov 2004;3:711.

777. Wong C. et al. Estimation of clinical trial success rates*. Biostatistics 2019;20:273.

778. Krzyszczyk P. et al. The growing role of precision and personalized medicine for cancer treatment. Technology (Singap World Sci) 2018;6:79.

779. Howard D. et al. New Anticancer Drugs Associated With Large Increases In Costs And Life Expectancy. Health Aff (Millwood) 2016;35:1581.

780. Saluja R. et al. Examining Trends in Cost and Clinical Benefit of Novel Anticancer Drugs Over Time. J Oncol Pract 2018;14:e280.

781. Likic R. Sustainability of costs of novel biologicals*. Br J Clin Pharmacol 2020;86:1233.

782. Ramsey S. et al. Washington State cancer patients found to be at greater risk for bankruptcy than people without a cancer diagnosis. Health Aff (Millwood) 2013;32:1143.

783. Barry B, Hardin R. Rational Man and Irrational Society? An Introduction and Sourcebook. Beverly Hills: Sage; 1982.

784. Hwang T. et al. Failure of Investigational Drugs in Late-Stage Clinical Development and Publication of Trial Results. JAMA Intern Med 2016;176:1826.

785. Cheng S. et al. A sense of urgency: Evaluating the link between clinical trial development time and the accrual performance of NCI-CTEP sponsored studies*. Clin Cancer Res 2010;16:5557.

786. Autobahn. Wikipedia https://en.wikipedia.org/wiki/Autobahn; accessed 2020/11/02.

787. Houghlen M. Is the Autobahn safer than U.S. highways? Motor Biscuit 2019; https://www.motorbiscuit.com/is-the-autobahn-safer-than-u-s-highways/; accessed 2020/11/2.

788. Carpenter DP. The political economy of FDA drug review*. Health Aff (Millwood) 2004;23:52.

789. Crimp D. Before Occupy: How AIDS Activists Seized Control of the FDA in 1988. The Atlantic Dec 6, 2011; https://www.theatlantic.com/health/archive/2011/12/before-occupy-how-aids-activists-seized-control-of-the-fda-in-1988/249302/; accessed 2020/11/04.

790. Carpenter D. Groups, the media, agency waiting costs, and FDA drug approval. Am J Political Science 2002;46:490.

791. Annas G. Faith (healing), hope and charity at the FDA*. Villanova Law Rev 1989;34:771.

792. Deaths, causes of death and life expectancy, 2016; Table 13-10-0394-01 Leading causes of death, total population, by age group. Statistics Canada https://www150.statcan.gc.ca/n1/daily-quotidien/180628/dq180628b-eng.htm. accessed 2019/04/28.

793. Yabroff K. et al. Economic burden of cancer in the United States: estimates, projections, and future research. Cancer Epidemiol Biomarkers Prev 2011;20:2006.

794. Stewart D. et al. Let's fight cancer like we're fighting COVID-19. National Newswatch 2020.

795. Binder L. We're expediting COVID-19 treatments. Why can't we do the same for cancer? Ottawa Citizen 2020.

796. Freireich EJ. Seventh David A Karnofsky Memorial Lecture-1976*. Clin Cancer Res 1997;3:2711.

797. Morgan-Linnell S. et al. U.S. Food and Drug Administration inspections of clinical investigators: overview of results from 1977 to 2009. Clin Cancer Res 2014;20:3364.

798. Kuhn TS. Structure of Scientific Revolutions 2nd Edition. Chicago: Univsersity of Chicago Press; 1970.

799. International Conference on Harmonisation of Technical Requirements for Registration of Pharmaceuticals for Human Use (ICH) adopts Consolidated Guideline on Good Clinical Practice in the Conduct of Clinical Trials on Medicinal Products for Human Use. Int Dig Health Legis 1997;48:231.

800. Wipke-Tevis D. et al. Impact of the Health Insurance Portability and Accountability Act on participant recruitment and retention. West J Nurs Res 2008;30:39.

801. Goss E. et al. The impact of the privacy rule on cancer research*. J Clin Oncol 2009;27:4014.

802. Nahra K. Privacy and Security Impacts of the 21st Century Cures Legislation https://iapp.org/news/a/privacy-and-security-impacts-of-the-21st-century-cures-legislation/ accessed 2020/12/01.

803. Hendricks-Sturrup R. 21st Century Cures Act Final Rule: Key Health Data Privacy Considerations. 2020 https://fpf.org/2020/11/02/21st-century-cures-act-final-rule-key-health-data-privacy-considerations/ accessed 2020/12/01.

804. Majumder M. et al. Sharing data under the 21st Century Cures Act. Genet Med 2017;19:1289.

805. Esteve A. The business of personal data*. International Data Privacy Law 2017;7:36.

806. Newell D. et al. The Cancer Research UK experience of pre-clinical toxicology studies to support early clinical trials with novel cancer therapies. Eur J Cancer 2004;40:899.

807. Dilts D. et al. Processes to activate phase III clinical trials in a Cooperative Oncology Group: the Case of Cancer and Leukemia Group B. J Clin Oncol 2006;24:4553.

808. Business process management. Wikipedia https://en.wikipedia.org/wiki/Business_process_management. accessed 2022/01/27.

809. Watters J. et al. Transforming the Activation of Clinical Trials. Clin Pharmacol Ther 2018;103:43.

810. Humphreys K. et al. The cost of institutional review board procedures in multicenter observational research. Ann Intern Med 2003;139:77.

811. Hall D. et al. Time required for institutional review board review at one Veterans Affairs medical center. JAMA Surg 2015;150:103.

812. Chaddah M. The Ontario cancer research ethics board*. Curr Oncol 2008;15:49.

813. Lynam E. et al. A Patient Focused Solution for Enrolling Clinical Trials in Rare and Selective Cancer Indications: A Landscape of Haystacks and Needles. Drug Inf J 2012;46:472.

814. Sateren W. et al. How sociodemographics, presence of oncology specialists, and hospital cancer programs affect accrual to cancer treatment trials. J Clin Oncol 2002;20:2109.

815. Comis R. et al. Public attitudes toward participation in cancer clinical trials. J Clin Oncol 2003;21:830.

816. Malalasekera A. et al. How long is too long?*. Eur Respir Rev 2018;27.

817. Cheng J. et al. Mandatory Research Biopsy Requirements*. Front Oncol 2019;9:968.

818. Villano J. et al. Incidence of brain metastasis at initial presentation of lung cancer. Neuro Oncol 2015;17:122.

819. Jin S. et al. Re-Evaluating Eligibility Criteria for Oncology Clinical Trials: Analysis of Investigational New Drug Applications in 2015. J Clin Oncol 2017;35:3745.

820. Laccetti A. et al. Effect of prior cancer on outcomes in advanced lung cancer*. JNCI 2015;107.

821. Ruckdeschel J. et al. A randomized trial of the four most active regimens for metastatic non-small-cell lung cancer. J Clin Oncol 1986;4:14.

822. Baka S. Randomized phase II study of two gemcitabine schedules for patients with impaired performance status and advanced non-small-cell lung cancer*. J Clin Oncol 2005;23:2136.

823. Cullen M. et al. Mitomycin, ifosfamide, and cisplatin in unresectable non-small-cell lung cancer: effects on survival and quality of life. J Clin Oncol 1999;17:3188.

824. Gridelli C. et al. Treatment of advanced non-small-cell lung cancer patients with ECOG performance status 2: results of an European Experts Panel. Ann Oncol 2004;15:419.

825. Kurzrock R. et al. Compliance in early-phase cancer clinical trials*. Oncologist 2013;18:308.

826. Herbst R. et al. Practical Considerations Relating to Routine Clinical Biomarker Testing for Non-small Cell Lung Cancer: Focus on Testing for RET Fusions. Front Med (Lausanne) 2020;7:562480.

827. Clinical Laboratory Improvement Amendments. FDA https://www.fda.gov/medical-devices/ivd-regulatory-assistance/clinical-laboratory-improvement-amendments-clia; accessed 2020/12/8.

828. Paik S. et al. Real-world performance of HER2 testing*. J Natl Cancer Inst 2002;94:852.

829. Kurzrock R. et al. A cancer trial scandal and its regulatory backlash. Nat Biotechnol 2014;32:27.

830. Schilsky R. et al. Development and use of integral assays in clinical trials. Clin Cancer Res 2012;18:1540.

831. Meshinchi S. et al. Lessons learned from the investigational device exemption review of Children's Oncology Group trial AAML1031. Clin Cancer Res 2012;18:1547.

832. Lazzari M. Developing a standard protocol for the introduction of new testing into a clinical laboratory. Laboratory Medicine 2009;40:389.

833. Cappuzzo F. et al. Epidermal growth factor receptor gene and protein and gefitinib sensitivity in non-small-cell lung cancer. J Natl Cancer Inst 2005;97:643.

834. Hirsch F. et al. Molecular predictors of outcome with gefitinib in a phase III placebo-controlled study in advanced non-small-cell lung cancer. J Clin Oncol 2006;24:5034.

835. Sholl L. et al. EGFR mutation is a better predictor of response to tyrosine kinase inhibitors in NSCLC than FISH, CISH, and immunohistochemistry*. Am J Clin Pathol 2010;133:922.

836. Douillard J. et al. Molecular predictors of outcome with gefitinib and docetaxel in previously treated NSCLC: data from the randomized phase III INTEREST trial*. J Clin Oncol 2010;28:744.

837. Getz K. et al. The Impact of Protocol Amendments on Clinical Trial Performance and Cost. Ther Innov Regul Sci 2016;50:436.

838. Fogel D. Factors associated with clinical trials that fail and opportunities for improving the likelihood of success: A review. Contemp Clin Trials Commun 2018;11:156.

839. Lou N. Contract Research Organizations: Outsourced Trials, Outsized Problems ASH Clinical News 2018: https://www.ashclinicalnews.org/spotlight/contract-research-organizations-outsourced-trials-outsized-problems/ accessed 2020/12/15.

840. Roberts D. et al. Contract research organizations in oncology clinical research: Challenges and opportunities. Cancer 2016;122:1476.

841. O'Leary E. et al. Data collection in cancer clinical trials*. Clin Trials 2013;10:624.

842. McLean J. et al. Drug-testing rules broken by Canadian researchers. Toronto Star 2014. https://www.thestar.com/news/canada/2014/09/16/drugtesting_rules_broken_by_canadian_researchers.html; accessed 2020/12/16.

843. Rand A. Atlas Shrugged. New York: Random House; 1957.

844. Legal Issues in Collaborations. Facilitating Collaborations to Develop Combination Investigational Cancer Therapies: Workshop Summary: National Academy of Sciences; 2012. https://www.nap.edu/read/13262/chapter/6. Accessed 2020/12/19.

845. Lai J. et al. Drivers of Start-Up Delays in Global Randomized Trials*. Ther Innov Regul Sci 2020.

846. Tran T. et al. Collaboration in Action: Measuring and Improving Contracting Performance in the University of California Contracting Network. Res Manag Rev 2017;22:28.

847. Sertkaya A. et al. Office of the Asistant Secretary for Planning and Evaluation. U.S. Department of Health and Human Services. Examination of Clincal Trial Costs and Barriers for Drug Development. July 24, 2014. https://aspe.hhs.gov/reports/examination-clinical-trial-costs-barriers-drug-development-0. Accessed 2021/09/26.

848. Pink D. Drive: the surprising truth about what motivates us.: Canaongate Books; 2011.

849. Ramagopalan S. et al. Can real-world data really replace randomised clinical trials? BMC Med 2020;18:13.

850. Fordyce C. et al. Trends in clinical trial investigator workforce and turnover: An analysis of the U.S. FDA 1572 BMIS database. Contemp Clin Trials Commun 2019;15:100380.

851. Vickers A. Do we want more cancer patients on clinical trials if so, what are the barriers to greater accrual. Trials 2008;9:31.

852. Ratelle L. Montreal Researcher Commits Suicide. The Chronicle of Higher Education 1994. https://www.chronicle.com/article/montreal-researcher-commits-suicide/. Accessed 2020/12/20.

853. Downing N. et al. Regulatory review of novel therapeutics*. N Engl J Med 2012;366:2284.

854. Batta A. et al. Trends in FDA drug approvals over last 2 decades: An observational study. J Family Med Prim Care 2020;9:105.

855. The Drug Review and Approval Process in Canada. SPharm 2018/06/02. https://spharm-inc.com/the-drug-review-and-approval-process-in-canada-an-eguide/. Accessed 2021/09/19.

856. Pfizer and Biontech to submit emergency use authorization request today to the U.S. FDA for COVID-19 vaccine. Pfizer 2020/11/20. https://www.pfizer.com/news/press-release/press-release-detail/pfizer-and-biontech-submit-emergency-use-authorization. Accessed 2021/09/25.

857. Samuel N. et al. Cross-comparison of cancer drug approvals at three international regulatory agencies. Curr Oncol 2016;23:e454.

858. Canadians wait more than 450 days longer for access to new medicines than Americans and Europeans. Fraser Institute 2021/05/13. https://www.fraserinstitute.org/sites/default/files/timely-access-to-new-pharmaceuticals-in-canada-US-and-EU-newsrelease.pdf. accessed 2021/06/23.

859. Mezher M. FDA user fee table*. Regulatory Focus Jan 4, 2021. https://www.raps.org/news-and-articles/news-articles/2020/7/fda-fy2021-user-fee-table, Accessed 2021/09/25.

860. Fees payable to the European Medicines Agency. EMA 2021/07/20. https://www.ema.europa.eu/en/human-regulatory/overview/fees-payable-european-medicines-agency. Accessed 2021/09/25.

861. Fees for Examination of a Submission: Drugs for Human Use. Government of Canada 2021. https://www.canada.ca/en/health-canada/services/drugs-health-products/funding-fees/fees-respect-human-drugs-medical-devices/pharmaceutical-submission-application-review-funding-fees-drugs-health-products.html. accessed 2021/09/19.

862. Project Orbis. FDA 2019/09/17. https://www.fda.gov/about-fda/oncology-center-excellence/project-orbis. Accessed 2021/09/23.

863. First Year of Project Orbis leads to 38 approvals of cancer therapies, including 8 by Health Canada. JDSUPRA 2021/01/13. https://www.jdsupra.com/legalnews/first-year-of-project-orbis-leads-to-38-9013775/. Accessed 2021/09/23.

864. Kantarjian H. et al. Why are cancer drugs so expensive*. Mayo Clin Proc 2015;90:500.

865. Siddiqui M. et al. The high cost of cancer drugs*. Mayo Clin Proc 2012;87:935.

866. Meeting the demand for lower drug prices. National Committee to Preserve Social Security & Medicare 2019. https://www.ncpssm.org/documents/

medicare-policy-papers/meeting-demand-for-lower-drug-prices-in-medicare/. Accessed 2020/12/31.

867. Revised PMPRB Guidelines. Overview of key changes. Public Webinar. Patented Medicine Prices Review Board 2020/07/08. https://www.canada.ca/content/dam/pmprb-cepmb/documents/consultations/draft-guidelines/2020/PMPRB-Public-Webinar-July8-2020.pdf. Accessed 2021/09/23.

868. The root cause of unavailability and delay to innovative medicines: Reducing the time before patients have access to innovative medicines. EFPIA 2020/06. https://www.efpia.eu/media/554527/root-causes-unvailability-delay-cra-final-300620.pdf. Accessed 2021/09/25.

869. Procedures for the CADTH pan-Canadian Oncology Drug Review. CADTH 2020/06. https://www.cadth.ca/sites/default/files/pcodr/pCODR%27s%20Drug%20Review%20Process/pcodr-procedures.pdf. accessed 2021/09/19.

870. CADTH reimbursement recommendation: Osimertinib (Tagrisso). Sept 2021. https://cadth.ca/sites/default/files/DRR/2021/PC0246%20Tagrisso%20-%20Draft%20CADTH%20Recommendation%20For%20posting%20September%202%2C%202021.pdf. accessed 2021/12/25.

871. Neumann PJ, Cohen JT, Weinstein MC. Updating cost-effectiveness--the curious resilience of the $50,000-per-QALY threshold. N Engl J Med 2014;371:796-7.

872. CPI Inflation Calculator. Value of $10 from 1975 to 2021. 2021. https://www.in2013dollars.com/us/inflation/1975?amount=10. accessed 2021/12/20.

873. Blome C, Augustin M. Measuring change in quality of life: bias in prospective and retrospective evaluation. Value Health 2015;18:110-5.

874. Khan I, Crott R, Bashir Z. Economic evaluation of cancer drugs. Boca Raton, FL: CRC Press; 2020.

875. Devlin NJ, Lorgelly PK. QALYs as a measure of value in cancer. J Cancer Policy 2017;11:19-25.

876. pan-Canadian Pharmaceutical Alliance. https://www.pcpacanada.ca/. accessed 2021/09/21.

877. Chapter 3. Section 3.09. Ministry of Health and Long-Term Care. Ontario Public Drug Programs. Annual Report 2017 of the Office of the Auditor General or Ontario 2017/Fall. https://www.auditor.on.ca/en/content/annualreports/arreports/en17/v1_309en17.pdf. Accessed 2021/09/26.

878. Pharmacutical sales/capita, US$ exchange rate Organization for Economic Co-operation and Development https://stats.oecd.org/index.aspx?DataSetCode=HEALTH_PHMC. accessed 2021/09/21.

879. Annual Report 2019. Patented Medicine Prices Review Board https://www.canada.ca/en/patented-medicine-prices-review/services/annual-reports/annual-report-2019.html. Accessed 2021/09/23.

880. Protecting Canadians from Excessive Drug Prices: Consulting on Proposed Amendments to the Patented Medicines Regulations. Government of Canada 2017/05/16. https://www.canada.ca/en/health-canada/programs/consultation-regulations-patented-medicine/document.html. Accessed 2021/09/22.

881. PMPRB Guidelines. Government of Canada https://www.canada.ca/en/patented-medicine-prices-review/services/legislation/about-guidelines/guidelines.html#app-e. accessed 2021/09/21.

882. Rawson N, Lawrence D. New Patented Medicine Regulations in Canada: Updated Case Study. Canadian Health Policy Jan 2020. https://www.canadianhealthpolicy.com/products/new-patented-medicine-regulations-in-canada--updated-case-study---en-fr-.html. Accessed 2021/09/22.

883. Danzon P. et al. The impact of price regulation on the launch delay of new drugs--evidence from twenty-five major markets in the 1990s. Health Econ 2005;14:269.

884. Kanavos P. et al. Does external reference pricing deliver what it promises? Evidence on its impact at national level. Eur J Health Econ 2020;21:129.

885. Spicer O, Grootendorst P. An empirical examination of the Patented Medicine Prices Review Board price control amendments on drug launches in Canada. 2020. https://www.canadiancentreforhealtheconomics.ca/wp-content/uploads/2020/08/Spicer-Grootendorst-2020.pdf. accessed 2021/06/24.

886. Stewart D. et al. Stiff price controls will hurt Canadians' access to groundbreaking drugs. Ottawa Citizen June 15, 2021. https://ottawacitizen.com/opinion/stewart-and-bradford-stiff-price-controls-will-hurt-canadians-access-to-ground-breaking-drugs. Accessed 2021/09/21.

887. Deshaies R. Research etc. Health Canada Pricing Reform Research Report. Jan 21, 2021. https://lifesciencesontario.ca/wp-content/uploads/2021/01/Impact-of-Health-Canada-Pricing-Reform-FINAL-Report-Jan-21-2021.pdf. accessed 2021/09/19.

888. New Medicine Launches: Canada in a Global Context. Life Sciences Ontario 2020/06/22. https://lifesciencesontario.ca/wp-content/uploads/2020/06/EN LSO Global-Launch-Benchmarking Webinar-June22-20 Final.pdf. accessed 2021/09/21.

889. Pinnow C. Lessons from the pandemic: we need increased collaboration between the biopharmaceutical sector and government. National Newswatch Nov 25, 2020. https://www.nationalnewswatch.com/2020/11/25/lessons-from-the-pandemic-a-call-to-action-for-increased-collaboration-between-the-canadian-biopharmaceutical-sector-and-government-for-all-innovative-products/. Accessed 2021/09/21.

890. Regulations amending the Regulations Amending the Patented Medicines Regulations (Additional Factors and Information Reporting Requirements), No. 3: SOR/2021-162. Canada Gazette, Part II 2021;155.

891. Rawson N. Clinical Trials in Canada Decrease: A Sign of Uncertainty Regarding Changes to the PMPRB? Canadian Health Policy April 2020. https://www.canadianhealthpolicy.com/products/clinical-trials-in-canada-decrease--a-sign-of-uncertainty-regarding-changes-to-the-pmprb-.html?buy type=. accessed 2021/12/31.

892. Rawson N, Skinner B. Time for the PMPRB to go. Hill Times June 18, 2021. https://www.macdonaldlaurier.ca/time-pmprb-go/. accessed 2021/12/25.

893. Labrie Y. Evidence that regulating pharmaceutical prices negatively affects R&D and access to new medicines. Canadian Health Policy June 2020. https://www.canadianhealthpolicy.com/products/evidence-that-regulating-pharmaceutical-prices-negatively-affects-r-d-and-access-to-new-medicines-.html. accessed 2021/09/19.

894. Experts in Chronic Myeloid Leukemia. The price of drugs for chronic myeloid leukemia (CML) is a reflection of the unsustainable prices of cancer drugs*. Blood 2013;121:4439.

895. Tefferi A. et al. In Support of a Patient-Driven Initiative and Petition to Lower the High Price of Cancer Drugs. Mayo Clin Proc 2015;90:996.

896. Schrag D. The price tag on progress*. N Engl J Med 2004;351:317.

897. Sharma S. et al. Oral chemotherapeutic agents for colorectal cancer. Oncologist 2000;5:99.

898. Cancer drug costs for a month of treatment at initial Food and Drug Administration approval. 2015. https://www.mskcc.org/sites/default/files/node/25097/documents/120915-drug-costs-table.pdf. Accessed 2020/12/29.

899. Light D. et al. Market spiral pricing of cancer drugs. Cancer 2013;119:3900.

900. Dolgin E. Bringing down the cost of cancer treatment. Nature 2018;555:S26.

901. Carrera P. et al. The financial burden and distress of patients with cancer: Understanding and stepping-up action on the financial toxicity of cancer treatment. CA Cancer J Clin 2018;68:153.

902. Ocana A. et al. Toward Value-Based Pricing to Boost Cancer Research and Innovation. Cancer Res 2016;76:3127.

903. Himmelstein D. et al. Medical bankruptcy in the United States, 2007*. Am J Med 2009;122:741.

904. Kantarjian H. et al. High cancer drug prices in the United States*. J Oncol Pract 2014;10:e208.

905. Afatinib. Committee to Evaluate Drugs 2016. http://www.health.gov.on.ca/en/pro/programs/drugs/ced/pdf/giotrif.pdf. Accessed 2020/12/30.

906. Kliff S. The true story of America's sky-high prescription drug prices. Vox 2018. https://www.vox.com/science-and-health/2016/11/30/12945756/prescription-drug-prices-explained. Accessed 2020/12/31.

907. Wouters O. Lobbying Expenditures and Campaign Contributions by the Pharmaceutical and Health Product Industry in the United States, 1999-2018. JAMA Intern Med 2020;180:688.

908. Rogers J. America needs to stop subsidizing Europe and Canada's prescription drugs. Real Clear Politics 2019. https://www.realclearpolitics.com/articles/2019/05/02/america_needs_to_stop_subsidizing_europe_and_canadas_prescription_drugs.html. Accessed 2020/12/31.

909. Ledley F. et al. Profitability of Large Pharmaceutical Companies*. JAMA 2020;323:834.

910. Hernandez I. et al. Drug Shortages in the United States*. JAMA 2020;323:819.

911. Tanne JH. US branded drug makers pay to prevent generic competition. BMJ 2008;336:1266.

912. Evergreening. Wikipedia https://en.wikipedia.org/wiki/Evergreening. Accessed 2021/01/01.

913. Rajkumar S. The high cost of prescription drugs: causes and solutions. Blood Ca J 2020;10:71.

914. Cohen D. Cancer drugs: high price, uncertain value. BMJ 2017;359:j4543.

915. Hillner B. et al. Efficacy does not necessarily translate to cost effectiveness: a case study in the challenges associated with 21st-century cancer drug pricing. J Clin Oncol 2009;27:2111.

916. Davis C. et al. Availability of evidence of benefits on overall survival and quality of life of cancer drugs approved by European Medicines Agency: retrospective cohort study of drug approvals 2009-13. BMJ 2017;359:j4530.

917. The myth of perverse physician incentives*. Community Oncology Alliance 2018. https://communityoncology.org/wp-content/uploads/2018/12/COA-Myth-of-Perverse-Incentives-Analysis-FINAL.pdf. accessed 2021/01/01.

918. Savage P. et al. Cancer Drugs: An International Comparison of Postlicensing Price Inflation. J Oncol Pract 2017;13:e538.

919. Cohen J. The curious case of Gleevec pricing. Forbes 2018. https://www.forbes.com/sites/joshuacohen/2018/09/12/the-curious-case-of-gleevec-pricing/?sh=7ea1ca1b54a3. Accessed 2020/12/29.

920. Human Drug Imports. US Food and Drug Administration https://www.fda.gov/drugs/guidance-compliance-regulatory-information/human-drug-imports. Accessed 2021/01/02.

921. Luhby T. Trump signs order pushing to allow drugs to be imported from Canada. CTV News 2020. https://www.ctvnews.ca/health/trump-signs-order-pushing-to-allow-drugs-to-be-imported-from-canada-1.5039482. Accessed 2021/01/02.

922. Neustaeter B. Canada bans exports of some prescription drugs as U.S. readies for imports. CTV News 2020. https://www.ctvnews.ca/health/canada-bans-exports-of-some-prescription-drugs-as-u-s-readies-for-imports-1.5209209. Accessed 2021/01/02.

923. Jommi C. et al. Implementation of Value-based Pricing for Medicines. Clin Ther 2020;42:15.

924. Van Harten W. et al. Responsible pricing in value-based assessment of cancer drugs*. Ecancermedicalscience 2017;11:ed71.

925. Carr J. Wage and price controls: panacea for inflation or prescription for disaster?: The Fraser Institute; 1976. https://www.fraserinstitute.org/sites/default/files/wage-and-price-controls.pdf. accessed 2021/12/01.

926. Reynolds N. Brilliant economist blew it on inflation. The Globe and Mail 2006. https://www.theglobeandmail.com/news/national/brilliant-economist-blew-it-on-inflation/article25677355/; accessed 2020/12/27.

927. Staudohar P. Effects of wage and price controls in Canada: 1975-1978. Industrial Relations 1979;34:674.

928. Feldman R. The perils of value-based pricing for prescription drugs. The Washington Post Apr 11, 2019. https://www.washingtonpost.com/outlook/2019/04/11/perils-value-based-pricing-prescription-drugs/ accessed 2020/12/23.

929. Final Economic Guidance Report. Atezolizumab (Tecentriq) for Small Cell Lung Cancer. CADTH pan-Canadian Oncology Drug Review 2020/01/30. https://www.cadth.ca/sites/default/files/pcodr/Reviews2020/10156AtezolizumabSCLC_fnEGR_NOREDACT-ABBREV_Post_30Jan2020_final.pdf. Accessed 2021/09/26.

930. Serebrin J. Big pharma turns to smaller firms of innovation. Globe and Mail 2013. https://www.theglobeandmail.com/report-on-business/small-business/sb-money/big-pharma-turns-to-smaller-firms-for-innovation/article12751137/. Accessed 2021/01/05.

931. Knaus C. et al. Pharmaceutical industry donates millions to both Australian political parties. The Guardian 2018. https://www.theguardian.com/business/2018/sep/25/pharmaceutical-industry-donates-millions-to-both-australian-political-parties. Accessed 2021/01/13.

932. Federal political financing in Canada. Wikipedia https://en.wikipedia.org/wiki/Federal_political_financing_in_Canada. Accessed 2021/09/26.

933. Campaign finance in the United States. Wikipedia https://en.wikipedia.org/wiki/Campaign_finance_in_the_United_States. Accessed 2021/01/05.

934. Clemons M. et al. Feasibility of using a pragmatic trials model to compare two primary febrile neutropenia prophylaxis regimens (ciprofloxacin versus G-CSF) in patients receiving docetaxel-cyclophosphamide chemotherapy for breast cancer (REaCT-TC). Support Care Cancer 2019;27:1345.

935. Basulaiman B. et al. Creating a pragmatic trials program for breast cancer patients: Rethinking Clinical Trials (REaCT). Breast Cancer Res Treat 2019;177:93.

936. List of countries by incarceration rate. Wikipedia https://en.wikipedia.org/wiki/List_of_countries_by_incarceration_rate. accessed 2021/05/23.

937. Private prisons. Wikipedia https://en.wikipedia.org/wiki/Private_prison#:~:text=Private%20prisons%20are%20operated%20in,further%20contracts%20with%20private%20prisons. accessed 2021/05/23.

938. Kamal R. et al. How does health spending in the US compare to other countries? Peterson-KFF Health System Tracker 2020. https://www.healthsystemtracker.org/chart-collection/health-spending-u-s-compare-countries/#item-spendingcomparison_gdp-per-capita-and-health-consumption-spending-per-capita-2019. Accessed 2021/01/20.

939. Kamal R. at al. How do healthcare prices and use in the U.S. compare to other countries? Peterson-KFF Health System Tracker 2018. https://www.healthsystemtracker.org/chart-collection/how-do-healthcare-prices-and-use-in-the-u-s-compare-to-other-countries/#item-start. Accessed 2021/01/20.

940. Shanosky N. et al. How do U.S. healthcare resources compare to other countries?. Peterson-KFF: Health System Tracker Aug. 12, 2020, https://www.

healthsystemtracker.org/chart-collection/u-s-health-care-resources-compare-countries/#item-start; accessed 2021/05/22

941. Hospital beds. Organization for Economic Co-operation and Development https://data.oecd.org/healtheqt/hospital-beds.htm#indicator-chart. Accessed 2021/06/03.

942. Stewart C. Computer tomography scanner density by country 2019. Statista Dec 3, 2020. https://www.statista.com/statistics/266539/distribution-of-equipment-for-computer-tomography/ accessed 2021/05/23.

943. Magnetic resonance imaging (MRI) units. Organization for Economic Co-operation and Development https://data.oecd.org/healtheqt/magnetic-resonance-imaging-mri-units.htm. accessed 2021/06/03.

944. Health Care Resources. Organization for Economic Co-operation and Development https://stats.oecd.org/Index.aspx?DataSetCode=HEALTH_REAC. accessed 2021/06/03.

945. Radiotherapy equipment. Organization for Economic Co-operation and Development https://data.oecd.org/healtheqt/radiotherapy-equipment.htm#indicator-chart. accessed 2021/06/03.

946. Price S. Largest Health Insurance Companies of 2021. ValuePenguin Apr 22, 2021. https://www.valuepenguin.com/largest-health-insurance-companies#:~:text=In%20the%20United%20States%2C%20there,companies%20that%20offer%20medical%20coverage. accessed 2021/05/23.

947. Himmelstein D. et al. Health Care Administrative Costs in the United States and Canada, 2017. Ann Intern Med 2020;172:134.

948. Canada. Office of the United States Trade Representative https://ustr.gov/countries-regions/americas/canada#:~:text=U.S.%2DCanada%20Trade%20Facts&text=U.S.%20goods%20and%20services%20trade,was%20%242.4%20billion%20in%202019. accessed 2021/05/23.

949. Sharpe A. Why are Americans more productive than Canadians? International Productivity Monitor 2003;6:19.

950. Downey KA. A heftier dose to swallow: Rising cost of health care in U.S. gives other developed countries an edge in keeping jobs. Washington Post 2004.

951. Guardado JR. Policy Research Perspectives: Medical Liability Claim Frequency Among U.S. Physicians. American Medical Association 2017. https://www.ama-assn.org/sites/ama-assn.org/files/corp/media-browser/public/government/advocacy/policy-research-perspective-medical-liability-claim-frequency.pdf. accessed 2021/05/23

952. Medical Malpractice Statistics and Facts. National Trial Law https://nationaltriallaw.com/medical-malpractice-statistics/. accessed 2021/05/23.

953. Milne V. et al. Is Canada's medical malpractice system working? healthydebate 2014. https://healthydebate.ca/2014/11/topic/cmpa-medical-malpractice/ accessed 2021/05/23.

954. Skinner BJ. The Medical Bankruptcy Myth Fraser Institute https://www.fraserinstitute.org/article/medical-bankruptcy-myth. Accessed 2021/06/28.

955. Medical Malpractice Liability: Canada. Library of Congress https://www.loc.gov/law/help/medical-malpractice-liability/canada.php. accessed 2021/05/23.

956. Life Expectancy of the World Population. Worldometer https://www.worldometers.info/demographics/life-expectancy/. accessed 2021/05/23.

957. Roser M. Why is life expectancy in the US lower than in other rich countries? Our World in Data 2020. https://ourworldindata.org/us-life-expectancy-low. accessed 2021/05/23.

958. Kamal R. et al. Access & Coverage. Peterson-KFF Health System Tracker 2020. https://www.healthsystemtracker.org/indicator/access-affordability/percent-insured/. Accessed 2021/01/20.

959. Kamal R. et al. Access & Coverage: Delay of needed care. Peterson-KFF Health System Tracker 2020. https://www.healthsystemtracker.org/indicator/access-affordability/delay-needed-care/. Accessed 2021/01/.20.

960. Kamal R. et al. Length of Life: Mortality Rate. Peterson-KFF Health System Tracker 2020. https://www.healthsystemtracker.org/indicator/health-well-being/mortality-rate/. Accessed 2021/01/20.

961. Kamal R. et al. Length of Life: Premature death. Peterson-KFF Health System Tracker 2020. https://www.healthsystemtracker.org/indicator/health-well-being/self-reported-health/. Accessed 2021/01/20.

962. Perkel C. Canada lags behind peers in doctors per capita, but average in physician visits. National Post Jan 31, 2020, https://nationalpost.com/pmn/news-pmn/canada-news-pmn-canada-lags-behind-peers-in-doctors-per-capita-but-average-in-physician-visits: accessed 2021/05/22.

963. COVID-19 coronavirus pandemic. Worldometer https://www.worldometers.info/coronavirus/. accessed 2021/05/23.

964. Dizioli A. et al. Health Insurance as a Productive Factor. Semantic Scholar 2012. https://www.semanticscholar.org/paper/Health-Insurance-as-a-Productive-Factor-Dizioli-Pinheiro/dc1e26c44340b081ca4ff1a412ac6ef1b87151e6?p2df. accessed 2021/05/23.

965. List of countries by firearm-related death rates. Wikipedia https://en.wikipedia.org/wiki/List_of_countries_by_firearm-related_death_rate. accessed 2021/05/23.

966. Kouyoumdjian F. et al. Do people who experience incarceration age more quickly? Exploratory analyses using retrospective cohort data on mortality from Ontario, Canada. PLoS One 2017;12:e0175837.

967. Li J. et al. Physician response to pay-for-performance*. Health Econ 2014;23:962.

968. Kamal R. et al. Patient Experience. Peterson-KFF Health System Tracker 2020. https://www.healthsystemtracker.org/indicator/access-affordability/4578-2/. Accessed 2021/01/20.

969. Health Care Wait Times by Country 2021. accessed 2021/06/21. World Population Review https://worldpopulationreview.com/country-rankings/health-care-wait-times-by-country.

970. Barua B, Moir M. Waiting Your Turn: Wait Times for Health Care in Canada, 2019 Report. Fraser Institute https://www.fraserinstitute.org/studies/waiting-your-turn-wait-times-for-health-care-in-canada-2019. accessed 2021/05/24.

971. Joint replacement wait times. Canadian Institute for Health Information https:// yourhealthsystem.cihi.ca/hsp/inbrief.#!/indicators/004/joint-replacement-wait-times/;mapC1;mapLevel2;/. accessed 2021/05/24.

972. Wait times for priotity procedures in Canada. Canadian Institute for Health Information 2020/07/09. https://www.cihi.ca/en/wait-times-for-priorit y-procedures-in-canada. accessed 2021/05/24.

973. System Performance. Wait Times of Diagnostic Imaging. Health Quality Ontario 2021/03. https://www.hqontario.ca/System-Performance/Wait-Times-for-Diagnostic-Imaging?utm_source=Ontario. ca&utm_medium=Referral&utm_campaign=WT%20Referral. accessed 2021/05/24.

974. Health spending. Organization for Economic Co-operation and Development https://data.oecd.org/healthres/health-spending.htm. accessed 2021/06/02.

975. List of countries by past life expectancy. Wikipedia https://en.wikipedia.org/wiki/ List_of_countries_by_past_life_expectancy. accessed 2021/06/01.

976. Canada's ranking on global life expectancy scale expected to drop by 2040: study. CTV News 2018/10/16. https://www.ctvnews.ca/health/canada-s-ranking-on-global-life-expectancy-scale-expected-to-drop-by-2040-study-1.4136851#:~:text=In%202016%2C%20Canada%20ranked%2017th,life%20ex-pectancy%20of%2081.6%20years.&text=Canada%20is%20currently%20 among%20the,according%20to%20a%20new%20study. Accessed 2021/06/01.

977. Siegel R. et al. Cancer Statistics, 2021. CA Cancer J Clin 2021;71:7.

978. Cancer-specific stats 2020. Canadian Cancer Society and Government of Canada https://cdn.cancer.ca/-/media/files/research/cancer-statistics/2020-statistics/canadian-cancer-statistics/2020-resources/ res-cancerstatistics-canadiancancerstatistics-2020_cancer-specific-stats. pdf?rev=a672053690e9493d921fe73453a749b4&hash=4F194657DA8 3EFC15ADB92A8E5ACBC5E. accessed 2021/06/02.

979. Boyd C. et al. Associations between community income and cancer survival in Ontario, Canada, and the United States. J Clin Oncol 1999;17:2244.

980. Kadiyala S. et al. Are United States and Canadian cancer screening rates consistent with guideline information regarding the age of screening initiation? Int J Qual Health Care 2011;23:611.

981. Bryan S. et al. Cancer in Canada: Stage at diagnosis. 2018. accessed 2021/06/23. Statistics Canada https://www150.statcan.gc.ca/n1/pub/82-003-x/2018012/ article/00003-eng.htm.

982. Khorana A. et al. Time to initial cancer treatment in the United States and association with survival over time: An observational study. PLoS One 2019;14:e0213209.

983. Huang J. et al. Does delay in starting treatment affect the outcomes of radiotherapy? A systematic review. J Clin Oncol 2003;21:555.

984. Raphael M. et al. The relationship between time to initiation of adjuvant chemotherapy and survival in breast cancer: a systematic review and meta-analysis. Breast Cancer Res Treat 2016;160:17.

985. General/Clinical Pathology Profile. Canadian Medical Association https://www. cma.ca/sites/default/files/2019-01/general-pathology-e.pdf. accessed 2021/06/03.

986. The Canadian medical imaging inventory, 2017. CADTH 2017, https://cadth.ca/ canadian-medical-imaging-inventory-2017. Accessed 2021/05/22.

987. Tolbert J. et al. Key facts about the uninsured population. Kaiser Family Foundation Nov 6, 2020 https://www.kff.org/uninsured/issue-brief/ key-facts-about-the-uninsured-population/. accessed 2021/06/27.

988. Love J. 14 Times Science Fiction Writers Predicted the Future. Ranker June 14, 2019. https://www.ranker.com/list/sci-fi-predictions-that-came-true/jordan-love. Accessed 2021/09/28.

989. Landes DS. The Wealth and Poverty of Nations: Why Some Are So Rich and Some So Poor: W. W. Norton; 1999.

990. Snowden D. Organizing a children's party. YouTube https://www.youtube.com/ watch?v=Miwb92eZaJg. Accessed 2021/10/06.

991. Kirkwood M. et al. The State of Oncology Practice in America, 2018: Results of the ASCO Practice Census Survey. J Oncol Pract 2018;14:e412.

992. Yang W. et al. Projected supply of and demand for oncologists and radiation oncologists through 2025: an aging, better-insured population will result in shortage. J Oncol Pract 2014;10:39.

993. Canadian specialty profile. Canadian Medical Association 2021. https://www. cma.ca/canadian-specialty-profiles. Accessed 2021/10/02.

994. Fundytus A. et al. Medical oncology workload in Canada*. Curr Oncol 2018;25:206.

995. Sussman J. et al. A Quality Initiative of the Program in Evidence-Based Care (PEBC), Cancer Care Ontario (CCO): Models of Care for Cancer Survivorship. Cancer Care Ontario 2017/03/28. https://www.cancercareontario.ca/sites/ ccocancercare/files/guidelines/summary/pebc26-1v2s.pdf. Accessed 2021/10/07.

996. Gyawali B. et al. Oncology training programmes for general practitioners: a scoping review. Ecancermedicalscience 2021;15:1241.

997. Cohen B. Society's valuation of life saving in radiation protection and other contexts. Health Phys 1980;38:33.

998. Cohen J. et al. Does preventive care save money?*. N Engl J Med 2008;358:661.

999. Historical trends in smoking prevalence. University of Waterloo https:// uwaterloo.ca/tobacco-use-canada/adult-tobacco-use/smoking-canada/ historical-trends-smoking-prevalence. accessed 2021/11/20.

1000. Canadian Cancer Statistics 2021. Canadian Cancer Society 2021. https:// cdn.cancer.ca/-/media/files/research/cancer-statistics/2021-statistics/2021-pdf-en-final.pdf?rev=2b9d2be7a2d34c1dab6a01c6b0a6a32d&hash=01DE85401DBF0217F8B64F2B7DF43986. Accessed 2021/11/20.

1001. Raaschou-Nielsen O. et al. Air pollution and lung cancer incidence in 17 European cohorts*. Lancet Oncol 2013;14:813.

1002. Rahman S. et al. Nicotinic receptor modulation to treat alcohol and drug dependence. Front Neurosci 2014;8:426.

1003. Montan P. et al. Pharmacologic therapy of obesity*. Ann Transl Med 2019;7:393.

1004. Romero-Laorden N. et al. Inherited mutations in DNA repair genes and cancer risk. Curr Probl Cancer 2017;41:251.

1005. Ceyhan-Birsoy O. et al. Paired Tumor-Normal Sequencing Provides Insights into TP53-Related Cancer Spectrum in Li-Fraumeni Patients. J Natl Cancer Inst 2021 Jul 7:djab117.

1006. Liu W. et al. Applications and challenges of CRISPR-Cas gene-editing to disease treatment in clinics. Precis Clin Med 2021;4:179.

1007. Ravindran S. Got mutation? 'Base editors' fix genomes one nucleotide at a time. Nature 2019;575:553.

1008. Cring M. et al. Gene therapy and gene correction: targets, progress, and challenges for treating human diseases. Gene Ther 2020 Oct 9. doi: 10.1038/s41434-020-00197-8.

1009. Drago D. et al. Global regulatory progress in delivering on the promise of gene therapies for unmet medical needs. Mol Ther Methods Clin Dev 2021;21:524.

1010. Lu C. et al. Phase I clinical trial of systemically administered TUSC2(FUS1)-nanoparticles mediating functional gene transfer in humans. PLoS One 2012;7:e34833.

1011. Bender E. Regulating the gene-therapy revolution. Nature 2018;564:S20.

1012. Milholland B. et al. Differences between germline and somatic mutation rates in humans and mice. Nat Commun 2017;8:15183.

1013. Aylon Y. et al. New plays in the p53 theater. Curr Opin Genet Dev 2011;21:86.

1014. Rivlin N. et al. Mutations in the p53 Tumor Suppressor Gene*. Genes Cancer 2011;2:466.

1015. Li S. et al. Development of synthetic lethality in cancer: molecular and cellular classification. Signal Transduct Target Ther 2020;5:241.

1016. Dedes K. et al. Synthetic lethality of PARP inhibition in cancers lacking BRCA1 and BRCA2 mutations. Cell Cycle 2011;10:1192.

1017. Sachdev E. et al. PARP Inhibition in Cancer*. Target Oncol 2019;14:657.

1018. Kurzrock R. et al. Click chemistry, 3D-printing, and omics*. Oncotarget 2016;7:2155.

1019. Adashek J. et al. The paradox of cancer genes in non-malignant conditions: implications for precision medicine. Genome Med 2020;12:16.

1020. Ascierto P. et al. The role of BRAF V600 mutation in melanoma. J Transl Med 2012;10:85.

1021. Pignon J. et al. Lung adjuvant cisplatin evaluation: a pooled analysis by the LACE Collaborative Group. J Clin Oncol 2008;26:3552.

1022. Moding E. et al. Circulating Tumor DNA Dynamics Predict Benefit from Consolidation Immunotherapy in Locally Advanced Non-Small Cell Lung Cancer. Nat Cancer 2020;1:176.

1023. Kurtz D. et al. Enhanced detection of minimal residual disease by targeted sequencing of phased variants in circulating tumor DNA. Nat Biotechnol 2021 Jul 22. doi: 10.1038/s41587-021-00981-w.

1024. Tutt A. et al. Oral poly(ADP-ribose) polymerase inhibitor olaparib in patients with BRCA1 or BRCA2 mutations and advanced breast cancer: a proof-of-concept trial. Lancet 2010;376:235.

1025. Dieras V. et al. Veliparib with carboplatin and paclitaxel in BRCA-mutated advanced breast cancer (BROCADE3)*. Lancet Oncol 2020;21:1269.

1026. Gonzalez-Martin A. et al. Niraparib in Patients with Newly Diagnosed Advanced Ovarian Cancer. N Engl J Med 2019;381:2391.

1027. de Bono J. et al. Olaparib for Metastatic Castration-Resistant Prostate Cancer. N Engl J Med 2020;382:2091.

1028. Golan T. et al. Maintenance Olaparib for Germline BRCA-Mutated Metastatic Pancreatic Cancer. N Engl J Med 2019;381:317.

1029. Rooney M. et al. Genomics of squamous cell lung cancer. Oncologist 2013;18:707.

1030. Kim K. et al. Recent progress in mapping the emerging landscape of the small-cell lung cancer genome. Exp Mol Med 2019;51:1.

1031. Leonetti A. et al. Resistance mechanisms to osimertinib in EGFR-mutated non-small cell lung cancer. Br J Cancer 2019;121:725.

1032. Haura E. et al. JNJ-61186372 (JNJ-372), an EGFR-cMet bispecific antibody, in EGFR-driven advanced non-small cell lung cancer (NSCLC). J Clin Oncol 2019;37:9009.

1033. Larkin J. er al. Five-Year Survival with Combined Nivolumab and Ipilimumab in Advanced Melanoma. N Engl J Med 2019;381:1535.

1034. Hellmann M. et al. Nivolumab plus Ipilimumab in Lung Cancer with a High Tumor Mutational Burden. N Engl J Med 2018;378:2093.

1035. Motzer R. et al. Nivolumab plus Ipilimumab versus Sunitinib in Advanced Renal-Cell Carcinoma. N Engl J Med 2018;378:1277.

1036. Larkin J. et al. Combined Nivolumab and Ipilimumab or Monotherapy in Untreated Melanoma. N Engl J Med 2015;373:23.

1037. Rodriguez-Abreu D. et al. Primary analysis of a randomized, double-blind, phase II study of the anti-TIGIT antibody tiragolumab (tira) plus atezolizumab (atezo) versus placebo plus atezo as first-line (1L) treatment in patients with PD-L1-selected NSCLC (CITYSCAPE). J Clin Oncol 2020;38:9503.

1038. Ahn M. et al. Vibostolimab, an anti-TIGIT antibody, as monotherapy and in combination with pembrolizumab in anti-PD-1/PD-L1-refractory NSCLC. Ann Oncol 2020;31 (suppl 4):Abst 1400P.

1039. Kordbacheh T. et al. Radiotherapy and anti-PD-1/PD-L1 combinations in lung cancer: building better translational research platforms. Ann Oncol 2018;29:301.

1040. Theelen W. et al. Pembrolizumab with or without radiotherapy for metastatic non-small-cell lung cancer: a pooled analysis of two randomised trials. Lancet Respir Med 2021;9:467.

1041. Chesney J. et al. Randomized, Open-Label Phase II Study Evaluating the Efficacy and Safety of Talimogene Laherparepvec in Combination With Ipilimumab Versus Ipilimumab Alone in Patients With Advanced, Unresectable Melanoma. J Clin Oncol 2018;36:1658.

1042. Zitvogel L. et al. The microbiome in cancer immunotherapy*. Science 2018;359:1366.

1043. Einsele H. et al. The BiTE (bispecific T-cell engager) platform: Development and future potential of a targeted immuno-oncology therapy across tumor types. Cancer 2020;126:3192.

1044. Owonikoko T. et al. A phase I study of AMG 757, half-life extended bispecific T-cell engager (BITE) immune therapy against DLL3, in SCLC. J Thorac Oncol 2021;16:S126.

1045. Sterner R. et al. CAR-T cell therapy*. Blood Cancer J 2021;11:69.

1046. Modi S. et al. Trastuzumab Deruxtecan in Previously Treated HER2-Positive Breast Cancer. N Engl J Med 2020;382:610.

1047. Stewart D. et al. Impact of decitabine on immunohistochemistry expression of the putative tumor suppressor genes FHIT, WWOX, FUS1 and PTEN in clinical tumor samples. Clin Epigenetics 2014;6:13.

1048. Reik W. Stability and flexibility of epigenetic gene regulation in mammalian development. Nature 2007;447:425.

1049. Issa J. et al. Targeting DNA methylation. Clin Cancer Res 2009;15:3938.

1050. Bolden J. et al. Anticancer activities of histone deacetylase inhibitors. Nat Rev Drug Discov 2006;5:769.

1051. Al-Hajj M. et al. Self-renewal and solid tumor stem cells. Oncogene 2004;23:7274.

1052. Taylor W. et al. Small-Molecule Ferroptotic Agents with Potential to Selectively Target Cancer Stem Cells. Sci Rep 2019;9:5926.

1053. Walcher L. et al. Cancer Stem Cells-Origins and Biomarkers*. Front Immunol 2020;11:1280.

1054. Clarke M. Clinical & Therapeutic Implications of Cancer Stem Cells. N Engl J Med 2019;380:2237.

1055. Pu X. et al. Inflammation-related genetic variations and survival in patients with advanced non-small cell lung cancer receiving first-line chemotherapy. Clin Pharmacol Ther 2014;96:360.

1056. Marsh S. et al. Irinotecan pharmacogenomics. Pharmacogenomics 2010;11:1003.

1057. Scartozzi M. et al. 5-Fluorouracil pharmacogenomics: still rocking after all these years? Pharmacogenomics 2011;12:251.

1058. Molla R. Genetic testing is an inexact science with real consequences. Vox 2019. https://www.vox.com/recode/2019/12/13/20978024/genetic-testing-dna-consequences-23andme-ancestry. Accessed 2021/10/31.

1059. Stewart D. et al. Extensive small cell lung cancer (SCLC) chemotherapy: a population kinetics assessment. Proc Am Society Clin Oncol 2020:abstract # e21101.